America's Best

IndustryWeek's *Guide to World-Class Manufacturing Plants*

THEODORE B. KINNI

John Wiley & Sons, Inc.

New York ▪ *Chichester* ▪ *Brisbane* ▪ *Toronto* ▪ *Singapore* ▪ *Weinheim*

This text is printed on acid-free paper.

Published by John Wiley & Sons, Inc.

This publication is designed to provide accurate and authoritative information in regard to the subject matter covered. It is sold with the understanding that the publisher is not engaged in rendering legal, accounting, or other professional services. If legal advice or other expert assistance is required, the services of a competent professional person should be sought.

Library of Congress Cataloging-in-Publication Data:

Kinni, Theodore B., 1956–
America's best : IndustryWeek's guide to world-class manufacturing plants / Theodore B. Kinni.
p. cm.
Includes index.
ISBN 0-471-16002-4 (cloth : alk. paper)
1. Manufacturers—United States—Case studies. I. Industry week. II. Title.
TS149.K53 1997
670'.25'73—dc20 96-27595
CIP

Printed in the United States of America

10 9 8 7 6 5 4 3 2 1

CONTENTS

PREFACE

In 1990, when *IndustryWeek* profiled its first class of America's Best Plants award winners, a unique reference began to grow. Without undue fanfare, Senior Editor John Sheridan's findings, interviews, and support materials describing the operational details of 12 of the most impressive manufacturing operations in the United States found their way into neatly labeled files.

When the fourth class of Best Plants was announced in 1993, the value of the Best Plants database, which now included a wealth of information including Best Plants applications, follow-up questionnaires, reams of supporting documents, and the findings of the entire team of *IW* editors dispatched across the country to record firsthand accounts of the finalist's plants, was beginning to become clear. That year, in addition to the plant profiles, Sheridan collected the lessons of the Best Plants in the first of what has become an annual feature in *IndustryWeek.* He also supervised the first annual Best Plants statistical profile, a benchmarking tool offered for manufacturers who wanted to measure their performance against the best. Conferences, which featured Best Plants managers telling their own stories, added to the mix.

By the time the 1995 winners were announced, the Best Plants database had become the kind of info treasure trove a business writer dreams about. When the doors swung open on the collected findings of 62 of America's Best Plants, the core of a special book was revealed. No unproved theory, no unsupported stories, no unspoken agendas—just the facts. Here was the stuff that works! And that is exactly what this book contains.

America's Best is based on the collective lessons of *IndustryWeek*'s America's Best Plants award winners. The components of successful manufacturing described here are taken directly from their experiences. It is a unique reference that reveals an integrated system of success (none of the Best Plants believe in silver bullets), innumerable individual solutions, a statistical profile, and a self-assessment to see how you measure up to the best.

This book is also a celebration of manufacturing excellence. *IndustryWeek* has been proud to honor the Best Plants. They have set the example and expanded the perimeters of world-class manufacturing. They have also been generous with their time and knowledge to a sometimes unbelievable extent. *America's Best* is one more thing, too: It is a challenge. *Help set the pace,* it whispers, *join the front-runners and build your own Best Plant.* If you undertake that challenge, we all win.

What's in It for You?

America's Best is presented in two main sections. Section 1 is a ten-chapter exploration of the common lessons of the Best Plants and is organized by chapter into nine components of manufacturing excellence as demonstrated by the Best Plants winners. Chapter 1 is an overview of these components. It serves a dual purpose: to summarize the organizational structure of the section and to describe the interconnectedness of the topics. To gain a complete portrait of world-class manufacturing, the rest of Section 1 can be read in sequence. Or readers who are further along the Best Plants path or who need immediate help on a specific topic, such as employee involvement or corporate citizenship, can turn directly to the relevant chapter.

Section 2 of the book is composed of three parts. The first collects in alphabetical order the individual profiles of each of the 62 Best Plants winners from 1990 to 1995. Based mainly on the findings of investigative trips made by *IW* editors in the year in which each plant won the America's Best Plants award, the profiles highlight unique aspects and offer a more complete portrait of the winners' plants. In contrast to the commonalties of the plants presented in Section 1, the profiles celebrate the diversity of the plants and effectively demonstrate that each journey to excellence must be customized to fit the unique culture and structure of the company.

Nine categories of achievement have been identified, and several appear in each profile. Each category is one of the nine subject chapters in the first section of the book and permits the reader to cross-reference at a glance the individual achievements of the Best Plants described in the profiles by subject.

The individual profiles conclude with contact information for each of the plants. The company name, address, telephone number, major products, and employee count are included for general reader queries. Requests for information other than benchmarking should be addressed to their logical departments. The contact name at each of the plants is intended for use solely by benchmarkers. In order to provide the most accurate information possible, all contacts were updated as closely as possible to the book's publication.

The second part of Section 2 is a statistical profile of the 1995 Best Plants class, including the 15 runner-up plants. Whenever possible, it also includes side-by-side comparisons from the 1993 and 1994 Best Plants classes. Abstracted and calculated by *IW* staff from the Best Plants applications and follow-up questionnaires, this profile provides useful benchmarks for all of the nine subject areas and hard numerical measures representing Best Plants' performance.

The third part of Section 2 contains a self-assessment based on *IW*'s Best Plants application. To discover how your plant measures up against the best, complete this assessment and compare it to the Best Plants Statistical Profile. Even an informal gap analysis will give you insights into those areas and chapters that should yield the highest return to your operation.

A note: This book is about manufacturing excellence and that is a moving target, or more accurately, 62 moving targets. The Best Plants are part of companies that are bought and sold; their employees move; their product lines shift; and their workforces and spaces expand and contract regularly. To steal a phrase: Change is the only constant in the physical details of these plants. In order to hit the moving targets that are Best Plants and minimize awkward phrases such as "former plant manager, who was promoted to vice president of manufacturing at ABC Corporation last year and was appointed COO at XYZ Incorporated this year, Joe Wise . . . ," I've frozen the progress of the Best Plants in the year the award was won. The contact information, which reflects the most recent responses from the plants, is the one exception, but even that will change with time.

Acknowledgments

In a very real sense, I am one of many coauthors of this book. Every member of the staff of *IndustryWeek* is also an author. It was their time and effort that provided much of the information in this book. John Sheridan, widely acknowledged as the Father of America's Best Plants, is the heart of *IW*'s Best Plants control center and, further, deserves a lion's share of the credit for the material on which this book is based. Close behind are the *IW* editors, past and present, whose names appear at the end of each of the Individual Plant Profiles in Section 2. They are the eyes and ears of Best Plants.

You need only read a single Best Plants application to know that each of the winning companies also deserves an author credit. They have opened their doors and generously shared their knowledge at a time when competitive advantage is a commodity in short supply.

This book is the first product of a strategic alliance in which each partner plays an essential role. Many thanks to *IW*'s management team, in particular David Altany, who nourished the seed of an idea, and Carl Marino, who made the commitment. Thanks also to John Wiley & Sons, in particular Jim Childs, who recognized the value, and Jeanne Glasser and Joseph Mills, who were instrumental in realizing the finished book. And, as always, thanks to John Willig—to describe him as simply a literary agent does no justice to his formidable abilities as a project developer, content editor, book marketer, and coach.

Thanks to the manufacturing experts who served on *America's Best* evaluation panel: Richard Schonberger, President, Schonberger & Associates; Alfonzo Hall, Plant Manager of General Motors' Cadillac Luxury Car Division's Pittsburgh Plant; Robert Hall of AME; and Bill Hansen of MIT. Their advice helped make this a better book. Many thanks also to Ford Motor Company CEO Alex Trotman for sharing his thoughts in the Afterword.

And finally, on the home front, love and thanks to Donna, whose organizational talents are world class and whose encouragement and patience made it possible for me to become part of the America's Best Plants team.

Theodore B. Kinni
Director, *IW* Books

PUTTING THE PIECES TOGETHER

Sometimes good ideas, like good vintages, take a few years to age into full-bodied excellence. In the case of this book, the maturation process took even longer. One hundred and fourteen years after its founding, *IndustryWeek* is finally publishing its first book detailing the best practices of world-class manufacturing companies.

As you're about to find out, the century-plus wait was well worth it.

IndustryWeek's mission has remained remarkably consistent since the founding of its great-grandfather publication, *Trade Review,* in 1882. The words or phrasing may change—after all, what editor can resist the urge to revise the words of predecessors—but the basic premise hasn't altered: to provide executives with the most current, most reliable, and most useful information possible, all with an eye toward helping them manage their companies more productively, more profitably, and more humanely.

As might be expected, that mission was carried out differently in 1882 than it is in 1996. *Trade Review* was a primitive journal that kept its audience, the brawny capitalists of the American Midwest, up-to-date on the burgeoning commerce of the late 1800s. As a new century approached, *Trade Review* broadened its reach by becoming *Iron Trade Review,* a publication for managers behind the industry that was building a nation. Thirty years later, changes in technology and *Iron*'s readership called for yet another rebirth, this time as *Steel.* The new magazine proved almost as durable as the alloy for which it was named; it operated continuously until 1970, when the publication's final metamorphosis turned *Steel* into *IndustryWeek.*

Even *IndustryWeek,* 26 years old and counting, has evolved since its inception. Originally conceived as a weekly news digest for U.S. industry, *IndustryWeek* eventually became instead a twice-monthly management magazine for senior executives in the international wealth-creating sector, which includes manufacturing and its supporting industries. Most recently, *IndustryWeek* has expanded its reach across a broad range of media and executive audiences, including forays into television, on-line services, and executive-level management development conferences.

And yet, exciting and successful as these ventures are, there's truth in the old adage that every journalist believes he has a great book inside him. The editors at *IndustryWeek* are no different: The overwhelming response to a number of our signature editorial events, including the America's Best Plants awards, has long convinced us that an *IndustryWeek* series of management books could fill an important need among our readers. With the invaluable help of author Ted Kinni and publisher John Wiley & Sons, the fulfillment of that belief—114 years in the making—now rests in your hands.

I hope you find *America's Best* as useful and entertaining—but not quite as long to complete—as it was for us to create.

John R. Brandt
Editor-in-Chief
IndustryWeek

Ever since the first issue of *IndustryWeek* rolled off the presses of Penton Publishing in January of 1970, the magazine and its tenacious, and often overworked, editorial staff have pursued a constant mission: to produce a publication that its readers would find both insightful and useful—and, perhaps, something more than that. Then, as now, our readers consisted primarily of executives and managers responsible for America's manufacturing industries—the wealth-generating bedrock of the economy. Over the years, the editors coined a variety of slogans or catchphrases to describe the magazine's mission in life. Early on, it was "A Voice for American Industry." More recently, it became "The Management Resource."

For more than a quarter century, the editors sought to guide and even inspire readers to become the kind of executives and leaders who know how to make the most of the resources at their command, including the intellectual capabilities of the workforce, thus creating a stimulating and rewarding work environment while steering their companies toward higher plateaus of success.

Analysis of management issues and executive styles has always been a staple of *IndustryWeek*. And because a majority of its readers labor in the world of manufacturing, we have long placed special emphasis on chronicling significant developments in manufacturing, including the emergence of just-in-time methods, the Total Quality movement, and the long-overdue recognition that empowered employees can be a powerful antidote to bloated corporate bureaucracy.

Since its inception, the magazine has closely tracked the overall "state of industry." As the U.S. manufacturing sector approached what some now regard as its historical nadir in the mid-1970s, *IndustryWeek* issued a clarion call for a broad-based effort to turn things around. Alarmed by the nation's rapidly eroding balance of trade, in 1971 we published a lengthy "impact report" which asked the unsettling question: IS THERE STILL TIME TO SAVE U.S. INDUSTRY? The cover illustration for that issue depicted the sands of time slipping through the neck of an hourglass.

Many American companies, lulled into complacency by their past success, ignored the wake-up call for another decade or longer. But, fortunately, there *was* still time. Nonetheless, uneasiness persisted as futurists discussed the implications of a "postindustrial society." Through the 1980s, many business journalists, even some on the *IW* staff, lamented the "deindustrialization" of America and the loss of jobs and growth markets to overseas competitors. But *IndustryWeek* continued to hammer out a message that many people seemed to have forgotten during the years when the marketing and financial wizards called most of the shots in American executive suites. It was a message that the Japanese had taken to heart—that core competence in manufacturing is a key source of competitive strength. In 1984, for example, *IW* published a feature article that reinforced the theme that manufacturing expertise should be viewed as a "strategic weapon."

When manufacturing mavens such as Richard J. Schonberger and Robert W. Hall began writing books that detailed their findings about the "secrets" of Japanese manufacturing (including concepts and techniques the Japanese had originally learned from Henry Ford), *IW* relayed their observations to its readers. When the Association for Manufacturing Excellence (AME) began to encourage U.S. executives to share their manufacturing know-how, *IW* reported on that organization's efforts and on major themes emerging from AME workshops and seminars. Along the way, the magazine kept its readers informed about key manufacturing technologies, supply-chain management issues, and world-class manufacturing practices. In 1990, it published a six-part series on "World Class Manufacturing," which helped crystallize awareness that this was not merely a Madison Avenue buzz phrase, but a blueprint for action.

One article in that series was a collection of profiles on the "gurus" of world-class manufacturing, including the late Dr. Shigeo Shingo, the Japanese engineering genius who played a major role in developing "quick-changeover" techniques and other elements of the Toyota Production System—the prototype of lean, flexible, world-class manufacturing. Throughout the 1980s, Dr. Shingo had visited the United States as a consultant, sharing his knowledge and insights with American manufacturers. At a conference in Indianapolis in 1990, an *IW* editor interviewed Dr. Shingo during his final trip to the United States. Near the end of the interview, the editor thanked Dr. Shingo and commended him on his efforts to help revive the flagging U.S. manufacturing sector. "You know," the editor said, "one day they may call you the savior of U.S. manufacturing."

"No," the elderly savant replied. "But if you tell your readers to follow my advice, *you* may become the savior of American manufacturing."

The remark provoked a chuckle, and evoked a rather formidable challenge. But Dr. Shingo's words, as well as encouragement from people such as Ken Stork, a former Motorola executive and past president of AME, did stimulate further thinking about what we might do to keep the ball rolling and accelerate the improvement momentum already evident within the U.S. manufacturing sector. Aware that numerous companies were making significant strides in quality improvement, streamlining production practices with JIT and cellular manufacturing (the plant-floor version of "reengineering"), increasing flexibility and responsiveness, building supply-chain partnerships, and harnessing the knowledge and brainpower of the workforce, *IW* decided to turn its editorial spotlight on "role model" manufacturers and showcase examples of plants that had begun to put all the pieces together. After all, what better way to encourage widespread adoption of world-class manufacturing principles and management practices than to illustrate how they have spurred success in real-world situations?

Thus in 1990, *IW* launched what was initially conceived as a "manufacturing excellence" awards program. At the insistence of then editor in chief Chuck Day, who had been looking for a signature campaign to capture the fancy of the magazine's manufacturing readership, the premiere special report carried the headline AMERICA'S BEST PLANTS—a trademark that has been used ever since.

At least one staff editor was slightly apprehensive about the implications of the phrase. With approximately 350,000 manufacturing facilities in the United States, how could we be sure that those we honored were truly the very best? With little hesitation, Chuck replied: "They're the best, if *we say* they are!" Magazine editors often think, or at least talk, like that. But Chuck fully expected that *IW* would make a rigorous effort to identify leading-edge plants that were representative of the nation's best, even if he did not yet realize how intensive the judging and selection process would ultimately become. And he understood that the descriptive term "America's Best" would elevate the level of interest in our program—much the way that the nation's sportswriters fuel interest in college football by picking the nation's "Top 25" gridiron teams.

Still unanswered, however, was the question of *how* we would approach the task. Unlike the Malcolm Baldrige National Quality Award program, we didn't have a $2 million Congressional appropriation to finance our search-and-selection process. (The first year, it took a bit of arm-twisting to persuade a budget-conscious publisher to spring for $600 worth of plaques for the winners.) No new staff was to be added. It was something the editors would have to do in their "spare" time.

The inaugural search lasted from early spring through late August, as we picked the brains of leading manufacturing consultants as well as executives affiliated with AME and other organizations, asking for their recommendations. After visiting a number of the candidate facilities and conducting in-depth phone interviews with others, a dozen exemplary plants were chosen. The first report on "America's Best Plants" appeared in *IW*'s October 15, 1990, issue.

We were rather proud that we'd pulled it off. And shortly afterward it became clear to me that we'd created quite a stir when, en route to another appointment, I decided to stop at the Ford Motor assembly plant in Wixom, Michigan, to personally present a commemorative plaque to the plant manager. I expected a cordial, but low-key reception, and perhaps a photo shoot for the plant newsletter. Instead, I was startled to be greeted by television cameras and a crew waiting to film the presentation ceremony for broadcast on Ford's worldwide TV network. But true appreciation of the impact of the "Best Plants" issue didn't sink in until the following spring. At an AME seminar in Vail, Colorado, someone in the audience mentioned the October report, which prompted another manufacturing executive to reply: "That was a great issue! As I read the stories on each of the winners, I made a checklist of all the things that they were doing and then compared that list with what we were doing or not doing in our plant. It gave me so many ideas that I drafted a whole new manufacturing strategy for our company!"

We knew then that we were on the right track—and that our search for "America's Best" plants had to become an annual event. Also, in keeping with the winners' emphasis on continuous improvement, we decided to explore ways to improve our process.

We suspected that if we limited our field of view to plants recommended by consultants and other experts, we'd end up with similar winners' lists every year and perhaps overlook a host of pretty dramatic, but little-heralded successes across the industrial landscape. Moreover, we had a strong hunch that the folks in the manufacturing trenches who were on their way to becoming world class, if not already there, had some rather intriguing stories to tell. So in 1991, we began to publicize an open call for nominations. The announcements—under the heading IS YOUR PLANT ONE OF AMERICA'S BEST?—included a simple nomination form.

We wanted to hear from people close to the action, the plant managers, quality managers, manufacturing VPs, who are most knowledgeable about the capabilities and performance improvements of the nation's most progressive production operations—from steelmakers and ice-cream manufacturers in the Midwest to electronics and other high-tech organizations in Silicon Valley.

The really good ones, we figured, know who they are. For one thing, they tend to actively benchmark "best practices" employed elsewhere. But a nagging question haunted us: *Would* they come forward and tell us about themselves and share some of the secrets of their success or would they, for competitive reasons, clam up and keep a low profile? We also wondered whether they would take the time to respond to a request for detailed information—the kind of information that we felt our judges would need to make the fairest possible decisions based on an evaluation of comparable data.

An anecdote shared by one of our first-year winners was somehow reassuring. Hank Rogers, quality manager at the Dana Corporation valve manufacturing plant in Minneapolis, recounted a conversation he'd overheard during a benchmarking trip to Japan in the late 1980s. While touring a Japanese plant, a member of the U.S. contingent asked one of the hosts: "Aren't you worried that we'll take the things you've shown us back to the United States and use them to compete against you?" And the Japanese executive replied: "No. We may *show* you what we do, but you won't be able to make it work in *your* plants." The hidden message was that adversarial relationships in American factories would undermine efforts to promote the kind of teamwork necessary to make significant improvements. But Americans *do* know something about teamwork—and they've shown a willingness to learn even more.

Fortunately, from the perspective of the "Best Plants" program, winning teams also appreciate, and deserve, recognition. And the *IndustryWeek* award has come to be highly prized within the manufacturing community, not only for the recognition it conveys, but also for the impact it can have in elevating employee morale and focusing employee effort.

Starting in 1991, the *IndustryWeek* "America's Best Plants" competition adopted a formal selection process that requires candidates to fill out questionnaires dealing with a wide range of topics, including: quality processes, employee involvement, manufacturing flexibility, supplier partnership, customer focus, community involvement, environmental and safety practices, use of technology, and evidence of bottom-line results such as productivity gains and marketplace success. In the years since, we've continued to refine and expand the Best Plants entry form, the latest version of which is included in this book so readers may assess their own standing in the community of world-class manufacturers. It asks for quantitative performance measurements, as well as descriptions of manufacturing approaches, management practices, and improvement initiatives. The questionnaires, along with supporting statements and other materials, are evaluated by a panel of editors in a two-stage judging process. Once the field is narrowed to 25 finalists, each of the finalists then receives an individualized follow-up questionnaire which poses "plant-specific" questions—seeking additional information, clarification of statements made in the initial entry, and documentation of performance data. After reviewing these submissions, the judges select 10 winners, typically in late July, pending final validation during site visits in August. In recent years, several outside experts have participated in evaluating the leading candidates to augment the judgment of the editors.

Admittedly and despite the fact that the entries include quite a bit of quantitative data, the evaluation process is partly subjective. We recognize that numbers can be deceiving, especially when attempting to compare, say, a steelmaking plant to a printed circuit board operation. For example, because some production processes are more complex than others, or require much tighter tolerances, a comparison of first-pass quality yields or order lead times will tend to favor certain types of industries over others. Thus, to the extent possible, the judges attempt to make allowances for differences in the nature of the products or production processes involved. This, in itself, injects an unavoidable subjective element.

If there is any bias to the selection process, it is probably a tendency to weight evidence of dramatic improvement over absolute performance levels. The simple explanation is that improvement rates, such as percent reduction in manufacturing cycle time, provide the most *commonly applicable* measurements among diverse operations.

Before the selections are confirmed, an *IW* editor visits each of the 10 plants to validate that the facilities are, in fact, deserving of recognition. Almost always, we have found the plants and the teams that operate them more impressive in person than they appeared on paper. The experience of getting a firsthand look, as well as a sense of the zeal and commitment-driving improvement efforts, is highly gratifying. It provides reassurance that American managers and workers have lost neither their pride in achievement nor their competitive spirit. And the often contagious enthusiasm for the pursuit of excellence evokes a sense of almost limitless possibility.

The editors of *IndustryWeek* relish the opportunity to cultivate such enthusiasm; and nowhere is it more evident than during the on-site ceremonies that occur when representatives of the magazine visit the winning facilities each fall to present the America's Best Plants awards. Typically, the celebrations are attended by top corporate brass, as well as state and local dignitaries; but their real purpose is to recognize and applaud the achievements of highly motivated managers and employees, and to encourage them to persist in their efforts. Aware that reward and recognition programs can be great morale builders, the management teams often give employees mementos of the occasion, such as baseball caps, T-shirts, coffee mugs, or jackets emblazoned with the "America's Best Plants" logo.

Sometimes, the winners pull out all the stops. Several years ago, a Westinghouse Electric plant in Cleveland erected two circus-size tents on its property, served a hot buffet lunch to some 3,000 employees and guests, hired a Dixieland band to entertain, and brought in U.S. Sen. John Glenn as an added keynote speaker. In 1995, the Siemens Automotive plant in Newport News, Virginia, secured a local high school marching band to kick off its event by marching through the plant, its trumpets and drums rallying workers to fall in step en route to the awards presentation site. And at Rockwell International's defense avionics plant in El Paso, a dramatic highlight was a flyover by a B-1 bomber which tipped its wings in a special salute to the workforce as the ceremony was about to get under way.

As *IW*'s Washington-based Senior Editor Bill Miller once put it, the atmosphere at the "Best Plants" awards-presentation ceremonies is reminiscent of the postgame celebrations in the locker room of the victorious Super Bowl champions.

With time, the fears we may have had about the willingness of outstanding manufacturing plants to share their success stories and to serve as "role models" for others in the common pursuit of excellence have been largely allayed. Sure, there are some excellent candidates out there who still prefer to keep their secrets to themselves. But the number of nominations and submissions of completed entries has grown steadily.

In business, success does breed success, often by inspiring others to strive harder. Anecdotal evidence suggests that *IW*'s "Best Plants" program, by giving visibility to pacesetting efforts, has had just such an effect. One of the best illustrations of that is the experience of the Lockheed Martin Government Electronic Systems plant in Moorestown, New Jersey, which in the late 1980s and early 1990s was troubled by loss of business stemming from defense-budget cuts and by heavy layoffs related to the outsourcing of work. When management and union leaders finally agreed to cooperate in an effort to turn things around, one of their first steps was to jointly attend *IW*'s 1992 "Best Plants" conference in Chicago, where they heard presentations by the 1991 winners. That experience was an eye-opener; and subsequent benchmarking visits to two of the plants helped to solidify a union/management partnership that rather quickly produced impressive results. Within two years, quality and productivity had improved so dramatically that once-outsourced work was being brought back into the Moorestown facility. And in 1994, the *IndustryWeek* judges were pleased to name the New Jersey plant as one of "America's Best."

Because *IW* wants to recognize a variety of outstanding manufacturing organizations, the eligibility rules for the "Best Plants" competition prohibit a winning facility from entering again for a period of five years, but other plants operated by the same company remain eligible. To date, seven highly regarded companies—Ford, General Electric, Varian Associates, The Timken Company, Johnson & Johnson Medical, General Motors, and Emerson Electric—have produced multiple winners.

In a number of companies with a strong commitment to manufacturing excellence, corporate management has encouraged its various plant teams to enter the competition as a means of benchmarking their progress and motivating further improvement. Even nonwinning plants report that they've found the exercise worthwhile as a way to track their performance and establish new goals. In some cases, reviewing the data on improvements over a five-year period has been an eye-opening experience. More than one participant has told us: "Until we collected and studied the information you asked for, we didn't realize how much we'd actually improved."

The Sony Electronics plant in San Diego was one of the 10 winners in 1994. It entered the competition at the urging of Stephen Burke, a manufacturing executive who had recently accepted a transfer and responsibility for business planning at the San Diego complex. Reviewing the completed "Best Plants" entry form, he later confided, helped give him a head start in his new assignment. "Thanks to that exercise," he said, "I learned more about the San Diego operation in one month than I probably would have in six months on the job."

Over the years, the "Best Plants" program has grown to dimensions never envisioned when it was launched in 1990. It now includes a series of conferences each year at which the winners share their success stories in person. Several winners are also featured in a special edition of *IW*'s *Management Today* show on the CNBC cable-TV network. Each year, *IW* staffers sift through the entries submitted by the 25 finalists, extracting data to create a composite Statistical Profile, which manufacturing professionals have found useful as a benchmarking reference. (The latest version is included here.) In 1996, *IW* Associate Editor George Taninecz inaugurated a follow-up study of past winners to evaluate how they've progressed. Also in 1996, the eligibility rules were amended to include facilities in Mexico and Canada, making Best Plants a broad-based "North American" competition. And now, thanks to the talent and vision of Ted Kinni, we're proud to present *America's Best*—the book.

In essence, this book is the culmination of a great deal of work by many people, including the *IW* editorial staff members who've devoted hundreds of hours each year to judging the entries. Veterans on the judging panel include Senior Technology Editor John Teresko, Senior Editor Bill Miller, and Associate Editor Tim Stevens, each of whom has repeatedly volunteered his time and best intellectual effort. A number of others, including Editor-in-Chief John Brandt and Publisher Carl Marino, have also served stints as judges. For several years, Administrative Assistant Joanne Honohan has been compiling candidate lists and coordinating mailing activities. And Associate Editor George Taninecz has assumed much of the burden of preparing entry materials, creating the annual statistical profile, and other administrative responsibilities. Many other staff members have also made important contributions. Keeping the "Best Plants" program going and growing has truly been a team effort.

A final observation: Among the frustrations of a business journalist—or any journalist, for that matter—is that it is impossible to fully appreciate the impact that a publication has on the audience it serves. But we, the editors of *IndustryWeek,* like to believe that our magazine has played a meaningful role in elevating the status of American manufacturing. So when, in 1994 and 1995, the World Economic Forum in Switzerland declared that the United States had reclaimed its former status as the most competitive industrial economy on the planet, we took a quiet pride in having encouraged, supported—and, hopefully, accelerated—that comeback.

By drawing attention to the latent competitive strength that can be unleashed when leading-edge manufacturing practices are coupled with down-to-earth common sense, *IW*'s annual "America's Best Plants" competition has, we believe, helped to foster the nation's manufacturing renaissance. At the very least, it has spread the gospel of world-class manufacturing. Without exception, the winners have been plants whose insightful managers and motivated production workers have learned, through brilliance of insight as well as trial-and-error discovery, how to put all the pieces together.

We hope this book will help you and your organization to follow in their footsteps.

John H. Sheridan
Senior Editor
IndustryWeek

SECTION I

MANUFACTURING EXCELLENCE: THE LESSONS OF AMERICA'S BEST

CHAPTER 1

BECOMING A BEST PLANT

Trying to arrange the America's Best Plants award winners into a coherent whole by organizing them into groups based on number of employees, plant size, annual sales, or industry is like trying to complete a jigsaw puzzle with pieces from 62 different puzzle sets. The winners seem to manufacture everything under the sun; workforces range from under one hundred to thousands of people; physical plants vary from buildings not much larger than a luxury home to millions of square feet; and annual sales volumes run from a few million to billions of dollars. There is no way to put together the Best Plants puzzle using these traditional categories.

But take a closer look at the pieces. When Best Plants describe themselves, there are words that pop up over and over: words such as *customer, quality,* and *responsiveness.* When the puzzle pieces are arranged by these words, they start to fall into place. The Best Plants puzzle does connect. It is joined at a deeper level by operational intent, core strategies, and supporting competencies. When the plants are examined in this light, a window factory can begin to look a lot like a weapons systems manufacturer, a car company has a lot in common with an ice-cream maker. Their products vary and the plants aren't physically alike, but Best Plants do think and act in similar ways.

The first section of this book explores nine major areas of similarity between the Best Plants. They are organized into three core strategies—customer focus, quality, and agility—that form the foundation for Best Plants performance and six supporting competencies—employee involvement, supply management, technology, product development, environmental responsibility and safety, and corporate citizenship—that help plants achieve the goals defined in the three core strategies. Together, they comprise a holistic view of manufacturing excellence.

Together is an important word. Best Plants repeatedly demonstrate that there is no alchemist's formula for an instantaneous transformation to world-class manufacturing. Rather, that transformation is a long-term (read "years") evolutionary effort. It demands a tremendous commitment from managers and every other employee. It also requires an investment in organizational resources and often means radical (read "painful") change in the culture and structure of the workplace. If that sounds too tough, so be it. To understate the effort required to create a world-class organization would only understate the accomplishments of the Best Plants and mislead those who would undertake a similar journey.

Lest this seem too discouraging, it is important to note that while the journey to manufacturing excellence may be hard work, it is also good and worthy work. There will be plenty of milestones to celebrate along the way and, as with any other faithfully pursued improvement program, the ensuing rewards will eventually begin to arrive with greater and greater momentum. Many of the Best Plants have recorded difficult early years and then, as their efforts begin to mature, quick and seemingly easy successes. It is much like building a house: The tough work of erecting a square and plumb frame provides a superior platform for the eye-pleasing finish carpentry.

Three Core Strategies

The three core strategies of Best Plants are simultaneously philosophical values and tangible goals. As we shall see, when they are properly harnessed, they add value to the plants in myriad ways. Together, the core strategies form a foundational triumvirate upon which a plant can build a structure for excellence. The three core strategies of customer focus, quality, and agility appear in every Best Plants application and are essential for long-term success.

Customer focus, the subject of Chapter 2, has always been a primary requirement of business. After all, if you aren't satisfying your customers, it doesn't seem possible to stay in business. Yet recent history proves that the ideas that customers should occupy the very center of a manufacturing operation and that the customer's desires and needs must be the first litmus test of any venture are easily overlooked. In fact, there are one or two stories among the Best Plants that begin with a company that has lost its focus on customers. But none still make this mistake—at the Best Plants, the customer is king.

Quality, the subject of Chapter 3, is the second core strategy and one that manufacturers internationally have been grappling with for several decades. The concept of quality has many applications and offers many benefits. In *America's Best*, when we talk about quality, we are talking about the defect-free, waste-free workmanship that derives from product and process quality. There are many broader definitions of quality that are also valid. For example, a quality perspective can also be applied to strategy, service, suppliers, and product development. In this book, those applications of quality are included in the chapters devoted to those topics.

Agility, the subject of Chapter 4, is probably least recognized of the three core strategies. If you canvassed the Best Plants and asked each whether agility was a strategy they practiced, there would probably be a fair number of quizzical looks. Yet, once defined, every one of the plants would instantly understand themselves as agile organizations. Broadly, *agility* is the ability of a plant to quickly, efficiently, and effectively respond to change. When we talk about agility we mean it to encompass lean manufacturing and just-in-time supply practices pioneered by the Japanese, flexible production structures, fast, clean-slate process redesign as advocated in reengineering, mass customization production strategies, and more. All the Best Plants have been building agility into their operations for years; they just haven't been calling it by that name until recently.

Six Supporting Competencies

The six supporting competencies of Best Plants assist in the achievement of the three core strategies. All six do not appear in every plant, and each plant places a differing degree of emphasis on each of them, but all are practiced in a great majority of the plants. The supporting competencies recognize that a manufacturing plant is more than simply a production line that converts raw material into finished goods. It is a living system that has impact far beyond its walls. Best Plants are also the people who work in them and the vendors who supply them. They are technologies that support the operation and the new products they create. And they have an effect on the environment and communities in which they operate.

This system must be tuned and balanced. Best Plants show that systems can become much more than the simple sum of their parts. In fact, it may not be possible to achieve an intense customer focus, a high-quality product, and agile plant without utilizing and integrating the six supporting competencies. Can you, for example, fulfill your customers' needs and desires in a timely, profitable way without an empowered and involved workforce and supplier partnerships and up-to-date information technology? Most Best Plants would answer no.

Employee involvement, the subject of Chapter 5, is a well-practiced supporting competency at every one of the Best Plants. All are using team-based structures to some degree; many have implemented self-managed work teams that are fully responsible for day-to-day operations. People are recognized by Best Plants as *appreciating* assets, and Best Plants are not afraid to invest heavily in their employees. This emphasis on people is a tremendous help in achieving the Best Plants' core strategies. They are able to mobilize the power of the entire workforce on product quality, plant agility, and their customers.

With supply management, the subject of Chapter 6, Best Plants extend the tenets of employee involvement down the supply chain. These plants are looking far beyond the simple invoiced cost of raw materials and are considering the total cost of working with a particular supply partner. When they calculate the costs of inventory storage and handling, poor parts quality, and missed shipments, Best Plants realize that the actual purchase price of the materials used in their products was sometimes only a small component of their real cost. They also understand that their suppliers can and should play an important role in functions that were once considered proprietary, such as product design and customer service. As with employees, the cooperative involvement of suppliers helps Best Plants achieve their core strategies.

The ability to adapt and implement the latest technology to support the drive to agility, quality, and customer satisfaction is the supporting competency addressed in Chapter 7. Best Plants apply the latest technologies throughout their operations. Electronic communications networks connect Best Plants to suppliers and customers; real-time manufacturing data arrives before the product itself; automated production systems reduce process variation; and computerized design tools build quality and manufacturing efficiency into new products long before they physically exist. The proper use of technological advances enables the Best Plants to maintain their drive toward world-class status.

The fourth supporting competency, the design of new products and the continuous improvement of existing products, is detailed in Chapter 8. Best Plants fully understand that new products, short development cycles, and low product costs are critical factors in their long-term success. To achieve these goals, the plants utilize fast-paced design processes that harness the expertise of their employees, suppliers, and customers to define and create new products. They also design products with an eye toward total cost; Best Plants' products are designed with quality and manufacturing efficiency in mind.

Environmental responsibility and workplace safety are the joint topics of Chapter 9. Best Plants know that everyone who shares this earth is to some degree a stakeholder in their business. They also know that waste in any form is a cost that they cannot afford. As a result, the plants actively and creatively seek waste-reduction solutions. In doing so, Best Plants help maintain a healthy planet and, at the same time, save money. Their concern for the environmental impact of their actions translates directly into a concern for all people, including their employees. Best Plants know that the safety of their employees is one of their greatest responsibilities; they also know that the cost of ignoring that responsibility might well be greater than they can pay.

Finally, Best Plants recognize their responsibilities in the community at large. Corporate citizenship, the topic of the final chapter in Section 1, examines the role and impact Best Plants have on the local, regional, and national communities in which they operate. In many cases, these plants are the largest employers in their area and are depended on to support the social well-being of their communities. Best Plants do not shirk this responsibility.

Uniting the Nine Components

The primary components of Best Plants described above are not an innate body of knowledge that is mysteriously imprinted on the subconscious minds of a plant's employees the first day they arrive at work. Rather, they are unified statements and plans developed with the guidance of senior management. The discipline that drives this process is called *strategic planning*.

It is fitting that strategic planning should be labeled a discipline because it takes a good deal of discipline to spend the time and effort required to develop and deploy plans. Planning is not a "sexy" part of business, and it is natural to feel that using the time spent *planning* goals to actually *work toward* them would speed progress. This might in fact be true for the simple goals of individuals, and there *is* such a thing as overplanning. However, in an enterprise as complex as a manufacturing plant, *no* planning simply translates into no common direction, no goals, and no future. Best Plants got where they are by planning their route. They didn't always accomplish exactly what they planned, but they adjusted their plans and started again.

The planning process used by Best Plants often incorporates the thinking of many people, but it is always conducted under the direct auspices of senior management. The most important job of plant leaders is this setting of direction. Each of the following chapters in this section could start with a long passage on how this or that topic requires the commitment and support of senior leadership. Logically, however, it need only be noted here. If a plant's management develops and deploys its plans faithfully, the necessary amounts of commitment and support for each of the components of Best Plants success will already be built into the process.

A Common Intention

The first and best lesson yielded by the study of America's Best Plants is that reaching a world-class level of manufacturing performance is a conscious effort.

It wasn't luck or some fateful fluke that placed these companies at the top of their industries. It was a shared intention to excel that was adopted with commitment, rigorously planned, and unceasingly pursued. Best Plants' managements know what they are about. Their employees don't come to work every day and unthinkingly do the same thing they did the day before. Everyone is working toward a common end. They understand and take responsibility for their goals and they are extremely articulate about their values and direction. Best Plants establish and begin communicating a common intent by using a variety of vision, mission, and values statements.

Listen to John Morrison, director of Operations at Johnson & Johnson's 1993 award-winning El Paso, Texas, plant:

> *All world-class companies have a vision, sets of goals, priorities, and missions that define their purpose and guide the course of every workday. Johnson & Johnson Medical Inc. (JJMI) relies on the Johnson & Johnson Credo, now over 50 years old, and a JJMI Mission Statement: "Superior Responsiveness to Customer Requirements." [This is] supported by the JJMI Critical Success Factors: Customer Driven Quality, Fast and Flexible Processes, Lowest Cost, and Total Associate Involvement. . . . Additionally, we have our quality, safety, and environmental policies. Finally, we have departmental goals, policies, and SOPs. These principles are constantly reinforced to each JJMI associate. . . .*

Johnson & Johnson's credo is a statement of corporate and cultural values that acts as a guide for its employees. Given the incredible amount of change that has occurred over the past five decades, this statement remains surprisingly relevant. Customer service, quality, agility, employees, supply-chain relations, product innovation, safety, and citizenship—all these ideas are specified in the credo and all are components of today's Best Plants. Together, they provide the employees of the El Paso plant with a set of guidelines they can share.

A company need not have the long history of a Johnson & Johnson to establish a common intent. XEL Communications, an independent company for just over 10 years, offers an equally compelling statement:

> *XEL will become the leader* in our selected telecommunications markets through innovation in products and services. Every XEL product and service will be rated Number One by our customers.
>
> *XEL will set the standards* by which our competitors are judged. We will be the best, most innovative, responsive designer, manufacturer and

A Values Road Map: Johnson & Johnson's Corporate Credo

We believe our first responsibility is to the doctors, nurses and patients,
to mothers and all others who use our products and services.
In meeting their needs everything we do must be of high quality.
We must constantly strive to reduce our costs
in order to maintain reasonable prices.
Customers' orders must be serviced promptly and accurately.
Our suppliers and distributors must have an opportunity
to make a fair profit.

We are responsible to our employees,
the men and women who work with us throughout the world.
Everyone must be considered as an individual.
We must respect their dignity and recognize their merit.
They must have a sense of security in their jobs.
Compensation must be fair and adequate,
and working conditions clean, orderly and safe.
Employees must feel free to make suggestions and complaints.
There must be equal opportunity for employment, development
and advancement for those qualified.
We must provide competent management,
and their actions must be just and ethical.

We are responsible to the communities in which we live and work
and to the world community as well.
We must be good citizens—support good works and charities
and bear our fair share of taxes.
We must encourage civic improvements and better health and education.
We must maintain in good order
the property we are privileged to use,
protecting the environment and natural resources.

Our final responsibility is to our stockholders.
Business must make a sound profit.
We must experiment with new ideas.
Research must be carried on, innovative programs developed
and mistakes paid for.
New equipment must be purchased, new facilities provided
and new products launched.
Reserves must be created to provide for adverse times.
When we operate according to these principles,
the stockholders should realize a fair return.

provider of quality products and services as seen by customers, employees, competitors, and suppliers.

We will insist upon the highest quality from everyone in every task.

We will be an organization where each of us is a self-manager who will:

- initiate action, commit to, and act responsibly in achieving objectives
- be responsible for XEL's performance
- be responsible for the quality of individual and team output
- invite team members to contribute based on experience, knowledge and ability

We will:

- be ethical and honest in all relationships
- build an environment where creativity and risk taking is promoted
- provide challenging and satisfying work
- ensure a climate of dignity and respect for all
- rely on interdepartmental teamwork, communications and cooperative problem solving to attain common goals
- offer opportunities for professional and personal growth
- recognize and reward individual contribution and achievement
- provide tools and services to enhance productivity
- maintain a safe and healthy work environment

XEL will be profitable and will grow in order to provide both a return to our investors and rewards to our team members.

XEL will be an exciting and enjoyable place to work while we achieve success.

Like Johnson & Johnson's credo, XEL's vision statement provides a fixed beacon for its employees. It is also a work-in-progress. The company explains that two points in their vision—responsiveness to customers' product and delivery needs and the need to overcome internal functional barriers to communication and cooperation—were added at a later date. That brings up a good point: The establishment of corporate intent, especially statements of values, should be a relatively constant source of guidance that employees, suppliers, and customers can depend on, but it need not remain forever fixed. Intentions evolve with time and changing environments, and if the corporate intent no longer reflects the world in which its employees must operate, companies need to modify their guidelines to reflect these changes.

The development process for intention statements, such as visions, missions, and values, varies among Best Plants. Often, they are the product of off-site retreats attended by the plant's management or a cross-functional group that includes participants from all levels of the business. Regardless of how the statements originate, they must be presented to and accepted by employees in order to provide value to a company. It isn't unusual to find corporate visions and missions hanging in the reception areas of companies of all kinds, but finding these same statements in use on a day-to-day basis is much less common an experience.

At the Chanhassen, Minnesota–based Fisher-Rosemount plant, statements of intent were developed in June 1991 with the help of a 36-member management team during an off-site meeting. The team started with parent company Emerson Electric's five-point "Best Cost Producer" strategy which was drafted in the early 1980s and includes a commitment to quality and customers, focused manufacturing strategies that compete on process as well as design, employee involvement, continuous cost improvement, and a commitment to provide the resources needed to support the strategy. To this, the team added its own vision, mission, and values statements.

A three-phase approach was used in communicating these new statements to its employees. First, within 48 hours of their return, the managers who helped develop the statements introduced them to employees in small groups. Next, every employee received a memo from the company's general manager that included a pocket-size version of the statements for their personal reference. And finally, senior management presented the new statements during a meeting of the entire workforce. Rollout complete, discussions led by senior managers continue during quarterly all-employee meetings, and annual reviews of the statements are conducted by a broad cross section of the organization.

Even with this thoughtful approach, Rosemount's guiding principles needed a significant revision. "It became clear as these statements were being used in making decisions that a key element was missing. The original vision, mission, and values did not include shareholders as long-term business partners," explains Operations Manager Amy Johnson. "To have credibility with the organization, it was important for the management team to formally acknowledge that Rosemount Chanhassen is in business to make money for its shareholders.

The overall well-being of the company is best served when the needs of customers, suppliers, employees, and shareholders are balanced. This relationship has now been included in each of the vision, mission, and value statements."

At Siemens' 1995 Best Plants award-winning Newport News, Virginia, plant, the mission statement was jointly developed by teams of management and union employees, and almost all employees voluntarily signed the statement. "An interesting debate between the employees emerged during the drafting of the Mission Statement as to whether People or Quality should come first," remembers marketing manager Sam Byrd. "The result is that they appear at the same level on the Statement. The result of all of this collaboration was that the Mission Statement became the vehicle for gaining consensus among the employees as we moved towards our shared vision." (Established in 1988, the vision of the business was to move up two slots to become the world's second-largest independent fuel injector manufacturer—a goal they expected to reach in the same year the plant won its award.)

Internal debates over the content of statements of intent should be expected and welcomed. Best Plants accept them as an indication that people are truly and passionately interested in the results of the work. However, for those who despair of ever reaching a consensus, XEL's Malcolm Shaw offers some good advice: "You will never satisfy all the people all the time. Don't be tempted to delay change because 100 percent of the company and your colleagues are not committed to your vision. The grass really is greener on the other side, and even those who hate the idea of change will get used to it."

Turning Intentions into Reality

Statements of intent impose a system of beliefs and ambitious goals on a corporate culture, but they usually do not specify an exact route for achieving their goals. Best Plants move their dreams closer to reality by developing strategic plans, which are intimately connected to and integrated with intent statements, and by deploying those plans down to the daily operating goals of the operation. Strategic plans, which usually include both long-range goals and shorter-term tactical requirements, vary widely in time frame. In any case, they inform employees how the plant will move toward its vision. These plans are actionable. They represent the blueprint that the plant will follow to realize its intentions.

The establishment and deployment of the annual business plan at MEMC Electronic Materials' Saint Peters, Missouri, plant offers a fine illustration of the transition from intent to reality that the Best Plants make in order to achieve their dreams. Management at the MEMC plant reviews and modifies their ongoing plant strategy at an off-site meeting each November. The result is a goals document that is expressed in terms of the plant's 4Cs. The 4Cs stand for Customer, Culture, Cost, and Citizenship, and they represent the key drivers in the plant's effort to become the best silicon wafer producer in the world. They also give the facility a structure on which to hang its goals. When management decides how to apply the plant's resources and capital investment funds, they assign expenditures in terms of the 4C framework.

The deployment of each year's updated business plan is well structured. The engineers, operational supervisors, and coaches are introduced to it in a December meeting, which is hosted by the plant manager and department heads. Afterward, supervisors use the plan to guide the establishment of their local goals, and coaches use it to assist the plant's empowered work teams in setting their objectives. By this process, alignment and linkage to the strategic plan is ensured throughout the plant. Once set and accepted, the management staff reviews progress against the plan's goals, and the plant manager reviews the collective progress with all employees in each ensuing fiscal quarter.

Strategy deployment also has a day-to-day component at MEMC. The plant management uses bulletin boards, company newsletters, electronic message boards, signs, E-mail, and video networks to keep everyone on the same track. Particularly creative is the plant's network of fax machines, which ensures that teams receive news quickly. All the messages on the network are labeled with one of the 4Cs, so they can be easily understood in relation to the plant's strategic goals. Half-hour team meetings scheduled at the beginning of each shift offer another opportunity to reiterate the common direction.

The collective power created by a stable strategy is clear at Varian Associates. The company boasts two Best Plants award winners, one in 1992 and another in 1994. Both units separately reported building their plans upon the same common foundation: the corporate Operational Excellence (OE) program. OE is composed of five core elements: total customer focus; a commitment to quality; flexible, responsive factories; a fast product development process; and organizational excellence as typified by communication, teamwork, and training. (See Varian Oncology Systems' Individual Plant Profile for more background.)

The 1994 award-winning NMRI unit interweaves the OE initiatives into its strategic plan, which with the participation of all the business unit's employees is developed into an annual operating plan (AOP). The AOP is organized by function. In manufacturing, for instance, just-in-time (JIT) action teams each establish their own quarterly goals and objectives based on the manufacturing AOP and are overseen by a steering committee. Once the AOP is complete and accepted, management tracks progress toward its goals on a monthly basis by monitoring functional activity reports, reviews it quarterly at strategic planning meetings, and semiannually presents their results to corporate management.

The successful deployment of strategic plans requires that concrete goals be set. At Milwaukee Electric Tool Corporation (METCO), the planning process gets more specific as it flows from its vision to mission to financial goals and, finally, to operating objectives. The company's vision and mission specify its intent: a vision that includes "Excellence in Quality, Performance, Value," and eight mission goals ranging from innovative products to ethical behavior. Once the intent is stated, the company commits to quantitative goals and objectives that are designed to support the qualitative goals of its vision and mission.

METCO's statement of financial goals specifies the expected percentages in operating margins, sales growth, and return on capital. Operating objectives specify sales per employee, hours of training per employee, percentage of sales to be generated by new products, and so on. Even a percentage of net income to be invested "in social, cultural, educational, and charitable activities" is set. These goals are updated annually and presented in a single-page document to all employees.

Evolving a Strategic Plan

Creating world-class manufacturing operations is an evolutionary process. Fully implementing the nine components of world-class performance requires substantial efforts and, often, significant organizational change. Best Plants undertake these challenges in manageable chunks, and their strategic plans integrate new components as their capabilities grow.

The Best Plants' strategies are also integrated in the sense that eventually all the components of excellence reside there. Where in the past a plant's financial objectives often occupied a higher position in the priorities of management than perhaps quality or employee issues, all are often present in a single plan in today's plants. By putting all the pieces together, Best Plants are better able to develop the connections between each core strategy and the supporting competencies. More important still, they can begin to realize the synergies between these components.

One important consideration: Even given the strong common themes, especially the three core strategies, that all Best Plants share, one plant's individual journey to excellence is never exactly the same as another's. Business models, structures, and culture vary widely among the plants; therefore, plants must adapt ideas and concepts to fit their particular circumstances. In other words,

Best Plants don't buy success off-the-rack. Their strategic plans reflect the individual characteristics that are unique to their companies.

Honeywell's Industrial Automation & Control (IAC) unit described over a decade of strategic evolution in their Best Plants application package. In 1981, management adopted a product-focused team strategy, which they struggled with "as the team design contained some weaknesses." In 1983, a second team-based product effort was initiated. Building on the lessons of the first effort, this team recorded a 25 percent improvement in product quality and a 30 percent reduction in cycle time. With the success of this second pilot, work teams were established throughout the production area. "Completed in two phases over a three-year period, our manufacturing process and organization changed from being process-focused with a functional organization to a system of focused factories, where each work team is a business within the business," remembers manager Gale Kristof.

In 1985, IAC brought their suppliers into the strategic mix with the Supplier Alliance process. Under this initiative, 44 suppliers would eventually account for almost half of the plant's production materials. A supplier stocking plan that delivers material directly to the factory floor, a resident supplier program, and supplier participation in the product development process were also established.

In 1987, IAC's strategic evolution continued as the company established an all-salaried workforce that featured the same pay program, benefits, and work rules for all employees. And, in 1988, management began to integrate quality into their manufacturing process with a quality-at-the-source program that gave employees full responsibility for the quality of their work. The result, says Kristof, "a fourfold improvement in quality, a 70 percent reduction in cycle time, 98+ percent success in meeting customer delivery commitments, and cost savings totaling over $20 million. That same year, the plant established a dedicated customer-satisfaction organization, whose mission was "to provide a competitive advantage for IAC through continuous improvement in all facets of the business in order to ensure customer delight with all products and services."

As the plant's strategy expanded, so did its appetite for success. In 1990, a major strategic initiative to attain world-class status was introduced at the plant. Says Kristof, "At the time, we were already recognized by the industrial control customer base as having the highest-quality hardware of any industrial control systems manufacturer [based on a *Control* survey]. But being Number One in 1990 was not good enough. In an industry where the hurdle is always rising, we knew that improvement would be required to remain at the top. Senior manufacturing management thus began an active program to develop strategies to achieve 'world-class' status." Management designed and adopted a three-year strategy to attain world-class status. The strategy specified stretch goals that forced the company to rethink, instead of simply improve, its processes. (See Honeywell's Individual Plant Profile for more information.)

When IAC won its Best Plants award, Kristof was able to report:

> *Three years after initiating our World Class Manufacturing program, we see our manufacturing process in a much different way than ever before. Suppliers and customers are integral to our planning process. Teams map their process to understand it, eliminate non-value-adding activities to simplify it and develop fail-safing techniques for process stability to make it automatic. Managers cooperate as a self-managed management team, taking responsibility and accountability as a team for the success of the product line and plan the changes required to meet our aggressive quality, cycle time and materials investment goals. By having a vision of what we could become, creating a plan to achieve it, setting aggressive goals, developing our own process for achieving those goals, empowering people with the skills and knowledge to contribute, using the talents of the work teams, and making it safe for people to experiment and take risks, we have accelerated our progress to making world-class a reality and not just a slogan.*

The growth in strategy described above well illustrates the synergies among the components of the Best Plants methodology. At IAC, one phase of growth supports another. Streamlined product design leads to teamwork, teamwork to focused factories, focused factories to supply management, and so on. The end result is a fully implemented, world-class strategic plan.

Another producer of industrial control systems, 1992 award winner The Foxboro Company, also pursued its vision of a world-class operation in an evolutionary fashion. The company's strategic plan is built around a series of major policy deployments, the latest of which focused on three strategic goals: to provide the highest-quality hardware; to develop the fastest response to customer demands; and to develop and maintain the lowest cost/function ratio in the marketplace.

These deployments have resulted in the formation of an operating platform "that does not change but serves as the basis for our continual improvement programs," reports production manager V. S. Venkataraman. The Foxboro platform is based on continuous quality improvement, a paperless JIT schedule, employee involvement, supply alliances, and other world-class programs.

Baxter's North Cove Plant Builds a Picture-Perfect Plan

Baxter Healthcare's IV Systems plant was attracting attention for its strategic Quality Leadership Process (QLP) long before it earned its 1994 Best Plants award. University of North Carolina's Dr. Aleda Roth identified QLP as a model for TQM that is capable of generating world-class results and customers. Other Baxter units, as well as companies outside the industry, have studied the system and benchmarked their own strategic processes against it. QLP, however, did not spring fully formed from the imagination of North Cove's management team, nor was it embraced overnight.

"[QLP] institutionalized a new participative style of management for this facility," remembers IE manager John Martino, "giving understanding to the term *quality* and a new way for each one of us to feel a part of our facility. Everyone went through many hours of training in 1985–1987, and we were skeptical of another program that would shortly end. . . . The response to get involved was slow."

This first incarnation of QLP focused solely on product quality and established employee teams to identify and solve the causes of production defects. During the same period, but in separate efforts, the plant was establishing JIT pilot projects introducing kanbans and cellular production, operator-based TPM programs, and an integrated quality plan. Again, Martino reports support was slow to develop . . . until "these pilots were successful and the skepticism turned into, 'when can I get involved?' "

It wasn't until 1988 that management begin consolidating their strategies into a single plan and all these efforts were brought under the auspices of QLP. The next year, they went the final step and joined QLP to the plant's business plan. That action marked the formation of a comprehensive strategic plan and deployment process. QLP, which is depicted as a Parthenon-like building (and is detailed in the Individual Plant Profiles), now includes 11 manufacturing strategies that are rigorously pursued in an effort to reach the plant's Vision 2000 statement.

QLP includes a deployment process that starts with five-year and annual plans. The plans provide the practical details of how the plant's goals will be accomplished. "All strategy implementations," explains Martino, "require extensive training, management meeting with employees to explain why a change is necessary, allowing employees to participate in the design and implementation, and providing a means of feedback to everyone."

Gilbarco Incorporated's manufacturing strategy manager Jerry Smith concisely describes his company's incremental approach to strategy as a similar journey:

> *The original Gilbarco manufacturing strategy, named CRISP for Continuous Rapid Improvement in the System of Production, was presented to the entire organization in early 1986, marking the beginning of the journey to manufacturing excellence. The original strategy was somewhat narrow in focus and consisted of four points: Total Quality Control; Just-In-Time; Focused Factories; and Total Preventive Maintenance.*
>
> *In the spirit of continuous improvement, the CRISP focus was broadened to include design engineering in 1988, employee involvement in 1989, Total Quality Management in 1990, and Total Productive Maintenance in 1993. The six points of the strategy today are: Total Quality Management; Total Productive Maintenance; Focused Factories; Concurrent Engineering; Just-in-Time; and Total Employee Involvement.*

North Cove's final formative step, the integration of financial goals and operational objectives into a single strategic plan, is not practiced by all plants, but it does seem to confer a certain intensity of focus upon those plants that choose to work from one plan. Baldrige Award–winning Marlow Industries describes itself as a total quality–managed company dedicated to total customer satisfaction and continuous improvement because TQM has been the driving factor in its manufacturing operations. As manager Ed Burke explains, "To us, total quality means full involvement of our customers, employees, management, and suppliers. We have no quality plan, only a single business plan embodying all our goals. We do not distinguish between quality and business goals . . . they are in fact the same. We have greatly simplified and clarified our TQM vision and goals by involving our employees in developing annual five-year strategic business plans which effectively combine all quality, operational, financial, human resources, training, and capital goals into one focused business plan."

Summary

The raw materials of America's Best Plants are three core strategies and six supporting competencies. The core strategies—customer focus, quality, and agility—require the development of a customer-driven business that can react quickly to changing conditions with high-quality products. The six supporting competencies—employee involvement, supply management, technology, product development, environmental stewardship and workplace safety, and corporate citizenship—are the skills that Best Plants develop and utilize to help achieve their core strategies.

Best Plants achieved their status by creating a common intent: a series of shared goals that everyone in the plant can embrace and pursue. Their common intent is made real by the development of detailed strategic plans, and these plans are achieved by structured deployments. This process is a never ending journey—world-class performance is a moving target that requires constant attention and effort.

You usually get what you expect. Set high expectations.

—Baxter Healthcare IV Systems Division

CHAPTER 2

CUSTOMER FOCUS

America's Best Plants confirm one of the fundamental realities of contemporary business: The customer is king. Best Plants know that satisfying customers is their most essential mandate. In response, they create and implement visions and strategies that are powered by customer needs and demands; they build organizational structures that support an intensive focus on customers; they create a high degree of intimacy and numerous connecting bonds to their customers; and they measure and continuously improve their ability to serve customers. In a very literal sense, customers drive many of these operations.

This realization and response is a far cry from the realities of the factory floor even two decades ago. Before the Customer Revolution, production was king. A plant's value was measured by its capacity, and customers took what was offered. If costs rose, companies simply increased the price of the product. Many plants, especially those that had only indirect connections to design, sale, and use of the products they built, seemed to exist in a "customerless" environment. Their only job, it seemed, was to keep finished goods rolling off the production lines.

Today's reality is much different, and customer focus is the first of the three core strategies of Best Plants. The award winners have extended the label "customer" far beyond any traditional definition. No longer is the customer simply the immediate purchaser of the product. Customers are also colleagues waiting to assemble a part into the final product; they are suppliers awaiting timely payment of invoices; and they are the end users of a product that is sometimes created, marketed, and sold three or more levels removed from the original manufacturer. At some time or another, everyone is a customer in a Best Plant.

Customer Focus Is a Corporate Value

Establishing customer focus as a core strategy in an organization begins at the foundations of the operation. At Best Plants, the drive for world-class customer satisfaction and service levels is repeatedly identified in vision, mission, and values statements. Starting with these broad statements of intent, customer focus is deployed throughout the business. It is addressed in strategic plans, built into the structure of the organization, and finally filtered into the day-to-day goals of every working unit of the operation.

The Foxboro Company's 1992 Best Plants award-winning Intelligent Automation plant defines its customers simply and elegantly: A customer is anyone who receives a product, service, or information from you. That statement doubles as the operating philosophy of the Foxboro, Massachusetts–based plant and is nurtured and supported with quantitative measures and aggressive goals. The company continuously measures on-time delivery, repair response, and defect resolution to help pinpoint how well customer needs are being fulfilled.

In Aurora, Colorado, 1995 Best Plant winner XEL Communications describes the value placed on customers succinctly and, at the same time, in broad strokes:

> *Instilling an attitude of top-class customer service is mandatory. Every person on every team has to understand that customers are the people you see every time you look at a teammate. In fact, we are all simultaneously customers and suppliers. Our final product is not a top-quality printed circuit board. It is a customer who has bought the board, paid for it, used it, is delighted with every aspect of it, and who wants to give us another problem to solve.*

This maker of telecommunication components drives its customer service "attitude" throughout the operation. Reduced cycle times, JIT inventory control and manufacturing, product development, computer-based production aids (such as CAD, CIM, and SPC), supply-chain management, staff training, and planning efforts are all identified not as means to increase efficiency or profit, but as ways to solve internal and external customer problems. To be sure, increased operating efficiencies and profits are by-products of these efforts, but the main focus at XEL remains on the customer. In this way, the company ensures that the rationalization behind improvement decision making continues to originate directly from the needs of the customer.

In logical progression, XEL measures its improvement efforts from the customer's viewpoint. Improvements are documented and refined through the use of periodical customer-satisfaction surveys, visits by company officers to customer purchasing points, and event-driven customer service functions, such as the warranty service activity and 24-hour customer service hotlines. Interestingly, XEL's representatives across the country are often former employees of the companies they serve. The company makes a policy of actively recruiting them because they "are better at finding out and anticipating needs, and have the confidence of customer departments in short-circuiting problems as soon as, or before, they arise." The result, according to the company: "We maintain that by [these] means, and the fact that products and services which delight our customers are a concept which is broadcast as a corporate goal to all company employees, we have not lost a customer in the history of the company."

Sony's San Diego Manufacturing Center, a 1994 Best Plants award winner and a major manufacturer of picture tubes for televisions and computer monitors, plans its customer focus using a formal program called CS100. "CS" stands for Customer Satisfaction, the numeral one signifies the Customer Is First, the two zeros signify the goals of Zero Defects and Zero Complaints. CS100 includes a regularly scheduled day for all employees to meet with customers and review customer feedback. Every employee also receives regular updates on the plant's CS100 status.

The goal of opening as many lines of communication to customers as possible figures prominently in many of the Best Plants' customer-focus strategies. The Timken Company's 1994 award-winning Faircrest Steel Plant calls communication "the heart of our customer satisfaction program." In order to create a constant and proactive flow of information from customers, the plant establishes direct communication links to its customers at as many different levels to as possible. Field sales engineers become so involved in customers' operations that in addition to collecting and transmitting customer requirements back to Faircrest, they sometimes even serve as a communications link inside their customer's business. Information regarding product quality is transferred directly from the customer to the plant's quality control department.

Fisher-Rosemount's 1993 Best Plants award-winning plant, which manufactures pressure transmitters in Chanhassen, Minnesota, built three levels of customer communication into its operational plan in order to better understand and anticipate its customers' needs and desires. At the highest level—Strategic Customer Interaction—a formal interchange for the purposes of strategic planning and customer focus include: a minimum of four executive visits per year with customers, suppliers, and Rosemount sales offices; alliances in which teams from Rosemount and its customers meet quarterly to discuss short- and long-term issues; and outside-consultant assessment based on analyses of customer input and market data. The second level of contact—Daily Customer Interaction—includes all the communication between the plant and customers that occurs in the normal course of business. Here, various contact reports capture frequent customer feedback, which is tracked and reported to the plant's management. The final level—Customer-Initiated Contact—provides cus-

Marlow Industries Starts with the Customer

Marlow Industries made a name for itself well before earning its Best Plants trophy in 1993. This manufacturer of thermoelectric coolers, solid-state devices that control the temperature of electronic components and equipment, is one of only a handful of small companies to win the Malcolm Baldrige National Quality Award (1991). A well-known adherent of Total Quality Management, Marlow pays particular attention to TQM's emphasis on the customer.

"The key to our success has been our emphasis on customer satisfaction," explains assistant vice president Ed Burke. "This has been accomplished through long-term customer alliances and close, direct, custom engineering/manufacturing with exceptional service support. As an example, we have no U.S. field sales personnel. To ensure that we truly build a close, personal, long-term relationship with our customers, marketing, engineering, and manufacturing personnel work directly with the customer to establish and implement the specific requirements, as contrasted by just selling a product. This approach has developed strong, long-term partnering with our customers."

tomers with fast, easy access into the organization through its National Response Center, a 24-hour-per-day, 7-day-per-week toll-free telephone center.

Building In the Customer

Adopting a customer-centered mind-set as a corporate value is the first step in a customer-driven plant. Next, Best Plants build that value into their organizational structures. They often accomplish this by reorganizing their physical operation around customer groups and specific markets. Sometimes, the connecting bonds between plant and customer are further strengthened by the establishment of long-term, strategic partnerships.

Marlow identifies Total Customer Satisfaction as "the responsibility and goal" of every employee. Like other Best Plants, it defines customers in a broad sense, focusing on both internal and external customers. To achieve a close relationship with external customers, the operation was restructured into market segments that focus on specific customers and product markets. Internal customers, the plant's employees, are visited at least once a week by their suppliers to ensure that quality, cost, and delivery are at the best possible level. "Assuring quality to each internal customer," says Burke, "assures the quality of the final product."

In addition to the strategies and techniques already described, Marlow uses a wide variety of tools to reach out to customers. Annually, the company develops its own surveys to establish its customer satisfaction levels, the future requirements of customers, and a benchmark survey. Independent customer surveys are conducted every three years. Direct customer contact is undertaken by senior executives, market-segment teams, and employees at every level of the company. Customers also participate in Marlow's strategic planning and product design processes.

The effectiveness of Marlow's approach is illustrated by the fact that when the company applied for Best Plants status in 1993, it had not lost a major customer for competitive reasons in 10 years, had held or reduced prices for the past 5 years, and had continued to be the worldwide market-share leader. In fact, the company's sales revenues represent greater than 80 percent of the worldwide market for thermoelectric coolers.

For example, John Crane Belfab, a welded-bellows-seals manufacturer in Daytona Beach, Florida, who earned Best Plants status in 1993, abandoned functional silos and recast its operation into a collection of four autonomous profit and loss work centers, each dedicated to a single customer or one market. These minibusinesses are completely responsible for all business activities, including product development, inventory control, quality assurance, and final product shipment.

Belfab's partnerships with its key customers include commitments that range from multiyear fixed pricing schedules to life-of-product production responsibility. The company dedicates production cells to these relationships and even goes so far as to use customers' color schemes and logos in the decoration of the cells. In 1993, quality manager John Michaelos enthusiastically described the results of the partnerships: "Market reception has been incredible, resulting in a 320 percent revenue gain from two long-term partnership agreements. Our two most successful partnerships are [with] Applied Materials and Allied Signal Garrett Engine Division. Sales volume for each of these accounts was approximately $100,000 [annually] prior to partnership development, with sales increasing to over 3 million in 1992. Also, Belfab was awarded the highest level of achievement as supplier of the year for 1991 by Applied Materials for performance in quality, delivery, value-added engineering, cycle-time reduction, and responsiveness."

Exxon Chemical Company's 28-*million*-square-foot Baytown, Texas, chemical plant maintains a similar series of partnerships. Monsanto, a major chemical customer, is connected directly by pipeline to the Baytown plant. A joint Exxon-Monsanto continuous-improvement team comprised of operations and technical personnel meets regularly to manage the operation. Their activities range from the reduction of product lead time to joint development of a new product.

Baytown's rubber-producing operation, which supplies material to the major tire companies, has created a series of annual technical conferences in that highly competitive industry. "We have developed a relationship such that they freely share with us their development plans for new products, while we share our development plans for new materials," explains Baytown site manager R. C. Floyd. "We are the only supplier trusted with this access to future development plans. These meetings serve as the basis for ongoing joint developments directly between our technologists and the customers."

Baytown's success at building its customers into the structure of its operations is well illustrated by the recognition it has received from those customers. In the three-year period between 1989 and 1992, the company was awarded the highest form of supplier recognition available from the largest customers in each major product area. These are Goodyear (elastomers), Monsanto (chemicals), Chevron (paramins), and Kimberly-Clark (plastics). Additionally, the Baytown plant received the Du Pont Diamond Award for the part it played on a team that concurrently developed a new material, process, and package for the packaged-food industry.

The formation of production cells is a fairly common practice among the Best Plants winners, many of which report impressive increases in efficiency and productivity after adopting the structure. (See Chapter 4.) Perhaps the most valuable benefit of cell production is that it can eliminate the traditional barrier between plants and their customers. At Belfab, for example, team members in focused cells are connected by direct telephone lines to their customers. Even this simple connection effectively eliminates several levels of communication from the typical information transfer and reduces response time significantly.

Another 1993 winner, formerly Westinghouse, now Northrop Grumman Naval Systems (NS) in Cleveland, Ohio, also uses the work-cell teams to provide direct communication links between its customers and production employees. "We have installed a system we call *Fast Facts,*" says operations manager David Shih, "whereby a customer may call or fax directly to the factory personnel for purposes of inquiry, quality concerns, etc." What Belfab and NS have both discovered is that the stronger the link they can forge between their operations and their customers, the deeper, more stable, and more profitable the relationship with that customer becomes.

Reorganizing into cell-based production is not the only way to build customers into operations. Best Plants frequently use teams to intensify their focus on customer needs. When General Motor's 1995 Best Plants award-winning Delphi Energy & Engine Management Systems operation in Grand Rapids, Michigan, reorganized in 1990, a strong customer focus was adopted as a primary goal. To attain that goal, two sets of cross-functional teams were created. First are product teams, which are composed of representatives from the manufacturing, quality, engineering, production control, finance, and maintenance operations within the plant and which focus exclusively on the current production needs of their customers. Concurrently, a second set of product-focused teams search out the future needs of customers. These teams include a lead manager, engineering support, customers, and major component suppliers. They meet monthly to discuss issues such as cost reduction, future product programs, and quality improvement. The information developed within the two sets of teams is shared via selected key members who serve in both groups.

Few manufacturers would argue that a strong customer focus has a positive effect on profits, but fewer still have as direct and motivational a structural link as Lockheed Martin's 1994 Best Plant award-winning Government Electronic Systems (GES) facility in Moorestown, New Jersey. GES's customer satisfaction ratings (in this case, the satisfaction of the United States Navy) are tied directly to a quarterly Award Fee. The weapon systems plant's performance is reviewed against criteria that are established jointly with their customers. The criteria are weighted: The score is based 50 percent on technical performance, 20 percent on schedule performance, and 30 percent on management and cost performance. The facility is rated on a quarterly basis and a score is calculated. The result of the score directly determines the fee the company is paid for that period.

Tying profits to customer satisfaction has proven very effective for GES. In the 12 years the program has been in place, contract delivery dates have been

100 percent on-time. The plant has consistently scored over 95 percent and is rated "outstanding" by the Navy.

Creating the Opportunity for Customer Contact

Best Plants recognize that every customer contact is an opportunity to improve their products and services. So, with strategy and structure in place, they create as many opportunities as possible for interaction with their customers. Best Plants eschew traditional functional thinking; no longer is a member of the sales department the only employee a customer sees. Instead, Best Plants send employees at every level to visit customers; they invite customers into their plants and establish easy internal access for customers at as many contact points as possible.

The result is an ever growing number of customer contacts, which Best Plants utilize as an important source of input. Information gathered in these contacts is used to establish higher levels of customer satisfaction, to discover new products and markets, to raise the quality and lower the cost basis of current products, and to continuously improve their own operating efficiencies. In other words, every customer contact is a valuable learning and improvement opportunity for Best Plants.

Even a cursory examination of the Best Plants database makes it clear that the opportunity for customer contact is limited only by the imagination. Most plants create a battery of traditional and not-so-traditional contact points, which work together to create a full portrait of customers and the marketplace. A good example is Symbiosis Corporation, a Miami, Florida–based medical device manufacturer. Founded with a $5,000 stake and a patent for a disposable syringe in 1985, Symbiosis went on to create disposable instruments for use in the rapidly growing field of endoscopic (minimally invasive) surgery. Seven years later, in 1992, the company was acquired by American Home Products for $175 million.

Symbiosis sells its output to marketing partners who, in turn, resell the company's products to end users in the medical community. To stay abreast of customer needs and emerging trends, the company has developed a strong contact base with both sets of customers. The company's marketing partners, who are in direct contact with end users, provide essential information on customer needs and trends. In fact, it was Boston Scientific, which bought Symbiosis from American Home Products in early 1996, that provided the impetus for this Best Plant's initial push into endoscopic instruments when it brought the new market niche to its partner's attention.

The end users of Symbiosis products—the medical community—are reached through direct contact with physicians. The company invites doctors to focus sessions in the field, employees attend surgical procedures, and clinical feedback is gathered on Symbiosis products as well as competitor's products. Other medical products manufacturers among the Best Plants winners also reach out to end users. Johnson & Johnson Medical's El Paso, Texas–based 1993 award-winning plant established Surgical Asepsis Seminars, as well as plant tours and product focus groups, for the surgeons, nurses, and other medical professionals who use their products.

Closeness to the customer is not restricted to medical products manufacturers. Steelmaker Timken's Bucyrus, Ohio, plant holds monthly visitor days to which customers (and suppliers) are invited. All levels of employees participate in presentations, shop-floor demonstrations, and luncheon discussions. Timken's second award-winning plant in Canton, Ohio, arms its employees with a video camera on customer visits. These visits are taped and their findings are presented to employees throughout the plant on their return. Honeywell's Phoenix, Arizona–based Industrial Automation & Control facility hosts an annual User Group symposium "that gives customers a forum to provide feedback. This meeting," explains Worldwide Manufacturing Programs manager Gail Kristof, "has been specifically structured to help customers exchange ideas and also advise us what their future requirements will be."

Best Plants have learned that with the proper training, employees at all levels are valuable customer missionaries. Accordingly, the plants often train all employees in the art and science of customer service. At Redwood Falls, Minnesota–based Zytec Corporation, a power-supply manufacturer whose 1992 Best Plants award was just one in a long line of honors including the Baldrige National Quality Award, mandatory *Service America* training is provided to all employees. The company maintains a policy that all customer requests are to receive immediate response and every employee is empowered to spend $1,000 on the spot to ensure a customer's satisfaction.

At John Crane Belfab, where customers work directly with production-cell team members, training programs have been created in partnership with Daytona Beach Community College. The skills required for effective customer contact, such as problem resolution, listening and response, and eliciting customer comments, are taught. Once employees are trained, customer contact is encouraged. In 1992, for example, an entire partnership product team of 14 employees traveled across the country to visit a Santa Clara, California–based customer.

Until recently, the idea of sending an hourly worker to visit a customer would have been looked upon as sheer folly by most plant managers, and even today, unenlightened managers might consider such an action risky at best. In contrast, Best Plants make a policy of sending employees to visit customers. At Baxter Healthcare's award-winning IV Systems plant in North Cove, North Carolina, production workers visit the hospitals that purchase their intravenous solutions to observe how the company's products are used in the field. Industrial-automation-control manufacturer The Foxboro Company, a 1992 award winner, started a Customer Friend program in which one employee adopts a customer and follows that customer's product from on-site system start-up until it is operating in a steady state.

Ford Motor Company uses a similar structure in its Lansdale, Pennsylvania, electronics plant (NPEF). The plant established a Customer Liaison Program composed of a network of liaison engineers, each individually assigned to a major customer. Liaison engineers act as single points of contact and take a proactive approach to identifying and providing swift resolution to customer issues. Liaison engineers make over 150 visits annually to NPEF customers; all major customers also receive scheduled quarterly visits to observe NPEF products' performance in the customer's process, discuss open items, and review satisfaction. "Liaison engineers provide the customer with fresh eyes to observe their processes and our product applications as well as make recommendations on material handling practices, scrap reduction, installation procedures, and throughput improvement," explains plant manager Dudley Wass.

Customer focus teams are a common sight inside and outside Best Plants. Johnson & Johnson Medical's El Paso, Texas, plant created a Value Analysis team that contacts customers directly to evaluate and update their needs and record their complaints and suggestions. Among its activities in 1992 and 1993, the team visited hospitals, observed surgical procedures, conducted discussion panels, and sponsored a booth at the annual conference of the Association of Operating Room Nurses. The plant reports that the resulting design changes led to improvements valued at over $2.5 million per year. General Motors' 1991 award-winning Cadillac Division sends its teams on dealership visits to collect information from sales and service staff. In an effort to reach out to the final authorities on their cars, hourly employees also telephone new Cadillac owners to hear their likes and dislikes firsthand.

For those who might think that a trip to visit a customer is little more than a vacation junket at company expense, 1995 Best Plants award winner MEMC Electronic Materials might not put up too great an argument. This silicon wafer producer sends its quality teams on customer visits as a reward for exemplary performance, thereby combining essential information gathering with employee recognition. The Saint Peters, Missouri, plant sponsors over 50 of

Varian Associates in Cyberspace

As access to the electronic frontier of communication expands, Best Plants are pioneering new ways to reach their customers. At Varian Associates' 1994 award-winning Nuclear Magnetic Resonance Instruments (NMRI) division in Palo Alto, California, a global E-mail facility, the Internet and a WWW site (http://www.varian.com) link the company's global customer base to Varian's application chemists, development engineers, sales, service, and manufacturing staff members. NMRI's Virtual Customer Support Network includes an electronic bulletin board which all customers can use to share information. This communications link not only encourages the flow of information between the company and its customers, but between customers as well. A newsletter, *VNMR NEWS,* is distributed to all NMRI customers through this electronic network, and the plant routinely utilizes E-mail to distribute new-product-development information, software fixes, and upgrades between formal releases. Thus, customers have almost instant access to information on new products.

The electronic customer support network includes a User Library, which contains examples of product applications and experimental routines. All customers are entitled to use and add to the library, allowing Varian's customer base to share nonproprietary experimental capabilities with each other. Updates to the user library are distributed monthly through E-mail. "Frequently, customers have discovered a way to improve a program," says NMRI's Ed Sehrt. "Varian, in turn, will distribute their solution via the bulletin board."

Cyberspace allows more than simply a forum for shared information. Using the same electronic network, Varian scientists and engineers can remotely log on to a customer's instruments, troubleshoot problems, and analyze the results. This real-time virtual service gives new meaning to the idea of "on-site" service and, in many cases, has significantly reduced customer downtime and the time spent on service visits.

these customer visits each year. Electronic components manufacturer Coherent, Incorporated, an Auburn, California–based 1994 award winner, established a weekly air-shuttle service to encourage employees at all levels to visit customers. "The goal," says operations director Diane Breedlove, "is to stay in touch with rapidly changing technologies, develop strong customer communications, and achieve a seamless manufacturing link."

Practice Fast Response

One important benefit of an outpost in cyberspace is the speed with which it is possible to respond to customers. Fast response to customer needs is more than a hallmark of Best Plants; it is a necessity in today's markets. Customers don't simply measure a plant's response against its competitors. Instead, customers are coming to expect the same response times from their suppliers on the job that they can get at home from the producers, distributors, and retailers of consumer goods.

For example, MEMC Electronic Materials, whose Vision 2000 states that "data will be as important as the wafer(s)" it makes for its customer, has created an on-line database called Customer Information System (CIS). CIS is a Visual Basic application which collects data from over 20 separate plant databases and stores it in a repository for customer queries. Combining database integration and Gateway software, the program provides customers with an encyclopedia of information regarding their MEMC products. Customers use software developed and distributed by the company to dial in and retrieve data. They can even create their own customized reports using CIS information with any Windows software. Manufacturing manager Perry Lee describes CIS:

> *The system will provide a much needed service to our customers. Instead of providing a standard Certificate of Quality Conformance or Capability Report, our customer can see what they want to see when they want. In addition, our engineers will easily be able to characterize products and processes. In both instances more accurate, flexible and timely delivery of data improves our value to customers and the quality of the product developed by our engineers.*
>
> *The benefits are numerous. Data refinement and analysis are greatly enhanced. The product information related to specific shipments is accessible immediately on-line (the product data is available before the product shipment itself). Customers can more readily study factors related to their own product performance. We can also add specific messages to our customers and thus*

enhance direct, fast communication. Predefined, customized reports can be provided to our customers. Internally, we have more powerful information to guide improvement activities.

Several Best Plants have cut response times by establishing 24-hour, 7-day-per-week phone centers for their customers. Fisher-Rosemount gives customers a single point of contact entry for all of their product and service needs through the use of the National Response Center. (See the company's Individual Plant Profile in Section 2.) XEL Communications maintains a 24-hour, 800-number hotline for customers, as well as staffing human-attended contact numbers for emergency situations. To ensure that all customer needs are resolved, the company uses a formal process for tracking contacts. As soon as a customer problem comes to the company's attention, it is ranked according to priority level, and responsibility for its solution is assigned to a department and individual. All problems logged in to the system are reviewed daily at the vice-presidential level in the engineering, marketing, or manufacturing functions as assigned by the vice president of quality.

Greensboro, North Carolina–based Gilbarco also recognizes its customers' need for a 24-hour support network. In response, it established a telesupport operation for the purchasers of its gasoline pumps and point-of-sale systems. The center, which fields 23,000 calls per month, can troubleshoot problems over the phone and dispatch service staff to the customer's location. Like XEL, Gilbarco established a formal system for tracking the resolution of customer complaints. Responsibility for response and resolution is assigned, and all complaints are logged. The system has three urgency response levels: 24 hours, 7 days, or 21 days, and all complaints are regularly reviewed by the senior managers of the company.

Copeland Corporation's Sidney, Ohio, scroll compressor plant described the need for fast and effective customer service succinctly in its 1994 Best Plants application: "There is no such thing as being oversensitive to customer quality issues." Immediate response is the plant's goal and, depending on the severity of the issue, employees are dispatched to customer's facilities no later than the next day for domestic customers (and as fast as travel arrangements will allow for international customers). Whenever quality defects arise with a customer, the plant supplies the customer with written corrective action.

When plants are closer to their customers, response times can drop precipitously. At the Allen-Bradley Company, the award-winning EMS1 operation supplies printed circuit boards to its internal customers, the Industrial Control Group's final assembly operations. Working in the same 3-million-square-foot building allows EMS1 to maintain almost constant contact with its upstream customers. The result: a policy goal to resolve all customer issues within one hour.

Design Products with the Customer

New-product design and existing-product development are supporting competencies of Best Plants, and the strategies, techniques, and tools the plants use in this area are detailed in Chapter 8. However, there is one aspect of product design and development that is rightly part of the three core strategies that all the award-winning plants share. That aspect, the participation of customers in every level of design and development, flows directly from the intense customer focus of Best Plants.

It may appear that customer involvement in product design and development is an obvious requirement of business success, but in many cases that knowledge has been hard-won by companies. Nor is the realization that customers must inform product development enough. Customer input must be earned. The prerequisites, as detailed above, are placing the customer at the heart of your plant, creating a structure that supports customers, inviting the customer into the operation at every opportunity, and responding to their wishes and needs in a way that communicates the value of their participation. Only then do Best Plants ask their customers to take an active partnership role in the work of developing products.

Government contractors, such as Rockwell International's El Paso, Texas–based Autonetics Electronics Division, have long maintained model design partnerships with their customers. Rockwell's 1995 award-winning facility, which supplies electromechanical components, determines the needs of its customers using Integrated Product Development (IPD) methodology and a systems engineering approach. Customers play an active role throughout the IPD process, which utilizes many of the techniques used in QFD (Quality Function Deployment, a matrix approach to design that leans heavily on the customer's needs and desires). IPD depends on cross-functional teamwork between customer, employees of the El Paso plant, and suppliers to create designs that take into account the entire life cycle of a product. This collaborative effort injects multiple viewpoints in the product design process and helps ensure that the customer's needs are anticipated throughout the design, development, and manufacturing processes.

The plant's systems engineering approach is aimed at transforming customer needs into a description of performance parameters for each of its products. As described by project administration manager Don Teague, systems engineering consists of four primary activities: functional analysis, synthesis, evaluation and decision, and a description of the product elements. "It requires," explains Teague, "a *mutual* understanding and support between the customer and Rockwell to interpret the objectives and requirements of the product."

Another way the company brings customers into its development cycle is the Reach Out (RO) Program. "RO emphasizes the need for employees and suppliers to look beyond conventional measures of performance and perceive their successes and failures through their customers' eyes," says Teague. "Knowledge of their customers' needs and expectations is a prerequisite to satisfying them. The RO Program plan of action requires El Paso's Integrated Product Teams to (1) identify their customers, (2) obtain knowledge of their customers' true needs and expectations, and (3) formulate team strategies for meeting those needs. The objective of the RO Program is not to sit and wait for feedback from the customer, but to go after it with a sincere effort towards continuous improvement."

Cincinnati Milacron's vividly named Wolfpack strategy also uses cross-functional teams to pull customers into the design process. In fact, the two simply stated requirements that form the basis for the Wolfpack program—understanding the product needs of the customer and developing the skills to meet those needs—are both aimed at excellence in design and development. This 1994 award-winning Batavia, Ohio, plastics injection machinery producer asks customers to identify the capabilities that they want in new machines and to identify any as-yet unmet desires. The findings from these team sessions become the fundamental inputs for new product design efforts.

In the transition from design to production, customer input continues to be actively solicited by Wolfpack teams. Product reliability data is gathered from customers and shared at weekly meetings with department managers. Also, six-month performance evaluations follow shipment of every product and ask customers to provide detailed information regarding product performance and the plant's service efforts. "Customers' early involvement in product planning will assure that the final product will best fit the widest segment of the market," says manufacturing manager Martin Lakes. "It will also provide continuing input for the next generation of products that changing requirements will demand and which improved resins will allow."

Gilbarco also saves space for customers on its cross-functional product teams, which include members representing the company's marketing, sales, manufacturing, and engineering functions. Customers attend many of the team meetings, which, according to manager Jerry Smith, "results in new products and the enhancement of old ones." The company's Advantage™ line of fuel-dispensing equipment, which can be customized on-site with graphics, card readers, and cash acceptors, was born of a customer request. Gilbarco's G-SITE™ System, a POS system, was also designed to fulfill specific customer requirements. At the same time the company was earning its Best Plants award, a team effort with customer participation was developing a full-screen display for gasoline dispensers.

Established in 1948, Varian Associates' Nuclear Magnetic Resonance Instruments (NMRI) plant thrives by enlisting the support of its customers to detect the technological and scientific developments that will affect its products and industry. As the company explains, it is influential players in the scientific community who set the pace and direction of the company's product lines. These

MEMC's SCQT Tracks Down Customer Needs

MEMC Electronic Materials' Saint Peters, Missouri, plant has established a Strategic Customer Quality Team (SCQT) to understand customer needs and expectations. SCQT is a group of select employees that meets directly with MEMC wafer customers to collect and analyze customers' current requirements and future expectations.

Annually, the team surveys MEMC's top-15 strategic customers, such as IBM and Samsung, to try to identify the product and service performance characteristics that will be important for the coming three years (a period of time that often reflects two full life cycles for customer's products). SCQT team members plot their findings using Pareto analysis of the issues, project goals are defined, and the improvement work proceeds. Progress reviews are held quarterly to assess results and plan any additional efforts needed to meet their goals.

This team is now in its seventh year of gathering regular customer inputs and responding to them with quantifiable improvements. In 1992, for example, the SCQT team identified the flatness of the company's silicon wafers as its customer's most important product concern. The team attacked the project and, by 1994, had doubled the company's wafer flatness capability.

scientists communicate their needs and invite suppliers to develop or enhance existing products. NMRI also conducts joint development projects, primarily in probe development and applications software, with its customers. Customers are also invited to participate in alpha and beta site testing, and the company hires some of its own customers in the scientific community as consultants and enters into technology agreements with others.

Measure for Success

There is a fine old saw that says "what you measure is what you get," and it is as true with customer focus as it is with any other business or personal goal. Of course, accurately measuring customer satisfaction is not always a simple matter. Like manufacturing excellence, customer perceptions are a moving target. Best Plants keep customers in their sights by measuring their success from their customers' perspective on an ongoing basis. To make matters more complicated, customer satisfaction has both objective and subjective components. Best Plants attempt to capture both components by using a variety of methods to measure how well they are fulfilling their customers' expectations.

Varian's NMRI plant in Palo Alto, California, uses quarterly surveys of 50 randomly chosen customers to help measure product performance and quality of service and support. The surveys are designed to reveal any gaps between customer expectations and the service they actually received. The plant's Hallmark of Quality Survey is conducted for every unit sold and installed. This formal survey measures customer satisfaction by gauging performance at acceptance of order, upon completion of installation, and throughout the product's warranty period.

Continental General Tire's 1994 award-winning Mount Vernon, Illinois, plant created a Customer Enthusiasm Index that is based on quantifiable internal measures such as adjustment rate, fill rate, delivery rate, billing error rate, and abandon rates. Fleet testing is utilized to collect performance feedback directly from end users. Soft measures, more subjective measures, are recorded using seven separate surveys, which are conducted by the plant's Akron, Ohio, home office.

It is not unusual for plants to depend on their corporate headquarters when they are removed from direct customer contact. The products manufactured by Tennessee Eastman Company, a 1991 award winner, are sold through Eastman Chemical Company's (ECC) worldwide sales organization, and customer-satisfaction data is measured on a companywide basis. ECC uses an ongoing customer satisfaction survey and database to measure and record customer perceptions. Customer complaint rates and customer retention levels are also measured. The amount of resources invested in these efforts is impressive. In 1991, then president Bill Garwood reported that 85 percent of ECC sales volume was measured and 12,000 survey forms had been distributed. Among the results: 70 percent of the company's customer base rate ECC as their highest-rated supplier.

The creation of surveys is part art and part science. Further, self-designed and -conducted surveys are vulnerable to bias. Accordingly, some Best Plants use outside experts to guide their satisfaction measurement efforts. MEMC Electronic, for example, determines customer satisfaction on an ongoing basis through a third-party customer survey.

Gilbarco also employed a customer-satisfaction consulting firm to design its annual survey. The company interviewed employees and customers to identify critical customer requirements. The first survey included 440 customers and was conducted in 1993. Results from the survey were posted in Gilbarco's Hall of Measures, a public spot in the facility where performance information is shared with all employees. Determined to get the most bang for their survey buck, Gilbarco formed employee teams to address the satisfaction improvement opportunities identified by its survey.

Gilbarco's example illustrates an important lesson: Best Plants don't simply collect these measures for their records; they use them to identify areas in need of improvement and take action. Automotive parts supplier Nippondenso Manufacturing reports that customer satisfaction is a function of both constant measurement and corrective-action systems. Repeat problems are almost nonexistent, the company reported in 1991, and the focus on across-line action, the implementation of corrective actions throughout all production lines of improvements, enhances quality throughout the plant. The result: a customer claim rate of 1.2 parts per million.

Honeywell's Phoenix, Arizona–based Industrial Automation & Control unit has conducted annual domestic customer-satisfaction surveys since the late 1980s. These were expanded to worldwide customers in the early 1990s. The surveys help identify customer requirements and are translated into performance planning and evaluation (PP&E) objectives for each manager and supervisor. The results are also shared with all employees during the plant's Quality College sessions.

Best Plants use many of the same techniques to measure internal customer satisfaction. Timken's Bucyrus, Ohio, plant conducts internal surveys from which actionable improvements are initiated. Follow-up surveys are conducted to ensure their effectiveness. Varian's Oncology Systems unit, a 1992

Best Plants winner, gathers internal customer feedback on problem boards, which are used to communicate customer response to internal suppliers. These customer feedback boards offer employees a place to indicate problems caused by their internal suppliers. They not only communicate customer needs to supporting departments, they require action and a commitment date for problem resolution.

Summary

The first core strategy of Best Plants is aimed squarely at the most critical ingredient of any business: its customers. Best Plants demonstrate that successfully establishing the customer at the center of your company starts by including the customer in corporate statements of intent and strategic plans. An intense customer focus is then injected into the very structure of the company and the bonds between company and customer are enhanced by creating multiple opportunities for contact. These contacts strengthen a plant's ties to customers when fast, effective response results. Best Plants always include their customers as an integral part of their product design and development processes. Finally, our plants continuously measure and improve their progress toward creating a world-class level of customer satisfaction.

Think outside-in as your customer thinks.

—Honeywell Industrial Automation & Control

CHAPTER 3

QUALITY

Nowhere did the Quality Revolution of the late 1970s and 1980s have a greater impact than on manufacturing companies. In the United States, entire industries disappeared as consumers discovered and embraced defect-free, long-lasting imported products that were priced significantly below the products they had previously been offered.

Many industries, and many of America's blue-chip companies, came under intense profit and market-share pressure during this period. Xerox Corporation, a 1990 Best Plants award winner, found itself in the unenviable position of trying to figure out how to compete against Japanese companies that were somehow able to sell their copiers at prices *below* the manufacturing cost of Xerox's own products. How was this possible, asked bewildered managers? There were a variety of reasons, but one of the most important was encompassed in the word *quality*.

Quality, which was thought of by many American companies as simply an inspection function, had been completely redefined while they weren't watching. Their competitors were not only building high-quality products, they were building high-performance production lines and workforces capable of continuously improving both products and processes. More important, they weren't adding to the cost of products by adding quality. They were actually lowering their production costs at the same time. They were building quality into the structure and the culture of their companies and eliminating waste, returns, and non-value-added activities from their production processes.

It is difficult to overstate the extent of the corporate consciousness-raising that occurred when American companies belatedly discovered this new competitive weapon called quality. For close to a decade, the concept spread like a wildfire through every aspect of manufacturing. Quality became *the* byword in industry, and as the concept expanded into every operating arena, it often seemed the sole objective of plants. The Best Plants were no exception and, although quality is no longer the sole concern of the plants (the stakes on this gaming table are always rising), it is still an integral part of their operations. Quality is the second core strategy of Best Plants.

Establishing a Quality Mind-Set

The main objective of the core strategy of quality is to consistently provide the best possible product at its price point in the marketplace. The means by which this is accomplished are far-reaching. But first, quality must become a basic component of the structure and the culture of a plant. This is a primary lesson of Best Plants: Quality is not a program that can be simply imposed on an operation; instead, it is a way of operating that permeates a business and the thinking of its employees.

Piecemeal attempts to introduce quality and off-the-rack implementations usually result in disappointment. Take a lesson from Honeywell's 1993 award-winning IAC plant:

> *Our early experiences with quality circles were less than successful because they were seen by many participants as "just another program." We learned that simply copying methods endorsed by other companies will not lead to desired results. Even proven approaches with documented benefits must be adapted to fit a company's own needs and processes. One must study examples to find out not only what works, but why it works. . . . Quality programs cannot be effective unless they are an integral part of day-to-day operations. Employees [have to] understand that continuous quality improvement is a process, not a program. We learned that successful implementation requires ownership, commitment, coordination, and creativity.*

Best Plants start creating a quality process by including quality in their corporate-intent statements. Quality values become the basis for its place in the strategic hierarchy of the plant. For instance, Unisys Government Systems' 1993 award-winning Pueblo, Colorado, plant enumerates seven values that serve as "the basis of operations for managers, employees, and teams":

- Quality is the responsibility of every employee and team.
- Quality improvements result from management leadership.
- Quality should be viewed from the customer's perspective.
- The focus of quality improvement must be on each job, at each step in the process.
- No level of defect is acceptable.

- There must be a committment to continuous improvement and the seven-step process to continuous improvement.
- Quality improvement reduces costs.

These quality values are brought to life in the plant's Total Quality Process (TQP) plan, which was introduced in 1990 and is shepherded by senior executives. TQP is deployed through a three-level structure and is guided by a quality steering committee. "The TQP Staff Council champions TQP culture and ensures the TQP process is compatible with our business strategy and plan," explains internal quality consultant Rudolph Krasovec. "The council includes our senior executives and members of the Benchmarking Partnership Team (BPT), who are selected from staff organizations. The next level includes the BPT, whose charter is to coordinate all our benchmarking activity, our TQP Staff Facilitation Team, and the people who work on applications for quality awards. This level leads our TQP initiatives and focuses on the cultural aspects of ensuring that TQP is our standard operating procedure. The third level comprises our TQP process operations and program teams."

The plant further strengthens its quality mind-set by tying quality performance measures to compensation. The Performance Planning and Evaluation process bases promotion, compensation, and performance feedback on a series of objectives which include quality measures. Employee performance is evaluated based on their accomplishments against these objectives, and this evaluation results in a rating, which then becomes a deciding factor in compensation and promotion decisions.

Englehard Corporation's 1991 award-winning plant in Huntsville, Alabama, depends on its Quality Council to lead the establishment of a plantwide quality mind-set. The council defines the plant's quality mission and guiding principles. It uses a Quality Operating System (QOS) to translate quality goals into operational objectives.

QOS defines three areas of quality for the plant: customer focus, internal focus, and prevention focus. By developing monthly performance measures in each of these areas, the plant is able to target quality improvement opportunities and continuously raise quality levels. In *customer focus,* for instance, customer rejects and returns are logged by part number and the reason for return, and this information is summarized in Pareto and bar charts for analysis. QOS's *internal focus* measures in-process parameters, first-run capability, and costs of quality, such as internal and external failure costs, as percentages of the plant's total product costs. The *prevention focus* utilizes customer quality ratings, supplier ratings, and advanced quality planning methods such as FMEA control

Fisher-Rosemount's Early Warning, Quick-Correction Quality System

Fisher-Rosemount's Chanhassen, Minnesota, plant established its quality mind-set using a systematic approach that is best portrayed as a wheel-like structure. The hub of the wheel is made up of the plant's quality leadership teams: the Quality Steering Committee, which draws its members from senior management and team sponsors, and the TQM Implementation Team (TQM-IT), which consists of team leaders from middle management. The individual members of TQM-IT act as the spokes of the wheel joining the hub of the quality system to a series of cross-functional employee teams that form the outer rim of the wheel. The TQM-IT convenes regularly to review team performance and to ensure that the progress toward quality goals remains on track.

The teams that comprise the wheel's rim are each focused on a specific area of quality concern. Some examples are the Documentation & Systems team, Information & Analysis team, and Strategic Quality Planning team. Membership in this network of teams includes about 10 percent of the workforce at any one time. In order to give all employees direct exposure to quality, team members typically serve a one- to three-year term, with a number of members rotating on and off the teams each year. The team's leadership also rotates.

The Chanhassen plant's quality system is based on two working principles: (1) "Have an early warning system" to identify quality problems and (2) "Take quick corrective action" to eliminate their causes. To firmly establish these principles, the plant simultaneously used a bottom-up and a top-down implementation approach. Employee-staffed cross-functional improvement teams, which are called PIT Crews and involve 75 percent of the workforce,

plans and Design of Experiments to identify and address quality concerns before they result in product defects.

A properly established quality mind-set cascades down through the organization. At Copeland Corporation's Sidney Scroll plant, the company vision specifies "outstanding quality" with specific objectives, such as delivering perfect products, reducing failures 10-fold in 5 years, and developing a Baldrige-level were organized to install the plant's early warning system. After 12 weeks of SPC and JIT training for all employees, an andon-type traffic-light system was put in place at each of the plant's process operations. A red light, which can be triggered by any employee, stops the line when defects or unacceptable variations occur in a process. These stoppages, which most manufacturers once viewed solely in terms of lost time and production, are now transformed into valuable improvement opportunities.

To enlist and maintain management support, a collection of key quality measures and a quarterly review process were developed. Managerial steering committees were chartered for the major product lines. Each team is responsible for the development and implementation of its own improvement plans.

Sounds comprehensive, right? "When this new system was first implemented, it did not produce the intended results," explains manager Amy Johnson. "Operating measures did not improve as rapidly as expected. It was later found that although the PIT Crews had the power to stop the line, they were not using it to deal with quality issues. [They] were not turning on red lights for a very simple reason—their supervisors did not support the new system. The supervisors had not made the transition from their traditional role in directing work flow and decision making to being leaders and facilitators of the new system. The bottom-up and top-down approach had missed an important segment in the middle of the organization.

"This situation was remedied by providing production supervisors with additional training focused on developing their leadership and training skills. As the production supervisors became comfortable with their new role, more red lights were called, more opportunities for continuous improvement were found, and operating measures began to improve dramatically."

The lesson: Establishing a corporate quality mind-set is a task that includes every employee.

---

quality status. The plant's quality policy commits it to "providing a total quality environment with a principle of doing it right the first time."

In its effort to pull these intentions off the printed page and onto the plant floor, the management team has utilized a series of techniques, both simple and complex. For one, quality's place on the hierarchy of managerial priorities is simply, yet effectively, established by making it the first item on meeting agendas. The quality levels of the plant's processes are constantly monitored by a computerized SPC system, and a quality scorecard system displays daily results in all production operations. Every one of the plant's products, innovative air-conditioning and refrigeration compressors, undergoes rigorous quality tests that simulate more than 15 years of service before they are released for shipment.

The plant's quality message is constantly repeated to employees. From 9½ hours of quality classroom training for new employees to the development of specific Coordinator Quality Responsibilities (CQR), quality is an integral part of performance expectations. Employees also participate in problem-solving teams that are chartered to study process variability and design fail-safing, or poka-yoke, devices to improve the plant's process capabilities.

Making Quality Everyone's Job

As Fisher-Rosemount's highlight illustrates, when quality is built into the culture and structure of a plant, it becomes everyone's concern. Quality in Best Plants is no longer the sole responsibility of the quality assurance or quality control departments, which historically have tended to be reactive and inspection-based jobs, not proactive functions. In fact, Best Plants often believe it is a waste of resources to try to inspect in quality. Instead, they prefer to build quality into process and products.

Asked to describe the least-fruitful quality technique he can think of, Rudolph Krasovec of Unisys' Pueblo, Colorado, plant replies: "Traditional methods of in-house and customer inspection, because it is the most costly with the least payback. However, we are required by our government customers to do this. We have made some progress in this area. . . . Employee and customer representative cross-functional team improvement initiatives have reduced and eliminated some inspection operations, resulting in significant time and cost savings."

Plant manager John Morrison of Johnson & Johnson's El Paso, Texas, operation agrees: "One method that we used until 1990 was quality assurance inspection of work-in-process goods. This method was unsuccessful because it did not involve people, did not focus on the root cause, and did not focus on the problem. The production associates were not given a sense of ownership."

The El Paso plant eliminated inspection as a support function and established the practice of self-inspection. "Implementation of self-inspection," Morrison continues, "accomplished two things: It provided a sense of ownership and a method for corrective action, and also freed up the quality assurance technicians to perform more value-added tasks."

The realization that inspecting in quality is ineffective compared to building in quality is why many of the Best Plants have reduced or eliminated their dependence on traditional quality functions. Instead, the responsibility for quality has been spread over the entire workforce. Quality is a part of the job description of every employee in a Best Plant.

Among other things, this translates into quality at the source. "At the source" means that the responsibility and accountability for quality resides with the people who are actually producing the goods and services. For example, Edy's Grand Ice Cream plant in Fort Wayne, Indiana, which began operating in 1986, has never had a laboratory or quality control staff. Instead, all quality checks are performed by the machine operators. Operators have full control over their output, which includes withholding product of suspect quality or stopping the production lines at any time.

An integrated quality system (IQS), featuring quality at the source, was put into operation at Johnson & Johnson's 1991 award-winning Sherman, Texas, facility starting in 1982. On its full implementation, which was not completed until 1990, its features included: formal operator training in all aspects of the quality requirements of the products each builds; successful completion of a quality certification test; and total operator responsibility for quality inspections and documentation. With the establishment of the IQS, former quality assurance employees were able to relinquish their ongoing inspection responsibilities and create a program of process and product audits.

MEMC Electronic Materials reports realizing significant benefits from self-inspection programs. "Self-inspection, the integration of former QC tasks into manufacturing processes, has greatly improved productivity by reducing inventory as well as manpower," says manufacturing manager Perry Lee. "Capital has been allocated to automate many inspection tasks in the future." Making quality everyone's job has paid off for the Saint Peters, Missouri, plant. As quality problems are identified, Quality Improvement Meeting teams analyze and attack their causes. With the teams' help, the plant reduced over fifteen rework processes to just two.

Quality at the source means more than ferreting out defective products. It also means identifying and eliminating the causes of those defects so that they can be eliminated. Integrating quality into the daily responsibilities of production workers meshes smoothly with the overall Best Plants' movement toward employee involvement and team-based structures at this plant. Once defects are identified, the effort and creativity of the entire workforce can be brought to bear on resolving quality problems.

Automotive parts supplier Nippondenso Manufacturing, for example, requires all operating employees to inspect their own work. When a problem is discovered, the plant's quality circles are used to determine its cause. Thus, the problem resolution process "results in improved operator skills, better teamwork, reduced customer complaints, and an improved training-and-operations manual," according to the company. All the plant's employees, including those in the business office, take part in quality control circles. The results of their work are presented during a biannual competition that is judged by the company president and a panel of executive staff members.

Stone Construction Equipment, a member along with Nippondenso of the 1991 Best Plants' class, integrated quality into the production process in its move to cellular manufacturing. Final inspection was eliminated and each work cell was given full responsibility for process and product quality control. In this case, the employee-owned company reports that the team approach enhanced the employees' sense of pride in their products and gave them a way to quickly resolve quality problems. Building on that sense of ownership, each employee signs off on his or her completed products by personally signing a product tag that is shipped to customers as an assurance of quality.

Self-directed workforces require self-inspection by their very definition. It makes no sense to give employees responsibility for managing their own performance if they are not also responsible for the quality of the products. Quality is an integral part of employee performance. In Aurora, Colorado, XEL Communication's self-directed work teams are simply charged with achieving a six-sigma quality performance. The objective: "to eliminate defects by making each team responsible for its own quality, each individual responsible for his or her own defects, and each work station responsible for checking all prior stages before adding value." This translates into work teams that have full control of their production processes. All team members are empowered to halt the assembly process and make process changes as warranted by the circumstances.

Quality at the source requires a significant investment in time and employee training. As we saw earlier, Johnson & Johnson's Sherman, Texas, plant spent eight years fully implementing the concept. It is neither realistic nor responsible to ask employees to take control of the quality of their work without also giving them the training, the tools, and the responsibility they need to successfully undertake the challenge. Chesebrough-Pond's 1992 award-winning Jefferson City, Missouri, plant, for example, trains all production employees in decision making, problem solving, and quality tools before giving them the responsibility of monitoring and improving the quality of the plant's manufacturing processes.

At Milwaukee Electric Tool, where quality at the source was established as part of the plant's move to a JIT-based cellular manufacturing structure, all team members received extensive training to support this shift in responsibility. Further, the plant codifies all of its standard work tasks using operation method sheets (OMS). OMS are prepared and logged onto personal computers located in each cell for continuous reference by team members. For ease of reference, each is color coded for TQC and quality verification requirements.

Once trained, Best Plants employees also need a constant flow of feedback about quality of the products and processes for which they are responsible. TRW's Vehicle Safety Systems plant in Cookeville, Tennessee, includes its quality strategy under its Six Fundamentals of Manufacturing system. The six fundamentals are waste elimination, quality, delivery, effectiveness, safety, and housekeeping. They are measured within each work cell by the production technicians employed there and compared against targets that are set by the cell members. Their performance is communicated to the workforce as a whole on 4- by 6-foot status boards adjacent to the facility's main entrance. Each work cell also maintains an individual status board for its specific area.

At Rockwell International's El Paso, Texas, plant, three factorywide feedback systems were installed to assist employees in the evaluation of their own performance. They measure and distribute quality, test, cycle time, and SPC data directly to the factory floor. The first, a product-tracking system, collects information via bar-code terminals located throughout the production floor. The second is a kanban system that utilizes work-in-process inventory control carts. And the third is a chart-based performance measurement system that is posted at each team's location throughout the factory. This data is converted into run charts, Pareto diagrams, and process control charts in an ongoing effort to keep each team informed of its performance status. Employees also maintain Operator Management books in which they record measures of their own performance.

Building Quality into Processes

The next lesson embraced by the Best Plants is that product quality is highly dependent on process quality. It was the late Dr. W. Edwards Deming who brought this idea into the mainstream of manufacturing thought. He taught that the largest cause of defective products was poorly designed processes and that no matter how much pressure was brought to bear on the workforce, products could not be consistently and/or significantly improved without addressing the systemic causes of their defects.

Japanese manufacturers were the first to fully accept Deming's ideas, which were based in part on statistical concepts developed in the 1920s and 1930s by Walter Shewart at AT&T. Deming was largely ignored in the United States until the early 1980s, when his thinking found an influential corporate patron in two-time Best Plants award winner Ford Motor Company. Until his death in 1993, Deming taught Ford and, eventually, almost every other major U.S. manufacturer the importance of understanding process variation and the basic principles of SPC.

Whether they directly credit the famous consultant or not, Best Plants are Deming disciples. They understand that defect-free products are the result of tightly controlled processes. Best Plants map, measure, and continuously improve their processes. This process orientation is also intimately connected to the Best Plants predisposition to practice quality at the source. It is frontline production employees who are responsible for much of the Best Plants' success with process control and improvement—a circumstance that Deming would have thoroughly approved.

As Deming taught, a comprehensive understanding of process is the primary weapon in the war to eliminate the process variations that cause product defects. This is a lesson well understood by Rockwell International's El Paso, Texas, plant. Although it is primarily a short-run production facility, El Paso's 20-odd products all flow through a series of common processes before being distributed to product-specific areas for final assembly. (The common processes allow the plant to realize the full benefit of automation and yet still maintain the customer-focused capabilities of a job shop.) In an effort to completely understand its processes the El Paso plant maps all of its core manufacturing and support processes using tree diagrams and flowcharts. Effective working tools for documenting and analyzing processes step-by-step, the charts and diagrams are posted in team areas.

After process maps are complete, El Paso uses SPC techniques to measure its processes. For each process, variables and attributes data is collected. Variables data is converted to Cpk measures. The plant's attributes data are measured in defects per million opportunities (DPMO). All this process data provides the raw material for the plant's continuous-improvement efforts. The results: In 1995, El Paso reported Cpk measures above 2.0 on 85 percent of its processes. (A Cpk of 2.0 is roughly equivalent to a six-sigma quality level.) From 1990 to 1995, the plant reduced its defect rate from 1,200 DPMO to less than 50 DPMO, a 96 percent improvement.

Exxon Chemical Company's huge Baytown, Texas, plant describes SPC as one of three technologies that have had the greatest impact on the plant's performance. (The other two are empowered teams and JIT methodology.) Site manager R. C. Floyd explains:

> *We have found that statistical techniques are perfectly suited for continuous chemical processing. Because chemical reactions occur over time and are very sensitive to even minor variations in the process conditions or the presence of trace impurities, statistical techniques have allowed us to greatly improve process stability. In one case, we improved the purity of an elastomeric material to such an extent that we were able to target a new market. This material, which was originally created as a window sealant for commercial buildings, is now also sold as a chewing gum base. Another material, originally designed as a tire component, is now used for pharmaceutical applications.*

Baytown uses computer systems to continuously analyze its product and control its process. By 1994, more than 90 percent of the facility's output was produced using this technology. Two of Baytown's four divisions reported Cpk values greater than 2.0 and the average Cpk for all the facility's products totals 1.70.

Fisher-Rosemount's pressure transmitter plant credits SPC as the "guiding light" of its manufacturing process improvement efforts. Its goal is to bring all the plant's processes to a three-sigma quality level. To support this effort, employees are trained to prepare and analyze SPC charts. The operator closest to each process is given the responsibility for charting relevant statistical data, calculating control limits, and monitoring the process.

Although SPC may provide a gold mine of information for quality-improvement-hungry manufacturers, its usefulness can be overstated. Westinghouse Electric's Cleveland, Ohio, plant uses computerized and manual SPC extensively, but when it comes to short production runs, operations manager David Shih cautions, "It is a tedious, time-consuming process to plot controls charts manually, which provides little real benefit for either quality improvement or troubleshooting needs, especially for short-run production quantities where production is frequently over before process control charts can be produced."

Nor is SPC a magic elixir that can be simply swallowed whole. John Michaelos of John Crane Belfab explains: "Several years ago, SPC evolved into the quality process control method of choice. Belfab jumped on the bandwagon by sending all colleagues to extensive SPC training with the expectation of using statistical process control to ensure all product manufacturing compliance. As quickly as we jumped on this wagon, we fell off. We failed to realize the fundamental use and applications of SPC. We have now implemented . . . SPC in applications that have value and are continuously finding new areas of use."

Further, installing an SPC-based system doesn't actually solve a plant's quality problems; it merely identifies them. Once processes are charted, limits established, and variations recorded, Best Plants have to eliminate the causes of process variance using other quality tools and techniques. Poka-yoke, also called fail-safing and mistake-proofing, is one commonly practiced technique that Best Plants utilize to build quality into their processes.

The solutions developed through the use of poka-yoke are often elegantly simple and inexpensive to implement. One such solution might be irregularly shaped parts that can be installed only in the proper way; another could be a stop on a piece of machinery to prevent a component from moving beyond a certain point. The objective of these simple fixes is to make it physically impossible to create defects.

Baxter Healthcare's North Cove, North Carolina, IV operation effectively utilized a poka-yoke solution in the plant's sterilization process. In one of the final production tasks before shipment, a series of multishelved racks holding innumerable full IV bags are wheeled into a large sterilizer. Any error in the sterilization process results in a loss of all the product. Even more serious would be a mix-up of sterilized and unsterilized product. To eliminate that possibility, the plant's staff designed gates for the area that allow product to move in only in the proper direction. The gates eliminate the possibility of releasing unsterilized product for shipment.

Johnson & Johnson's Artcraft facility teaches employees poka-yoke techniques as part of the 42 annual training hours each receives. The facility boasted 30 active quality improvement teams in 1993, three of which were cross-functional and staffed by employees from both the El Paso plant and its sister operation in Mexico, Juarez Manufacturing. These teams eliminated cutting defects on several lines of surgical caps and gowns and also implemented the electronic log sheet in the sterilizer process, a poka-yoke solution that kept the sterilizer reject-free for a record 180 days in 1992 and 1993.

Artcraft pegs much of its success with mistake-proofing and other improvement techniques to employee involvement (an essential support of quality improvement and the subject of Chapter 5). "Production associates visit customers and supplier plants and maintain direct contact with their counterparts at these facilities," says operations director John Morrison. "They are involved in three different cross-functional teams with Juarez manufacturing dedicated to quality improvement and waste reduction. This has virtually eliminated rejects on selected headcovers, fenestrated cuts, and gown sleeves."

Honeywell's IAC business unit uses poka-yoke to help it achieve its stretch goal of a tenfold reduction in defects. Employee-initiated Corrective Action Teams use Pareto charts to identify repeat failures, then undertake root-cause analysis to uncover an optimal solution. "Since defect-free output can only be achieved through error prevention, one of our primary strategies was to utilize fail-safing techniques," explains manager Gale Kristof. "Fail-safing became one of the cornerstones of our team training and an integral part of all problem-solving sessions. To date, we have in excess of 100 documented improvements based on fail-safing from team members."

Poka-yoke is a well-established activity at TRW's Cookeville Vehicle Safety Systems plant, which calls mistake-proofing an operational "philosophy." A full-time mistake-proofing engineer coordinates strategic planning, audits, and training in the technique. For each of the facility's processes a mistake-proofing checklist is completed, evaluated, and audited. "TRW Cookeville, with all levels of management support, aggressively pursued a mistake-proofing philosophy which has become ingrained in the everyday thought processes throughout the organization," says section head Walter Marcum. "Our dedication to our mistake-proofing philosophy has resulted in low in-process and customer reject rates."

Poka-yoke-type solutions to quality problems aren't the only techniques employed to prevent defects. Wilson Sporting Goods' Humboldt, Tennessee, plant installed a long list of quality job aids including: a black light to ensure proper coverage in the painting process; size gauges at winding operations to ensure correct ball diameter; and scales in the packing operation to ensure every box of product is fully packed before shipment. As with poka-yoke solutions, the goal is to help employees in their ongoing quest to eliminate product defects by building quality into processes.

Building Process Quality

Building quality into processes further requires that Best Plants build process quality. This may sound like hairsplitting on the surface, but every piece of machinery and equipment that comprises a process must run as intended to reach high-quality levels on a consistent basis. Equipment wear and failure are also major causes of defective products.

Best Plants keep their equipment operating in perfect condition using a series of techniques that are grouped under the generic heading of Total Productive Maintenance (TPM). TPM is really an entire family of maintenance tasks, including predictive maintenance, preventive maintenance, equipment improvements, and even maintenance prevention—a strategy that attempts to eliminate the ongoing need for maintenance altogether. All these related techniques support a single goal: process quality.

Lockheed Martin's Moorestown, New Jersey, plant employs a dedicated equipment maintenance team to maximize process quality and equipment availability. An extensive preventive maintenance program covers all the plant's equipment. Emergency work orders, which are generated when machines are down, are tracked to profile equipment performance. Both sources of maintenance information are collected in a database that is used to profile equipment maintenance history for predictive purposes. For capital improvement planning, the plant's Manufacturing Engineering Group assesses all equipment for life expectancy and replacement timing.

SPX Corporation's Power Team takes a team approach to its TPM program. Each TPM team is made up of the machinist who operates the equipment, the area supervisor, the scheduler, and a member of the plant's maintenance staff. The team members work together to schedule all preventive maintenance; in case of breakdowns, every team member is involved in the resolution of the incident. The machinist is responsible for routine maintenance chores, such as greasing, oiling, and checking fluid levels. The machinist also takes a frontline role in the early identification of developing maintenance problems.

The plant's more sophisticated CNC equipment is serviced using laser alignment equipment and a personal computer. "This allows us to not only tweak the machines," explains plant manager Jim Schultz, "but also gives us an early warning for things like ball screw and spindle bearing replacement."

Power Team uses a homegrown computer program to track and schedule its preventive maintenance tasks. Active for over a decade, the program creates a weekly schedule of equipment maintenance, all of which is based on either calendar days or run hours. The program also keeps a history of parts used in the maintenance and repair of each machine and the hours spent working on it for the previous 12-month period.

As we have already seen with product quality, there is a strong trend toward maintenance at the source at Best Plants. For example, Engelhard Corporation's Huntsville, Alabama, production operators are responsible for a whole litany of maintenance tasks. Production workers clean, inspect, and replace coater carriers and double-diaphragm pneumatic pumps. They maintain the plant's outfeed conveyor system, fabricate transfer hoses for slurry transfer operations, and install blowout-prevention wires on predryer box inserts, a task that requires drilling and cutting operations.

Tennessee Eastman's TPM Is World Class

Tennessee Eastman Company's (TEC) Kingsport, Tennessee, facility began its TPM program in 1986. Since then, it has earned recognition as a leading practitioner in the field. This process manufacturer went so far as to develop its own unique approach to TPM, which it successfully markets to other companies. But it is the internal results of TPM that are really impressive.

When TEC applied for Best Plants status in 1991, it had trained 120 teams, representing approximately 90 percent of the manufacturing workforce, in the philosophy and techniques of TPM. These teams identified over 3,000 specific areas (or interfaces) where maintenance skills could be transferred to production workers, and they had already accomplished that transfer in 2,000 instances. The program's documented results: $24 million in savings.

TEC started its zone-based TPM effort in response to what program founder Bill Maggard calls a "major opportunity for waste reduction at the interface between operators and mechanics." (See the Individual Plant Profiles.) In early 1987, three pilot programs located in different business units at the site were authorized. By June, TEC had three success stories to relate, and approval for plantwide expansion was granted.

TEC's TPM implementation process is a useful generic approach to the discipline (and one that is strikingly similar to total quality implementations). Starting with basic informational introductions, the process calls for the establishment of a steering committee, which oversees one or more area-specific implementation teams. With input from the employees in each area, these teams chart the maintenance tasks and define them into zones of responsibility. The tasks are charted, reviewed, and approved by the steering committee; TPM training is developed and delivered; and employees are provided with the tools and supplies they will need to assume their new roles. Work begins, results are monitored and evaluated, and the effort expands. The goal, according to TEC, is to continuously improve equipment reliability and availability.

Further, Engelhard's production crews are active participants in the plant's housekeeping routines. They clean, prepare, and coat the plant's floors, maintain aisle striping, and prepare and paint their own equipment and workstations. This results in significant cost savings to the company as the dedicated maintenance staff is able to spend its time on more complex chores; the use of outside maintenance contractors is minimized; and uptime, operator knowledge of equipment, and the overall condition of the facility are all improved.

The age of a plant is not a significant factor when it comes to maintaining process quality. New plants face the same conditions as their elder brethren—process quality deteriorates very quickly if they ignore the establishment of clear TPM practices. Just a year old when it earned its 1991 America's Best Plants award, Steelcase's Context plant was already excited by its TPM results.

"Preventative maintenance, predictive mantenance, machine improvement, and autonomous maintenance by operators are all being addressed in the Context plant," says plant manager Dick Bierschbach. "Results are excellent. We are 50 percent complete in writing PM procedures on all our new equipment. Predictive maintenance is being done through use of vibration analysis, infrared surveys, lubricant analysis, etc. Machine improvement programs and total operator responsibility programs are being planned."

Continental General Tire's Mount Vernon, Illinois, plant, a more mature operation in years, reports that almost three-quarters of its work orders are now planned maintenance, and machine uptime runs steadily at 99 percent or higher rates. This plant also established a computer network to log and store its database on machine performance. It offers an excellent piece of advice for plants working on process quality: "Make maintenance proactive—not reactive."

Approaching Zero Defects

Even taking into account the waning of the quality mania of the 1980s, the pursuit of quality remains a serious business at the Best Plants. Once acceptable product defect levels seem ridiculously high by today's standards. In fact, modern quality goals require new standards of measure. Six-sigma quality, which translates into a defect rate of 3.4 parts per million (ppm) and was originally established by Motorola as an internal stretch goal, is the de facto standard at many of the Best Plants. Gilbarco's 4 × 94 program is a good example. Launched in 1990, 4 × 94 achieved a defect rate goal of 4 ppm in all business activities by 1994.

A few of the Best Plants have moved beyond six-sigma and are achieving close to perfect quality. Nippondenso, for example, reported a customer claim rate of a mere 1.2 ppm in 1991. Other plants have also recorded defect-free production for long periods of time. In their 1995 Best Plants application, General Motor's Grand Rapids, Michigan, valve-lifter plant reported 18 months of shipments to a major customer without a single defect. The plant concluded 1994 with a defective parts per million rate of 2. Two of four production teams worked at a level of less than 1 ppm. This is a level of performance that represents a 91 percent reduction from 1992 defect levels and, needless to say, the plant exceeded its customers' supplier goals in every instance.

Sony's San Diego Manufacturing Center is another award-winning plant that has set aggressive quality goals. In this case, the plant's Zero Defects (ZD) program focuses on perfect first-pass yields. Established in 1990, ZD serves as a cornerstone for the plant's focus on the continuous improvement of process yields and throughput. Since its inception, SDMC has registered a 65 percent increase in first-pass yields. The plant's goal is to cut losses from defects in half each year. Toward that end, each defect found is turned over to management for review and resolution. In color-television assembly, for example, the CRT quality adjustment was automated using a video camera and computer system. This new system enhanced adjustment accuracy and raised inspection productivity.

John Crane Belfab's "No Use As Is" nicely illustrates the Best Plants' quest for perfection. Explains quality manager John Michaelos: "In 1990, we implemented a program of No Use As Is. The Use As Is disposition is frequently used in industry to accept product that does not meet specification. This was a major step in our quality management program that did not come without pain. Use As Is requires our president's approval and he won't!"

At Milwaukee Electric Tool's Jackson, Mississippi, plant, the quality system also aims at a target of zero-defect process capability. Basic quality tools, such as control charts and root-cause analysis, are used to understand problems. Poka-yoke techniques, an andon system used to shut down the line and focus managerial attention when defects occur, and 100 percent conformance to a three-step (do-check-verify) production routine are the plant's weapons in the battle against defects.

Varian's 1992 award-winning Oncology Systems business, which logged 100 percent first-pass yields for some of its components, uses the Malcolm Baldrige National Quality Award criteria as the basis for comprehensive reviews of its operations. Where gaps between the criteria and reality on the plant floor are identified, employee-staffed Corrective Action Teams and Continuous Process Improvement Teams chart the process, benchmark performance, set new goals, and implement improvements. In its Best Plants application, the plant reported improvement actions completed or on-schedule for 50 of the 63 performance gaps and partial completion for 10 others.

Oncology Systems also established a "Five by Five" program, which concentrates on five of the company's components that in total comprise fully one-third of its warranty and scrap costs. A high-profile action team, which uses a Pareto-based problem-solving approach, is assigned to each component and charged with reducing component-failure costs by a factor of 5 in five years. These costs are monitored monthly, and management reports that the teams consistently exceed their goals. In 1992, the plant's component-failure costs were less than one-third of their 1989 levels.

Along with these programs, a formal fault-resolution process is in place at the Palo Alto, California, plant. It is a five-step problem-solving technique that, while allowing for an interim fix, mandates a permanent resolution to quality deficiencies. A correctly designed resolution identifies and removes repetitive problems related to process and/or design deficiencies permanently.

Summary

Quality, the second core strategy of Best Plants, is an integral element in their structure and culture. Best Plants implement and nurture a quality mind-set among all their employees by establishing quality-based corporate values, which are in turn supported by formal systems that expressly reflect the individual nature of each business unit. Quality in Best Plants is everyone's job, and every employee is responsible for the quality of his or her own work. The plants support quality at the source by providing their employees with the training, tools, and information they need to achieve high quality levels.

Best Plants understand that a great majority of defects are caused by out-of-control processes. They map, measure, and analyze their key manufacturing and support processes. They search out the root causes of defective work and take permanent action to eliminate those causes. Total productive maintenance, a crucial support of defective-free processes is religiously practiced at Best Plants, and workers take much of the responsibility for the smooth, defect-free operation of their machines and equipment. Through these activities and aggressive quality goals, Best Plants are approaching perfect first-pass yields and zero defects.

Quality is a strategy, not a program.

—*Marlow Industries, Incorporated*

CHAPTER 4

AGILITY

Once upon a time manufacturing was a fairly simple business. A plant's operating goals were clear: Maximize profits by running a single product for as long as possible; then, change over the lines and run the next product for as long as possible. As for customers, they could have any product, in any color they wanted . . . as long as they wanted the product that was in stock and the color they preferred was black, to steal a famous phrase from Henry Ford.

Economies of scale ruled the manufacturing world and everybody knew that mass production and the full utilization of plant capacity was the way to make money. The longer the machines ran and workers toiled without interruption, the lower the overall cost of the product. Why disrupt the economic balance of a plant by introducing new product lines? It was easier to build a new plant dedicated to running as much of the new product as possible. And, if there was no more room for finished goods in the warehouse, well, build another warehouse. Full capacity generated capital expansion.

As in the age-old truism, the seeds of destruction were sown along with the successes of scale economics. The same thinking that lowered costs with volume production also resulted in inflexible plants that could not be easily reconfigured, swollen raw material inventories, mountains of work-in-process, and huge finished goods inventories whose value in the marketplace often bore little relation to the numbers associated with them in corporate ledgers.

It was a manufacturing paradigm that worked just fine as long as everyone played the same game. In other words, it was a system teetering on the brink of total collapse. And collapse it did. Change did in this long-accepted manufacturing paradigm. Innovation, customer demands, and competition began to increase exponentially. The world started to turn faster and the plants couldn't keep up.

To catch up with their new environment, plants needed a new operating system, a system that was able to keep up with the ever increasing rate of change. That system is embodied in the word *agility*, the third and final core strategy of

Best Plants. Agility is broadly, and simply, defined among the Best Plants: It is the ability to thrive under conditions of constant and unpredictable change.

Logically, thriving in conditions of constant and unpredictable change is largely dependent on the ability to create constant and unpredictable solutions to the challenges and opportunities that change brings. Therefore, agility, like world-class performance, is a moving target. There is no step-by-step map to agility. Its strategy and tactics change with every new demand from the marketplace and every technological advance. Agile manufacturers practice the art of creative response.

With that said, there are some basic foundations of agility that many of the Best Plants share. They build flexible operating structures and staff them with workers who fully understand the meaning and purpose of agility. Best Plants run lean operations, which produce profits by manufacturing goods based on customer demand, not economies of scale. And Best Plants continuously reduce their process cycle times using the clean-slate redesign techniques of re-engineering and the same *kaizen* (or continuous incremental improvement) approach used to boost product quality. Add these practices together and a portrait of the agile manufacturer starts to appear.

Creating Flexible Structures

Agility requires a production structure that is easy to manipulate. The physical structure of a plant and the layout of the machinery and equipment must be flexible enough to quickly respond to the rise and fall of customer demand for products and the rapid replacement of old products with new ones. In a time when the entire life cycle of a new product is often measured in months, a Best Plant does not have the luxury of a retooling or capital expansion plan that in the past may have taken years to fully realize. Instead, the lines must be ready to run as soon as the product is available.

The transition to flexible structures is not an overnight process. It comes together, like the elements of Best Plants performance, in the way a jigsaw puzzle is completed. But it always starts, as it did at Honeywell's IAC plant in Phoenix, Arizona, with one inescapable conclusion. Says Gale Kristof:

> *We knew that one requirement for us to make significant improvements in quality, cycle time, and materials management would be to change the manufacturing process itself. To support this, aggressive goals were intentionally selected to cause all areas to rethink how they did business rather than just modify the existing methodology. Work teams used process mapping to identify*

unnecessary operations and implemented a kanban card system in conjunction with smaller lot sizes and a build-to-order philosophy. Based on these changes, our transporter equipment became a bottleneck and was removed. Additional space became available and work-in-process inventory dropped significantly. As better information tools were developed, we streamlined the kanban segment and moved closer to a continuous flow process. As continuous flow has been further refined, personnel from various work teams are moved to other teams requiring temporary assistance. Further cycle-time improvements were then realized by relocating some production cells closer to their source of supply and/or customers.

This cascade of change enabled IAC to create a short-cycle, build-to-order, continuous-flow manufacturing process. It is to accomplish this agile system that many Best Plants reorganize the infrastructure of their facilities. Kennametal Incorporated's metalworking manufacturing division, based in Solon, Ohio, reorganized into a more agile structure when two plants were combined into one in 1988. Management adopted a focused factory design, establishing five product-based units within the plant. The independent units are staffed with a dedicated team of employees, who produce their particular products from start to finish entirely within the unit and work toward their own performance objectives.

"This was an entirely new concept for us, which was implemented at the time we moved into the new building," remembers unit manager Richard Monday. "Focused manufacturing has since become a key factor in improving plant performance, employee morale, responsiveness to customers, and our competitiveness in the world marketplace."

Inside the focused-product structures, this 1991 award-winning tooling product manufacturer further created 11 dedicated manufacturing cells, which the company reports have resulted in additional improvements in productivity and quality, as well as reductions in lead time and manufacturing costs. The cells are staffed by self-managed teams who take responsibility for work scheduling, quality, housekeeping, routine maintenance tasks, and product flow. Each member is cross-trained to operate all the machines in the cell.

Siemens Automotive also uses a product-based focused factory structure, and it reports that this new system has been the most significant "practice" in improving the facility's flexibility. Each product group is a self-contained plant within a plant, with all the resources required for a complete manufacturing operation. This structure effectively eliminated functional barriers within the Newport News, Virginia, facility. Traditionally independent support staff, such

Super Sack Manufacturing Goes Cellular

Until 1989, Super Sack Manufacturing Corporation's operating structure was a very traditional, functionally organized, batch-based system. Manufacturing vice president David Kellenberger describes it as follows:

> Every bag that we made traveled through the whole facility. Our supervisors tried to balance the production flow as best they could. Our employees waited to be told where to move or what to do and tried to look busy. We built up buffer stock between each operation so that everyone would have work. If we did not have a few days work-in-process, we would not feel comfortable. (If a mistake was made, everyone would have work repairing the bags.) When we ran a new product, it meant that everyone in the plant had to learn about the new product at the same time. When we ran a small order, everyone in the plant had to spend time learning all of the intricacies about the order. After producing just a few pieces, they would have to move on to another order. Between each department, we had material handlers to navigate the parts and pieces to the next department, and lead persons to communicate the plans from the line supervisors to each department. And so on . . .

Today, this producer of flexible intermediate bulk containers commonly used in the chemical, agricultural, pharmaceutical, and food industries works in a totally different manner. Continues Kellenberger:

---

as materials-handling and maintenance personnel, now report directly to their assigned product group. Almost three-quarters of the plant's workforce is organized into teams, which maintain a high degree of control over their work areas. The new structure, in place since 1988, enables the company to respond rapidly to customer demands and production needs. More concretely, it has also been a major contributing factor to a *tripling* of the plant's production volume in five years, a period during which the workforce grew by only 20 percent.

Originally designed as a batch manufacturing structure that was organized by functional units, Milwaukee Electric Tool's Jackson, Mississippi, plant made its

Now each bag [travels only] through one team. This team becomes an expert on that particular bag and customer, so that quality, productivity, and customer service are improved. One employee passes the part to the next employee. Instead of maximizing machine utilization, we focus on maximizing employee effectiveness. Our incentive system recognizes team and plant performance, so that team members can concentrate on reducing non-value-added activities. If there is a problem, the one part is reworked. We do not have to repair every other part in the buffer stock, because we do not have the buffer stock.

Super Sack staffs its manufacturing cells with self-directed work teams. The cells pull all the manufacturing tasks that go into any one specific product together into one place and the teams, of 8 to 10 people each, take full responsibility for making a complete product. Among the many benefits of the new system: The company can concurrently manufacture a multitude of customer orders; short production runs are more economical; work-in-progress is vastly reduced; production defects are minimized; and work-flow inefficiencies and process bottlenecks are readily identifiable.

But best of all, says Kellenberger, "is the ability to respond to *other* customers' needs. Without a huge inventory or assembly line full of work, the teams are more easily quickly changed over to meet a rush requirement. Under the old system, efficiencies and output would suffer when rush orders came in. Now we can quickly adapt to meet any situation."

move to a 100 percent cellular production in late 1990. Its manufacturing cells are broadly organized into three areas: tool assembly, component parts manufacturing, and accessory manufacturing. Each measures its performance in takt time, based on a standard lot size of a single unit.

To achieve its highly flexible one-piece-flow, mixed-model strategy, the Jackson plant's cells utilize techniques such as setup reduction, continuous improvement (kaizen) supported by daily meetings and problem-solving teams, and a kanban-based inventory system. In Milwaukee's system, the members of the work cells plan, schedule, and expedite their production and maintain direct contact with vendors. The operating improvements recorded since the adoption of the cellular structure, which are included in the plant's Individual Profile, have been impressive.

Building to Demand

As the Best Plants well know, excess inventory, whether in raw materials, work-in-progress, or finished inventory, is a barrier to agility and a constant source of loss. One of the most valuable features of an agile manufacturing system is that it offers a way to drastically reduce, and in some cases virtually eliminate, inventory. That feature, which all the Best Plants have adopted to some degree, is part of the demand-based manufacturing philosophy labeled *just-in-time* (JIT).

JIT is part and parcel of the lean manufacturing philosophy pioneered by Eiji Toyota and Taiichi Ohno at the Toyota Motor Company. It first emerged in the early 1950s. Because of the unique environment Toyota faced at the time, a mass-production system was neither a realistic nor profitable alternative for the company. Instead, Toyota needed a system capable of turning out small lots of many products. It was this need, as yet undreamed of in the United States, that resulted in the agile, demand-based operating system that became famous as the Toyota Production System and that eventually turned conventional manufacturing on its head.

Lean manufacturing eliminates inventory and its accompanying waste by creating finished products on an as-needed basis. It "pulls" raw materials and work-in-progress through the process only on a demand order for finished goods. When this demand system is working smoothly, material is pulled into production on a just-in-time basis, eliminating the need for inventories. Many benefits accrue, but one of the most valuable (and one that wasn't truly understood until recently) is agility. The use of a build-on-demand system forces a plant to develop its ability to quickly adapt to change.

Because of the competitive advantages Toyota gained from its new production system, the automobile industry and its suppliers were among the first to face the new realities of a demand-based production system. So it shouldn't come as much of a surprise that the Best Plants award winners in this industry were already a good ways down the path toward agile organizations long before the word *agility* was regularly appearing in everyday business conversation.

Nippondenso, for instance, was reporting in 1991 that "our customers require shipments of our products many times per day. For example, we deliver to our major customer 16 times per day (every hour for two shifts) directly to line side. These hourly deliveries include many different products and models in each shipment and rarely involve full-pallet quantities."

Nippondenso supports customer demands for highly flexible, just-in-time supply with processes that have been designed to allow random production of multiple models and a kanban-triggered inventory system that includes the plant's suppliers. The results: There is an average of 2 hours of in-process inventory in the plant at any one time, and customer lead times on an emergency order have been reduced to 90 minutes.

As inventory levels decrease, Best Plants find that raw materials and parts require substantially less space and can be moved closer to their production areas, a shift that reduces cycle times. John Crane Belfab, which eliminated 8,000 square feet of storage, feeds its customer-focused work cells with a point-of-use inventory system. Their kanban replenishment system features specially designed parts kits shipped directly from the company's suppliers to the cell itself. The kits, which are shipped in demand-based quantities and usually occupy a single shelf or two, are stored in each work cell within a few feet of where they are used.

Belfab further reduces cycle time and builds flexibility with quick-change tooling that allows the plant's staff to more easily switch setups. The redesigned dies used in production welding operations represent reduction of setup time from 20 minutes to under 3 minutes. Point-of-use inventory and quick-change tooling are two of the techniques that have helped Belfab drive its average lead times down to 2 weeks in operations that used to require 16 weeks. Inventory turns, in 1993, had been increased to 16.6 from 1989's 5.4 turns. And best of all, says quality manager John Michaelos, "we quickly respond to customer needs."

Varian Associates' 1994 award-winning Nuclear Magnetic Resonance Instruments (NMRI) unit has established focused factories for the production of three major components of their spectrometers. Each is supplied through the same kind of kanban-based, point-of-use inventory system. Manager Ed Sehrt describes the company's reasons for installing a JIT production system:

> *One key objective in adopting JIT methods was to expose process control problems and force them to be resolved. A steady reduction in rework and scrap costs as a percentage of sales is a good indication that this has been successful.*
>
> *A second objective was to organize the production process so that parts and assemblies, both purchased and made in-house, were available on the shop*

floor when they are needed. Lot sizes of one were incorporated, and kanbans were established for shop processes and with our vendors. Inventory reduction became a natural consequence of this approach.

The final objective was to establish a climate of continuous improvement. All of manufacturing, work centers and support groups, were organized into JIT action teams. Presently, every team establishes quarterly goals and measurements based in three categories: quality, cycle time, and cost. A JIT Steering Committee oversees the progress of these teams and presents quarterly awards to the ones with the best records. The teams are empowered to manage themselves and provide their own ideas on where improvements can be made.

The Electronic Products division in Auburn, California, was the first in Coherent's corporate family to convert from traditional manufacturing to a JIT-based production system. The results of the switch are impressive: Response time dropped from 45 days to 3 days and product cycle time dropped from 5 days to 3 hours. Operations director Diane Breedlove compares the old and new systems:

Prior to the implementation of JIT manufacturing, the Printed Circuit Board (PCB) line was managed as a typical batch manufacturing operation. A planner scheduled work orders which were queued and picked off a pick list, along with other work orders from inventory in the stockroom. The work orders (kits) would then be moved to the PCB area to be audited and queued with other work orders.

Eventually, the work order would be unpacked by the operator, set up, and worked on in a batch. When the finished product was completed, it would be taken to the finished goods area in the warehouse, where it was logged in through a computer transaction. When the product was ready to ship, it would be moved to the Shipping Department for packaging and shipping.

After the PCB line was converted to a demand flow process, the work order requirement, warehousing, material moves from the warehouse to the manufacturing floor, auditing, queuing of the kits, and setup time to begin manufacturing were eliminated. The trigger to build a printed circuit board is now the consumption of the PCB, and it is done electronically.

In case the description isn't vivid enough, Breedlove charted the time differences between batch production and JIT for *IndustryWeek*'s Best Plants judges, as follows:

	Batch days	JIT days
Issuing the work order to the stockroom	3	0
Picking the material (kiting)	5	0
Moving material to the PCB area	1	0
Auditing and queuing the kit	5	0
Setup time by operator	1	0
Manufacture of the work order	5	0.5
Movement to the stockroom, queue time, and put away	2	0
Total	22	0.5

Days improvement in cycle time: 21.5
Percentage improvement: 21.5/22.0 = 98%

Creating Agile Workforces

For all the benefits of a flexible, lean production structure, it won't work very well without people who are capable of making it run smoothly. The creation of such a workforce is a major component of agility. Employees are an integral part of the conversion to a demand-based, fast-response production system.

Unfortunately, many Best Plants found that their traditional batch manufacturing systems did little to prepare employees for the responsibilities they were being asked to assume in an agile environment. The traditional system encouraged employees to become expert at a single, often minute, task in the production process. Their main responsibility was to repeat that task over and over, as quickly and accurately as possible, as directed by a supervisor. This performance scheme is as opposite to the needs of agile manufacturing systems as night is to day.

As John Morrison of Johnson & Johnson Medical's Artcraft plant pragmatically states, flexibility requires, "a team-oriented workforce that allows for maximum cross training, higher productivity, and less overtime." Toward that end, Artcraft maintains six job descriptions for 242 employees. Nine of every ten employees is cross-trained in two jobs, three out of four are qualified in three jobs, and half of the workforce is prepared to work at at least four different jobs. Adds Morrison: "It is not uncommon to see a mechanic helping package material or a production worker helping out in the warehouse." Artcraft's emphasis on cross-training is supported by 42 hours of training per employee per year.

When it comes to the need for a creative, flexible workforce, Morrison would find himself in perfect agreement with the management at Symbiosis. Symbiosis' customer-focused work units depend on and require top-notch employee skills. Each focused unit at the company includes a cross-functional staff which is responsible for every aspect of product design, quality, manufacturing, materials, and finance. Explains manufacturing vice president Scott Jahrmarkt:

> *Symbiosis is a high-tech, labor-intensive company. As a result, the success of the company depends heavily on the performance of the people producing the products. All of our production lines are U-shaped work cells containing team leaders and assemblers. Production throughput is monitored on an hourly basis by the operators who are responsible for the quality of the products they produce. The indirect personnel of the focused factory (buyers, planners, engineers, quality assurance, material handlers) reside physically within the areas of production as an aid or support to the manufacturing lines and to be on call should questions, comments, or problems arise. The responsibility of the teams is driven down to the lowest level, thus in essence creating a true factory within a factory that can be measured quite easily in terms of time, quality, service, and cost and/or profit.*

At John Crane Belfab, agility is supported by customer-focused work cells and a 100 percent self-directed work-team structure. The plant employs no middle managers, supervisors, secretaries, work leaders, or quality inspectors. To create an environment capable of nurturing work teams, the plant eliminated as many barriers between employees and management as it could. Eleven hourly pay grades and thirty-five job descriptions were reduced to just two job descriptions, and the manufacturing workforce is all salaried.

Belfab's integrated approach also includes skill-based compensation, flex-time scheduling as needed, and a particular emphasis on training programs. "We realized early in our development that the fundamental changes that had to occur would not be possible without educating the entire workforce," states John Michaelos.

Belfab partnered with Daytona Beach Community College to develop a comprehensive training and personal development program that is offered to all of the plant's employees. The Quality Through Empowerment (QTE) program is a three-year, ten-session course curriculum designed in joint analysis with the college to teach skills consistent with Belfab's quality values and long-term goals. Courses include SPC, blueprint reading, basic inspection, and empowerment. Says Michaelos:

> *The importance of training to develop an understanding that accountability and responsibility belong to each and every colleague cannot be understated. We have found that providing our colleagues the opportunity to fully participate in the decision-making process results in an environment that breeds innovation. Since our colleagues run the business units, their creativity focuses on improvements . . . how to build a part better, how to deliver it to the customer faster, and how to anticipate changes in the customer's needs.*

Gilbarco's 1989 switch to self-directed production teams and an agile cellular structure needed plenty of employee assistance to take hold. States manager Jerry Smith, "The area most affected by the manufacturing strategy . . . has been employee-related." To support the change, Gilbarco created 57 hours of training. Major training modules include 10 hours of JIT, 18 hours of quality, and 24 hours of interpersonal skills training.

Gilbarco also encourages and supports team activities such as basketball, softball, and bowling to help employees become more familiar with working as part of a group. Further, it purposely takes a team approach to all major problems, product development projects, and other activities. The company encourages continued learning with its on-site learning center, which is equipped with personal computers containing educational software covering more than 25 self-paced learning courses ranging from the kindergarten through high school levels, including GED preparation and other adult education courses. The center is open to employees and their families, and the employees are able to choose their own topics and learn at their own pace. Private cubicles are used to house the equipment, which is available 24 hours, seven days a week, and the company will also provide learning coaches on an as-needed basis. Employees use the equipment on their own time, and in 1993 nearly 200 employees were enrolled in the program.

Super Sack's cell-based strategy for agility also relies heavily on employees. But the company made a fairly common mistake when it initially underestimated the commitment in time and resources a team-based cellular structure would require. Warns manufacturing vice president David Kellenberger:

> *Self-directed work teams are not an overnight fix-all. A long-term approach is necessary. The term* self-direction *is misleading in the beginning. It is much*

Employees Drive Change at Delphi Energy & Engine Management Systems Plant

On several occasions in the past, General Motor's Energy & Engine Management Systems' Grand Rapids, Michigan, plant ran headlong into a brick wall while trying to instill the principles of lean manufacturing into its workforce. According to manager Peter Gallavin, "Previous attempts, as far back as the early 1980s, failed because the organization moved from information to action without two key ingredients, namely, employee understanding of the process and, most importantly, employee commitment."

When it reorganized into product-based focused factories in 1990, the plant's management was determined not to make the same mistake again. This time, employee education, training, and communication played an important role in the conversion. In a training program conducted under the auspices of the GM/UAW Quality Network, the entire workforce was taught the principles of synchronous manufacturing. During 1991, a 3½-day workshop, which focused on the elimination of waste, lead-time reduction, workplace organization, and visual control and pull systems, was conducted over 50 times. In all, 1,200 employees participated in the program. Additional workshops were conducted for various division and corporate staff and all of the plant's major suppliers.

Understanding that the hearts, as well as the minds, of employees had to be won over to the new system, groups of employees were taken on visioning trips to several manufacturing operations that had successfully adopted the faster, more flexible system. "The purpose of these trips was to proactively minimize any anxiety of key individual employees who were pivotal to the change process," explains Gallavin.

Employees involved in the change process received additional training in techniques such as problem solving, pull-type production systems, lead-time analysis, and poka-yoke. After training, the company launched 36 "synchronous implementation teams" focusing on wide-scale continuous improvement. These teams are empowered, cross-functional work groups made up of hourly employees and supported by an engineering facilitator. Their progress is reviewed on a monthly schedule by the plant manager.

With establishment of its focused factories, Delphi eliminated as many independent support functions as it could, dividing these previously shared resources among product groups as needed. It assigned specific colors to each product team alignment and repainted all of its operations throughout the entire 1.4-million-square-foot facility to build team identity. To further build coherence, flexible operator classifications were established within the product teams, but cross-team movement was prohibited. This restriction, says Gallavin, "enhanced our training activities since we can now focus our efforts on those job-skill requirements unique to each product team."

Delphi's emphasis on employee involvement in establishing its new production system created significant additional benefits for the company. With their added knowledge, hourly skilled-trades employees now assist the engineering staff in the establishment of specifications for all new equipment purchases. Job setters and operators assist in the machine qualification process, which is conducted at the vendor's facility prior to installation in the Delphi plant. Line employees also participate on the project team that is building proprietary machinery, equipment, and tooling in an attempt to maintain the plant's lead over competitors for as long a time as possible, and a machine rebuild team is redesigning and rebuilding excess and obsolete equipment to maximize the plant's return on assets and increase process capabilities.

Hourly employees also contributed significantly in the plant's shift to a pull system of production. Relates Gallavin:

> In January of 1992, the process started in earnest as each product team established implementation teams consisting of hourly manufacturing employees, supervisors, engineers, and materials management personnel. These groups put together the visual control boards, determined the number of signals in each pull loop (product buffer/queue), developed loop instructions for fork truck drivers and operators, and established the appropriate communications to ensure customer satisfaction. By April 1, 1992, pull loops had been established on over 500 processes, allowing use of the Materials Resource Planning (MRP) system to be discontinued and the movement of material through the plant to be managed totally through the pull system. Later in 1992, all suppliers were placed on a pull system. Transformation to the pull system was achieved by small groups of people working with their customers to develop a material-movement process that would work for them.

more work for a management staff for several years while team members are learning to take on new responsibilities and feel a sense of ownership for their job and to their team.

A huge commitment of time and resources for training must be made. One of our greatest mistakes at the beginning was our failure to recognize how important training was. We recognized that our new *teams would need training and support, but did not see that our teams of existing employees would need a great deal of training in communication, goal-setting and general team-building skills to make a successful transition. In fact, old teams need even more training than new ones.*

Once Super Sack recognized the need for advanced training, it initiated a new tradition of continuous learning with a High Performance Team Building training program for mature teams. In 1995, the company continued it with a new training program, this time in the area of conflict resolution.

Milwaukee Electric Tool also views employees as major contributors to the agility of the Jackson, Mississippi, plant. In its continuing effort to create an agile workforce, the company adopted a number of interrelated programs and strategies, including pay-for-knowledge compensation, self-directed work teams, a peer appraisal system, visual control boards, executive cell reviews, and kaizen "workshops."

"The change from a closely supervised, individually focused workforce to a team-focused, empowered workforce created a discomfort level with some employees," remembers plant manager Harry Peterson. "To help overcome this, we provided training to our facilitators to improve their skill level in dealing with employee concerns. We conducted regular monthly meetings in each cell so employees would be fully informed and have a chance to provide their input to METCO management. We have used employee teams to help deal with this."

The policies and practices that helped turn the Jackson plant into an agile operation were successfully implemented with a good deal of communication. Management took the time to thoroughly explain to all employees how each additional piece of the agility puzzle actually worked, specified the desired result, and further described how that result would support the needs of all the plant's stakeholders. "After this, detailed training was conducted where needed," says Peterson, "and appropriate feedback systems were established. This is capped off with a true open-door policy and a respect-for-all-employees policy at METCO."

Not all of the Best Plants are yet organized into self-directed teams, but they are all striving to create workforces that will help them support an agile operating structure. For example, Baxter Healthcare's IV Systems division, which has been somewhat constrained by its highly regulated industry, is nevertheless preparing its managers and supervisors for their new roles in a team-based future. "This plant is definitely relying continually on more self-directed work teams and shifting to a flatter organization structure," explains manager John Martino. "Several steps were taken in the planning process to ensure a smooth transition beginning with full and constant communication. During the last two years, we have begun an intensive supervisory training program with the initial session conducted for all first-line supervisors being a two-day coaching and feedback seminar. We have built upon this initial foundation with additional sessions for supervisors and managers focusing on managing change and conflict management. Team facilitation training is also occurring with current supervisors and other individuals as they take on leadership roles. In all, we have conducted over 7,000 hours of supervisory training in this two-year period [1992–1993]."

Continuing Reduction of Cycle Times

Implanting the culture and structure of an agile manufacturing system and fine-tuning the engine of employee involvement creates the working foundation for this core strategy. But, as noted at the start of this chapter, the journey is ongoing and ever changing. There is one constant, however, and that is time. Time *is* money in the agile world, and Best Plants continuously reduce the time needed to produce goods.

Northrop Grumman's (formerly Westinghouse Electric's) Naval Systems plant in Cleveland, Ohio, made waves among prime defense contractors when it became one of the first to adopt a comprehensive JIT program that included focused factories, work cells, and a demand-based pull system. Along with the switch came a deeper awareness of the importance of continuously reducing cycle time in all areas of the operation.

"The passionate drive of a continuous-improvement approach emphasizing cycle-time reduction ultimately resulted in dramatic process-yield improvement, schedule conformance, as well as cost reduction," says manager David Shih. "[Naval Systems] firmly believes that cycle time reflects the overall cost-schedule-quality performance of the entire organization [and is] a much better and more comprehensive measurement than the old productivity measurement."

The plant's emphasis on cycle time led to a formal mapping of all processes. Coordinated by the plant's Industrial Engineering staff, the process maps were created by the employees already working in the respective processes. The detailed flowcharts that resulted were used to identify opportunities and create an improvement implementation schedule. The results: a 70 percent reduction in overall process steps and cycle time reductions of 75 percent.

Ford Motor Company's North Penn Electronics Facility (NPEF) adopted business process reengineering (BPR) in its search for ever greater agility. "BPR," explains plant manager Dudley Wass, "is a methodology that identifies and abandons the current rules and assumptions underlying our manufacturing operations to produce dramatic breakthrough levels of improvement in yield, overall equipment effectiveness, manufacturing cycle time, and inventory reduction."

On winning its award in 1993, NPEF was analyzing two of its product lines using BPR methodology with the intention of transferring the results to the plant's other product lines upon project completion. Among the first results of BPR was an innovative information system known as FEMS, which offers, among other improvements, a 98 percent reduction in changeover time. (See the plant's Individual Profile.)

Another major agility breakthrough is the plant's integration of a memory-chip-programming operation into the production line. Programming the memory after insertion into the product modules eliminates an off-line batch programming operation that previously required two days to perform. Additional benefits, reports the plant, include a direct labor reduction from seven operators to two and a reduction in parts inventory.

"Lowering inventory is a long-term, continuous process," maintains John Morrison of Johnson & Johnson's Artcraft plant. In a typical Best Plants experience, Artcraft found that the switch to an agile JIT environment reveals new barriers to continuous-flow production as the change starts to take hold. These barriers require a steady flow of solutions. "Since we made the decision to lower inventories and 'go JIT' in 1990, we have found many of the rocks in the stream that appear along the way," continues Morrison. "We have dealt with some of the more unique situations, such as traffic across the international border, with equally unique solutions. Obtaining the highest-level security agreement with U.S. Customs has greatly reduced our transportation inventory between our plants in Juarez and the El Paso Sterilizer. This and many other innovative solutions have allowed us to reduce inventories and lead times, and have made cash available for capital improvements. This creates a cycle of continuous improvement that has kept us highly competitive."

The same team approach that serves the Best Plants so well in quality improvement often does double duty when building organizational agility. The mandate of Fisher-Rosemount's process improvement teams, or PIT Crews ("to continuously improve value to our customers as measured by quality, reliability, service and total cost") is broadly stated to empower employees to attempt a wide range of improvements. There are 150 teams involving roughly 75 percent of the plant's employees at any given time.

Highly localized, PIT Crews work on their own processes. But, in their totality, all their ideas add up to large gains for the plant. Between 1989 and 1992, for example, the PIT Crew for Model 3051 cut cycle time by 60 percent. Crews in Models 2088 and 2024 reduced cycle time 33 and 36 percent, respectively. In the circuit card assembly area, over 190 implemented PIT Crew suggestions added up to a tripling of productivity in material storage.

A plant's willingness to depend on its employees remains a key support of agility. The way in which Chesebrough-Pond's Jefferson City, Missouri, plant approached the construction of a new production line offers a good illustration. "The line was designed and installed locally," relates industrial engineer Leo McCarthy. "During the installation phase, line personnel who would eventually run the operation were involved in layout, actual installation, and testing. A significant objective was to ensure that a size changeover could occur within 30 minutes. Using this as an objective, the line personnel examined every opportunity for improvement and were empowered to make changes as appropriate. The result: Line changeovers occur in less than 15 minutes, making changeovers a nonissue."

Summary

Agility, the third core strategy of Best Plants, represents the ability to thrive in a business environment of constant and unpredictable change. Best Plants foster agile operations by creating flexible production systems that are capable of turning on a dime. These systems often include focused factories and manufacturing cells staffed by teams of employees. Agility is enhanced by pull-type production systems, which build goods based on demand. These systems utilize the principles of JIT and other lean manufacturing techniques to reduce inventory to its lowest possible levels.

Agile structures require an educated, committed workforce. Many of the Best Plants use self-directed work teams to staff work cells, enhance response times, decrease time spent on decision making, and to recapture the resources spent on managerial support. Finally, Best Plants continue to build their agility by enlisting the help of their skilled workforces in capturing the competitive advantage inherent in continuous reductions in process cycle times.

> The old batch processing methodology, even if executed with precision, cannot compete with new techniques using demand-pull, one-piece flow in an employee-empowered atmosphere.
>
> —*Milwaukee Electric Tool Corporation*

CHAPTER 5

EMPLOYEE INVOLVEMENT

Whenever the leaders of America's Best Plants discuss the source of their plants' successes, there is always one reason that comes first. Ask them the most important element in the pursuit of a world-class level of customer focus, and the same answer is tendered. Ask them the key to attaining perfect quality or how to build an agile organization, and you'll get the same answer again and again: "people." *People* are the power that drives Best Plants.

People have always been the driving force of manufacturing, but until the last decade or so, "people" was not always the answer you would get when you asked the questions posed in the previous paragraph. Instead, you might hear about automation or strategy or plain old managerial savvy. The way some managers told it, employees were actually a substantial barrier to business success. They didn't want to work; they demanded too much money and too many benefits; they just wouldn't do what they were told.

Well, one of the most satisfying aspects of studying the Best Plants is hearing over and over again exactly how foolish that thinking was and is. The sincerity with which Best Plants' management talks about the people at work in their plants is genuine and heartfelt. There is no doubt that they clearly believe that the participation, creativity, and knowledge of every employee was a decisive force in all their accomplishments. They also know that the development and maintenance of a skilled and involved workforce is a critical enabler of operational success. It is, in fact, the first supporting competency of Best Plants.

The development of a world-class workforce is a task on which Best Plants spend a great deal of time, effort, and resources. They don't silently value their employees; they communicate those feelings. Best Plants trust and respect people, treat them honestly and with equality, and maintain an atmosphere of open communication.

Best Plants invest in people. They understand that people are *appreciating* assets, that employees grow more valuable with time and experience. They spend a great deal of money developing the full potential of their employees through job training and continuing education.

In addition to their investment in employees, Best Plants create innumerable opportunities for people to contribute to the success of their companies. They encourage and obtain high levels of employee participation in a wide variety of formal and informal improvement efforts.

Finally, Best Plants recognize and reward employees for their contributions. From companywide celebrations of special achievements to financial participation in the profits and the savings that employees generate in their activities, the plants understand that positive feedback is the only way to keep employees involved. They also know that the levels of employee participation needed to create a world-class plant is a right that is not simply granted to an employer, but one that must be earned.

Opening Up to Employees

In the bad old days, communication between management and workers in the typical manufacturing plant was a pretty clean-cut affair. Management decided where it wanted to go and sent the marching orders down the chain of command. (If that has a martial ring to it, it's honestly come by—the traditional hierarchical model of organizations was borrowed from the military.) If workers wanted to communicate with management, they approached their direct superior. If they had a complaint, they could take that to their shop steward or take it home. And that was about all the communication you needed.

The communication links in America's Best Plants are a far cry from the rigid single line of communication of old. They must be radically different; after all, the pyramid of command and control no longer exists in many of these plants. With layer upon layer of middle managers and supervisory personnel eliminated and supervisor/employee ratios heading ever higher, communication becomes a critical issue.

Super Sack offers a good portrait of the communication issues that surface when the organizational structure of a company is radically changed. As the company reorganized its employees into self-directed work teams (SDWT) and eliminated the functional-distinctions boundaries of a batch-driven manufacturing process for a series of work cells, its communication structure was fundamentally disrupted. In its place, the company had to create a series of new links. Manufacturing vice president David Kellenberger explains the requirements of a team-based structure as follows:

Under real SDWT, supervisors are not used. We have a plant manager and a quality control manager, of course, but their job is to communicate customers' expectations to the teams and show the teams where and how they fit into the big corporate picture. It is not management's job to supervise, but to help communicate corporate expectations and assist and encourage employees in their establishment of team goals. Technical support staff is available to assist where needed. Each team has a team leader who is elected by the team for a three-month period. Their job is to conduct daily meetings, to make sure the team knows what they are to do, and to coordinate team activities and share team information. Each team also has a coach whose role is to motivate, reward positive behavior, assist teams in working through conflict, and be available for support. Coaches may be either staff or hourly members, but all have a primary job apart from their coaching. They may not have expertise in making products and are not responsible for organizing, staffing, directing, or controlling.

Super Sack initiated a daily plantwide meeting to create a regular two-way flow of information between plant management and all of the employee team leaders. This meeting includes a review of the previous day's work, customer feedback, an open discussion of issues facing the plant, and scheduling. "Good decisions are based upon good information," says Kellenberger. "Therefore, we operate on the premise: If it is important for us to know, it is important for you to know."

The elimination of all supervisory jobs created another important communication task at Super Sack. Eliminating the job description of supervisor did not eliminate the supervisors themselves. Instead, these people were now needed to act as the facilitators, coaches, and technical support resources of the new organization. From the beginning of the conversion, the company invested a good deal of effort in communicating this managerial role shift. Weekly facilitator training sessions were conducted for supervisors for the first few months and continued at greater intervals afterward. Four years into the switch to SDWTs, they are still used.

The face-to-face interaction encouraged by Super Sack's daily meetings is the most important form of corporate communication, according to TRW's Vehicle Safety Systems plant in Cookeville, Tennessee. This plant is not meeting-shy. It conducts daily team meetings, production meetings, continuous-improvement meetings, and quarterly plantwide meetings to provide opportunities for two-way employee communication. Skip-level meetings, which bring senior managers face-to-face with employees, are used to gather questions, comments, and suggestions. The most effective means of communication, says section head

Walter Marcum, "is spending time on the shop floor with the work-cell production technicians and support groups."

Face-to-face communication requires trust and understanding between people. At Exxon Chemical Company's Baytown, Texas, site, a unique Human Diversity Pilot project, which is integrated with the business needs of the plant, is helping to build structurally sound communication bridges between the plant's several thousand employees. The Diversity Pilot trained more than 60 percent of the site's employees in awareness of the cultural and other differences between people. This figure includes all the plant's supervisors, managers, and change agents.

The pilot is managed by the Diversity Pioneers, who act through four subcommittees: Communications, which includes a volunteer editorial team; Policies and Practices, which ensures that all people are treated fairly in the conduct of business; Training, which ensures that all people are equally prepared for the future; and Success Criteria, which measures how well the plant is progressing toward its diversity goals.

"The baseline for this measurement is an employee attitude survey which was first conducted in November 1991," explains site manager Ray Floyd. "Since then, we have retested these results in August 1992 and April 1993. On a five-point scale, the percent of our people who scored the company in one of the top-two categories has increased each eight months, and is now nearly three times higher than it was originally."

XEL Communications echoes the beliefs of many Best Plants when it says that verbal contact is the most efficient and effective form of communication. With a workforce of roughly 100 employees in 1986, informal verbal exchanges provided a viable way of communicating plant objectives and gathering feedback. Now, with three times as many employees, the plant retains its person-to-person information flow by thinking of the company as consisting of three "villages." The villages—core manufacturing, OEM manufacturing, and engineering/marketing functions—each consist of around 100 people. "Each has its own team handwriting," says manager Malcolm Shaw, "and one of our management challenges is to manage the interfaces without having to default to a traditional management model."

Copeland's Sidney Scroll plant calls a well-trained, highly motivated workforce "the greatest competitive weapon," and a key ingredient to developing that weapon is a sound communication process. Ongoing communications promote and reinforce the team culture at Sidney Scroll. Explains operations director Randy Rose:

Employees become more productive and more committed to their work when they understand the business. Detailed communications include information on Copeland's competitors, our markets, our financial performance, and our goals. Employees will respond appropriately to competitive pressures when economic communications relating to competitors, pricing, costs, quality, products, and other issues are clear and effective. The beneficiaries of this effort are not only employees themselves but also customers, suppliers, and shareholders.

Managers at Copeland's compressor manufacturing plant meet weekly with employees, keeping them updated on customer issues, concerns, and needs, and business process changes. Feedback from employees is also solicited before process changes are made.

Among the more traditional communication methods in use at the Sidney Scroll plant are daily teams meetings, plantwide quarterly meetings, and an annual state-of-the-business meeting which is hosted by the CEO and COO of the company. Company newsletters and magazines are published, and parent Emerson Electric conducts opinion surveys of all employees to determine employee attitudes, identify areas for improvement, and assist in the promotion of a team-based culture.

Open lines of communication also include the plant's trade union. Sidney Scroll management keeps union leadership informed of current management practices and organizational design plans. It also invites union leaders to attend process improvement training sessions and cites union support as an important reason for successful negotiations in creating broad job classifications and responsibilities (there are 5 job classifications at this plant versus as many as 60 in other Copeland plants). In employee-related areas, such as scheduling adjustments, the union representatives are asked for their input and offered alternative choices.

In the case of General Motors' Engine & Energy Management Systems plant, a union/management partnership called the Quality Network (QN) established the plant's entire production system strategy, including a formal communication plan. In 1992, a communication action strategy that specifies the types of information that need to be communicated to the workforce and the means and routing of each communication, was created and adopted as part of the QN initiative. From corporate feedback to the reasoning behind important decisions to community activities, the QN communication plan offers a structure for constant and consistent employee contact.

General Tire's Rules of Employee Involvement

From the first day of employee orientation, Continental General Tire's Mount Vernon, Illinois, plant is looking for opportunities to communicate with its employees. The plant's means of communication include management-by-walking-around, regular plantwide meetings, daily safety talks, newsletters, and even that classic informal communication network, the grapevine. Because the plant's management knows that actions can speak louder than words, employees are invited to participate in policy and benefit decisions, dismissal and disciplinary actions, and plant shutdown scheduling. Says Floyd Brookman, "Involving our employees is simply the way we do business."

Management-by-walking-around, a managerial tool popularized by author and consultant Tom Peters, has been raised to an art form by plant manager Jim Dolwick, who says, "I have a lot of friends on the factory floor who know I spend time in the plant. You have to smell it, touch it, and see it to be able to confront and correct problems."

Here are some of General Tire's rules of the road for employee involvement:

Job descriptions are guidelines, not confines. Thoroughly screen applicants and hire only the best. Then train, empower, inspire, and remove restrictive craft lines.

Investing in a Winning Workforce

Creating a workforce capable of successfully pursuing the three core strategies of Best Plants requires an organizational commitment to lifelong learning and a substantial investment in training and education. Listen to Harry Peterson at Milwaukee Electric Tool: "*Train train train.* Plan on a minimum of 100 hours training per employee per year. This would combine concept, classroom, and hands-on training. Kaizen workshops work and are excellent training tools."

And just in case Peterson didn't make the point clear enough, XEL Communication would add: "Training and education of the workforce are never suffi-

Promote and maintain education. Education should be a never-ending process and the plant offers 100% tuition reimbursement for all colleagues.

We never communicate enough. We actively encourage communications and work to dissolve real and perceived barriers. Many facilities include "open-door" policies in mission statements and employee handbooks, but it's just the way we do business in Mt. Vernon.

Success in any program requires management commitment at all levels. Employee involvement and STAR [a gainsharing-based continuous-improvement program] drive the success of our plant, yet neither would be viable without the firm commitment and determined effort of management in Akron, Ohio, and Mt. Vernon, Illinois.

Recognize and reward performance excellence. We present symbolic rewards for acts of valor or commemoration. An employee team designed a recognition grid to assure that everyone in the plant shared equal opportunities for recognition. Managers present various tokens and reward people responsible for setting performance records, promoting safety, personal milestones, birthdays, or any event they deem worthy of special recognition.

Promote and ensure job security. A vital part of our culture is knowing that our jobs are secure. Being the low-cost, high-value manufacturer of products within our company ensures that jobs will exist in our facility for as long as we maintain the edge.

cient, never adequate to fulfill the vision, and never complete. Training is never too expensive, only ignorance. And don't forget to have management attend the same training the workforce does."

XEL describes itself as a "brain-driven" organization and uses a volunteer-based Education Task Force to plan and administer its curriculum of work-related skills training. The task force's program includes basic and remedial academic skills, operational skills, SPC and computer-based training, as well as team, relationship-building, and communications skills.

In addition, XEL's human resources department created a partnership with the Community College of Aurora in a benchmark Workforce 2000 project, funded by grants awarded in 1989, which took effect in 1990. Since then, the company has been involved in four consecutive educational projects, two of them in partnership with the state of Colorado and two with federal agencies.

Educational partnerships are a common feature among the Best Plants. Allen-Bradley's 1992 award-winning EMS1 facility, for instance, created a Surface Mount program for its technicians and engineers that is conducted under the auspices of the Milwaukee School of Engineering. The program consists of seven courses in mathematics and electronics. MSOE instructors are brought to the plant to teach the courses; employees attend on their own time. They earn college credits for course work in algebra, trigonometry, electronic circuits, semiconductors, electronic packaging, digital circuits, and surface-mount technology. For those employees in the program, annual training totals over 150 hours per year.

Super Sack created its corporate university in conjunction with Grayson College of Denison, Texas. Super Sack University allows employees to earn an associate degree in manufacturing technology. Located in an area where the high school dropout rate is 30 percent, Super Sack also offers an on-site GED preparation course to its workers.

Nippondenso Manufacturing takes a hybrid approach to employee training, utilizing both internal and external training strategies. Internally, the company developed over 300 hours of management skill training, as well as an evaluation system that measures the effectiveness of training by recording participants' increase in knowledge, reaction to the training, and on-the-job skill use before and after training.

At the same time, the company maintains partnerships with local educational institutions, including community colleges, vocational schools, universities, and a specialized math and science education center. The plant's apprenticeship program depends on a nearby Regional Manufacturing Technology Center for advanced education in the skilled trades.

At some of the Best Plants, employee training is primarily an in-house function. Baldrige Award–winning Marlow Industries pursues its corporate goal of Total Customer Satisfaction through Total Employee Satisfaction. Employees, the company says, "are our most competitive asset." Training is one of the fundamental methods that the company uses to create Total Employee Satisfaction. Employee training starts on the first day of employment and continues throughout the year, totaling more than 80 hours per employee. Marlow's training is primarily developed and conducted internally and it covers all the disciplines in use within the company. It also includes Marlow Professional Qualification Training (PQS), in which all manufacturing personnel are trained to master their primary skill set and are also cross-trained to enhance organizational flexibility.

Symbiosis Corporation Provides Mentors for Hourly Employees

In 1994, Symbiosis instituted a unique employee-mentoring program. Unlike traditional mentoring practices, in which senior executives take younger managers under their wing and guide their climb up the corporate ladder, the Symbiosis effort is designed to assist nonexempt hourly employees.

The underlying objective, explains mentoring-program coordinator Diane Flennory, is to improve employee morale by supporting personal growth, promoting a sense of belonging, and providing opportunities. "We are making an investment in the morale and productivity of this organization," she says. "The mentors work with the 'mentees' to help them reach their full potential. It is a win-win situation. The employees win and the company wins."

A pilot program launched with 13 employees in the firm's receiving-and-inspection department was highly successful. Says Flennory, "Within two months, I saw attitudes improve, productivity increase, and morale improve. And I saw these employees going the extra mile for Symbiosis."

In 1995, the program was expanded to include 30 mentors, each assigned to work with one employee on a voluntary basis. The mentors carry various corporate credentials. They can be managers, engineers, or supervisors. One thing they all have in common is that they have been with the company "long enough to be grounded in the Symbiosis Way," says Scott Jahrmarkt, vice president for manufacturing.

Mentors are expected to spend at least eight one-hour sessions with their charges on company time. "They can do it wherever they want to—in the office, in the cafeteria, in the boardroom," Flennory points out. The employees enrolled in the program also attend workshops on subjects such as computer skills and personal finance.

The program format includes a mix of structured time and "free-flowing" time. Of every two hours that mentors spend with employees, one hour is devoted to discussion of the training topic. The other hour is considered free time—topics may include almost anything. "I think it's beneficial to the [employees]," says Jahrmarkt. "It can give them insight into what they might want to do in their careers. It is another form of educating our employees."

—John H. Sheridan

Rockwell International's El Paso, Texas, plant features an in-house Technical Learning Resource Center (TLRC) that covers over 4,500 square feet. The TLRC includes fully equipped technical training rooms, a training laboratory, two general-purpose classrooms, a library/conference area containing in-house and home study materials, and a computerized records center. Formal course work in technical processes and programs, as well as training in interpersonal skills, is provided by full-time instructors, who are certified to provide training in all Department of Defense specifications and standards. Training activities are integrated with and complement employee involvement programs.

Perhaps it's true that everything is bigger in Texas. The training facilities at Exxon's sprawling Baytown, Texas, site rival Rockwell's. Baytown maintains a large library of technical, business, and general information, as well as a newly established library of quality and improvement resources. The plant's learning center includes eight interactive video training stations. Eight more satellite locations featuring unit-specific materials are located within the site's various manufacturing divisions.

"Essentially, at any time, people can identify a skill or understanding that they need and receive customized, interactive training in that area immediately," says Ray Floyd. "We are developing the capability to deliver interactive video training through our IBM mainframe. Each employee on the site has access to that machine through terminals located at individual workstations. Our objective is for anyone, anywhere, who has a free half hour, to be able to receive interactive training that is specifically suited to his or her exact needs."

Baytown also maintains a Central Training Organization of twelve full-time professionals and a library staff of six people. Twenty-six additional full-time trainers conduct business-unit-specific training and are located throughout the site.

Interestingly, for all of this site's educational assets, Floyd would warn plants against mass training strategies. "Early in our quality journey, it was believed that the quality tools were so wonderful that great progress could be achieved simply by training everyone," remembers the site manager. "When this was done in a vacuum, nothing actually changed. Most people failed to use their new knowledge acquired from mass training when they returned to the job. Additionally, by the time other systems were in place to support quality improvement, many had forgotten and required retraining."

Creating Opportunities for Involvement

Best Plants fully utilize the creativity and talent of their world-class workforces by providing a wide variety of opportunities for employees to become involved in workplace improvement. Team-based improvement activities and individual programs, such as suggestion systems, are two of the primary means by which employee participation is encouraged and improvements captured.

For anyone who still believes that employee participation is unnecessary and that managers can do all the thinking in a world-class environment, take a quick lesson from Air Products & Chemicals' South Brunswick, New Jersey, plant's frank admission:

> *We have learned that employee suggestions and empowerment will lead to the most effective and efficient solutions to most issues. This is exemplified by a few of our past mistakes. In the first several years of the plant's operation, workers were not always consulted on process changes or designs of new systems. This resulted in a significant amount of rework on a few projects. For example, a loading ramp in the warehouse too narrow to operate a forklift on and baffles in a storage tank which served only to trap the product and later needed to be removed.*

Rockwell International's Defense Electronics plant in El Paso, Texas, reiterates the point when it says, "Without [an] organization that makes full use of each employee's skills and ideas for continuous improvement, world-class performance is unattainable."

One way the El Paso plant encourages employee participation is with its Opportunity for Improvement (OFI) program. Through OFI, employees submit ideas for improving workplace performance. Each idea is tracked for implementation, effectiveness as measured in savings, and response timeliness. Ideas that significantly contribute to improvements in quality or productivity are eligible for monetary awards, ranging from $25 to $10,000, depending on the value generated for the company.

Although formal suggestion systems predate today's high-involvement employee strategies by many decades, Best Plants adopt a good idea when they see one. Nippondenso Manufacturing's BEAM (Because Every Associate Matters) suggestion program, which won a best-suggestion-system prize in 1992, its first year of operation, is a good example. The Michigan chapter of the National Association of Suggestion Systems honored BEAM for results in:

- Number of suggestions submitted: 1,728
- Percentage of associates participating in the program: 50 percent of 850 eligible
- Percentage of suggestions adopted: 96 percent
- Resulting cost savings: $851,000

Chesebrough-Pond's personal product manufacturing plant in Jefferson City, Missouri, created a Continuous Quality Improvement Challenge (C-QUIC) process, through which any employee can initiate a suggestion for workplace improvement. Idea submission is purposely kept simple: A form, available in all departments, is completed listing the idea, the problem it addresses, the solution, and the submitter's name. The plant responds to every C-QUIC with a written thank-you and, within two weeks, a formal notification confirming the idea's acceptance or nonacceptance.

Idea implementations are the submitter's responsibility, unless the idea requires a larger effort, in which case it is assigned to one of the plant's Quality Action Teams. The plant records all challenge progress in its Total Quality database, and C-QUIC project activity and status listings are posted throughout the plant on a weekly basis.

Varian Associates' 1992 award-winning Oncology Systems business created a similar employee feedback and idea-gathering system known as the Employee Action Request, or EAR, system. The plant broadened the scope and usefulness of its system by allowing requests related to any area: technical (product or process), workplace, customer, or personnel. A request is made simply by filling out a one-page form. The forms are reviewed each week by a cross-functional team. Estimates of implementation costs and savings are calculated, their implementation is assigned, and a completion date fixed.

Redwood Falls, Minnesota–based Zytec Corporation updated the traditional suggestion system by turning it into an "idea-implementation system." Rather than gathering ideas, judging them, and then, often after a good deal of time and consideration, implementing them, Zytec's system concentrates on immediate solutions. In the plant's Implemented Improvement System, employees document their ideas on a standard form and take it to a manager for immediate authorization. Idea implementation is the employee's responsibility and, if additional help is required, that same employee manages the project. When fully implemented, the employee assesses the results and returns to management for review and sign-off.

Not surprisingly given the Best Plants' predilection for teamwork, many of the involvement opportunities offered to employees are team-based. Declares James Benson, general manager of The Timken Company's Bucyrus, Ohio, plant, "Associate involvement and empowerment is driven by the belief that people will be most effective in an operating environment that promotes the opportunity to participate, learn, and progress: a *team* environment. The objective of a team environment is to get everybody to think of ways to make the company better."

The Bucyrus plant creates opportunities for its employees to participate in improvement efforts using self-directed work teams, CEDAC (a group problem-solving tool), and a suggestion system that rewards employees for both participation and implementation. (See the Individual Profile for details.) In 1992, when the plant earned its Best Plants award, its goal was to involve 100 percent of workforce in at least one problem-solving project during the year. Progress toward that goal is tracked monthly and displayed prominently in a main aisle of the plant.

"Teams, teams, teams is the measure of total employee involvement," according to plant manager Tim McGraw of Johnson & Johnson Medical's Sherman, Texas, plant. On winning their Best Plants award in 1991, the plant was able to boast 275 teams involving 70 percent of employees. (Their goal is also 100 percent involvement.) Every employee is encouraged to spend one hour per week "off" the job, working on team-based improvement efforts. In 1990, these teams identified nonconformance costs in excess of $10 million and generated savings in the amount of $2.6 million.

Improvement activities are not the only opportunities for employee involvement at the Best Plants. They also involve employees in areas that might surprise the traditional manager, such as the resolution of disciplinary problems and even employee termination. At GE's Electrical Distribution & Control Division, a new process was needed to replace the traditional, ungainly, employee complaint procedure. In the old procedure, the immediate supervisor was the first stop in a five-step process whereby employees had to press a complaint level by level all the way up to the division headquarters. With each step, of course, their frustration grew.

Now, complaints are presented to a quasi-judicial panel consisting of three hourly and two salaried employees, whose final decision is binding. The panel is not empowered to change plant policies, but it does make factual determinations involving their applications. All plant employees are invited to enter the pool from which panel members are drawn, and participants are remunerated.

Finally, creating opportunities for employee participation also means removing the barriers to involvement. The Best Plants as a group are highly conscious of the two-edged nature of the continuous-improvement sword. They know that employees quickly recognize that a successful program of improvement means they may well be working themselves out of their jobs. In fact, the much-remarked-upon resistance of employees and trade unions to improvement programs has often been based on the well-founded fear that what is good for the companies where they work is not always good for them. Best Plants, while stopping short of comprehensive employment guarantees, understand these fears and, in no uncertain terms, assure employees that their help in building a world-class factory will not cost them their jobs.

States Harry Peterson of Milwaukee Electric Tool, "We have taken the position with all METCO full-time regular employees that they would not be subject to layoff as a result of improvements made through our efforts to achieve world-class status." Happily, at this Best Plant, as well as many others, the growth in productivity has been matched by growth in demand. The company does, however, actively avoid layoff scenarios by using product promotions to stimulate sales and by moving employees to other work cells as necessary.

Cincinnati Milacron's Batavia, Ohio, plant has given "considerable thought" to avoiding excess staff. To avoid layoffs, it maintains its permanent staff at an average business level. Spikes in business are handled with the help of temporary-personnel agencies and contract services. The plant also continuously seeks diversity by expanding into new products and markets as a way to moderate the effect of business downturns. And, finally, the company actively cross-trains employees to obtain maximum flexibility in work assignments.

At Sony's San Diego Manufacturing Center, where production per employee almost doubled in a three-year period, it is clearly communicated that no employee will be laid off as a result of the plant's improvement programs. Instead, management freezes new hires and uses cross training and flexible workforce programs to create new homes for displaced workers.

At Varian Associates' award-winning NMRI unit, where no layoffs have occurred as a result of flattening the organizational chart and consolidating functional departments, Ed Sehrt explains the company's strategy:

> *Sourcing strategies are planned well in advance with consideration of its impact to the workforce to ensure placement within the business unit. Vacancies from attrition are filled after first consideration is given to potential internal candidates, and external new hires are accepted with broader skills to match all potential positions. A similar approach is taken with administrative positions, making possible an easy transfer from one responsibility to another without losing efficiency.*
>
> *By cultivating a pool of highly skilled employees with a broad range of expertise, a highly flexible workforce permits temporary assignments to areas of periodic peak demand and serves as a tool for cross training. Thus, permanent reassignments become effective with little trauma to the employee.*

The employees at TRW's Vehicle Safety Systems plant in Cookeville, Tennessee, know that as their ideas improve the operation, many tasks could be eliminated. They also know that any employees who are displaced by methods improvements or reduced work-cell size will not be laid off. Instead, they are absorbed into other growth areas within the facility. This policy is stressed in all of the plant's continuous-improvement training. The company goes a step further and explains that when a position is not immediately available for a displaced employee, the employee will be trained, with pay, in programs such as the plant's Team Boot Camp, until a position becomes available.

Some welcome final advice on how to create a high-involvement workforce comes from Diane Breedlove of Coherent, Incorporated.

- Involve employees in the writing and implementation of policies and procedures. It is important that they know their opinions and best interests are being considered.
- Ensure employees have all the information they need to make good business decisions (i.e., profit and loss statements, budgetary data, return-on-investment information, and other information not typically shared with direct labor employees).
- Trust that employees will make good decisions.
- Allow employees to make mistakes without being criticized (as long as they are not damaging to the business), and encourage them to try other alternatives if their initial decisions prove unsuccessful.
- Challenge employees to continually set and achieve higher goals.
- Provide an open-door environment where employees feel comfortable about expressing ideas and asking questions.
- Recognize individuals and groups for their accomplishments (i.e., time off with pay, group and individual lunches, small gifts, thank-you cards, etc.).
- Be willing to offer incentives to employees for performing tasks, duties, and responsibilities typically handled by management (i.e., gainsharing, other incentive plans).
- Be truthful and honest with employees. Don't be afraid to tell them you don't know the answers to their questions or that something is confidential and cannot be discussed.

Recognize and Reward Employee Involvement

Recognition and reward are two sides of the same motivational coin, and Best Plants know that without both, their employee involvement efforts will slowly and surely wither away. Recognition is just as important as financial reward

Idea for Improvement at Baxter Healthcare

Baxter Healthcare's IV Systems Division in North Cove, North Carolina, is one the Best Plants that has not formally converted to a team-based production system. Nevertheless, a team culture permeates the facility, and almost all the myriad opportunities for employee involvement are team-based. Rightly so, as the plant is, according to manager John Martino, "one family with one future." He explains some of the plant's team improvement activities as follows:

> Self-directed teams range from formal quality working teams, which are established by charter and work toward stated goals with documented meetings and results, to ad hoc teams established for specific looks at various issues on a less formal basis. Focus groups are widely utilized to garner information about specific issues, processes and practices. They generate potential ideas for improvement and serve as sounding boards for proposed changes. Total Productive Maintenance working teams have been solidly entrenched throughout the operation, significantly reducing machine downtime by turning responsibilities over to operators. Safety committees are widely used in the plant for generation and implementation of preventative measures, as well as for accident investigation and problem resolution.

North Cove's employees are involved in a variety of participative and empowered activities, both formal and informal, and one of the fundamental supports of these programs is the plant's Idea for Improvement (IFI) process.

---

and the Best Plants are prompt to thank their employees in any number of creative ways.

At General Electric's Auburn, California, plant, for example, every December employees choose from a list of "awards" for each of the following four quarters. The program is tied directly to the attainment of plant goals in the areas of inventory, cost, productivity, and service. For 1995, the awards included

Each of the plant's nine quality improvement teams have IFI coordinators, who are mostly hourly employees empowered to form cross-functional resolution committees that address and implement improvements. The IFI systems support the achievement of annual cost savings of $11.6 million at the plant, a significant advantage in a price-driven marketplace in which this plant strives to remain the low-cost provider.

In a plant that projected a baseline of 859 improvement projects in 1994, employee accomplishments are numerous. They include:

First-pass-yield release team. A cross-functional team reduced product release times to a 24-hour window of opportunity from 1991's average of 58 hours.

Maintenance team. A group of employees redesigned the lubrication system for compressors, a single improvement that saved $350,000 by extending the service life of machinery to 19 months versus 6 months prior to change.

Maintenance team. Employees redesigned double work-saver truck beds to eliminate a recurring cause of downtime. The result: only one failure in 40 cumulative months of run time for five trucks.

Interplant shipping order team. A cross-functional team redesigned packing configurations that allow customers to have all materials available for their assembly. This also improved service requirements and documentation for the customer.

Plastic department reorganization teams. Teams of employees analyzed work flow and achieved a 30 percent reduction in head count.

Automatic printer/filler team. A cross-functional team implemented a process improvement that increased production by 55 percent, reduced line scrap by 70 percent, and reduced product defects by 35 percent over a two-year period.

pizzas or steaks (prepared and served by the salaried staff), T-shirts, and turkeys.

Unisys' Pueblo Operations created the Customer Satisfaction Team Olympics (CSTO) to promote participation in Total Quality Process teams and to recognize team members for their contributions. The unique aspect of the program is that it is self-administered, and anyone can nominate a team, including the team itself. The team members are empowered to decide in which award category their implemented improvement fits—bronze, silver, or gold—and the awards range from a team gift certificate for dinner to individual cash payments. On applying for the Best Plants status, the plant had given 41 team awards, including 7 at the gold level.

Fisher-Rosemount's PIT Crews serve the plant's improvement drive so well that a PIT Stop subcommittee was formed to "promote the PIT mechanism for continuous improvement by providing recognition and visibility, encouraging functional interchange and communication, and facilitating training and development." The PIT Stop stages events, such as a recognition and training picnic for PIT Crew leaders, a PIT Crew Olympic competition where teams compete against each other in 10 improvement-related events, the decoration of a Christmas tree with ornaments promoting the PIT Crew's "New Year's Resolutions," and a "Play PIT ball" program. In addition, PIT Crew success stories are publicized and celebrated in the plant's TQM newsletter, *On Target.*

Financial bonuses and profit-sharing plans are also a common form of reward among the Best Plants. At Super Sack, a generous bonus plan allows its members to earn 10 percent of all sales generated above the plant's break-even point. The teams know the selling price of the product they are manufacturing and determine their goals based upon the bonus level they wish to achieve. The bonus is paid twice yearly: just prior to summer shutdown and at the Christmas holidays. The plant also periodically offers special incentives that are intended to generate fun and create additional compensation for the teams.

At Tennessee Eastman's Kingsport, Tennessee, facility, all employees participate in Success Sharing, a companywide compensation plan in which employees may contribute up to 5 percent of their compensation and, in return, receive an annual (pro rata) lump sum based on Eastman Chemical Company's return on assets. Eastman Chemical Company employees also participate in the Eastman Kodak Company Wage Dividend, which is based upon Kodak performance.

Formal gainsharing plans are in force at a number of the Best Plants. For example, Rockwell's El Paso, Texas, manufacturing facility gives all employees a financial share in the facility's quality, cost, and schedule performance gains. The reward system requires a measurable gain over planned and approved goals. At the beginning of each fiscal-year quarter, the plant's Integrated Product Teams participate in setting performance goals for that period. Quality goals maintain the highest performance priority and account for 60 percent of the payout. At the completion of each quarter, the El Paso IPT's actual performance-to-goals results are tallied and a total facility performance measurement is calculated. Payouts are distributed in December of each year.

The Timken Company's Faircrest Steel Plant's Performance Plan is another gainsharing system that rewards employees for continuous improvements in quality, productivity, and cost reduction. "The old incentive systems paid by individual operations," explains senior human resources representative Tim Chapin. "Consequently, individual operations worked to optimize their area without regard for downstream operation processes or quality issues. Conversely, the Faircrest Performance Plan truly engenders teamwork because the entire plant is considered a continuous process and is on one incentive system."

There are a few Best Plants that are still very close to their entrepreneurial roots and one in particular, 1995 award-winning Symbiosis Corporation, has created the kind of recognition and reward story that business legends are made of. When the company founders sold Symbiosis to American Home Products, they thanked the workforce that helped them earn the $175 million purchase price by giving them 14 percent, or $24.5 million. Fifty-four key employees received payments ranging from $52,500 to over $1 million each. Eight employees hit the $1 million mark or better. In addition, every one of the plant's 700 employees received a $20 tax-free for every week they had worked for the company. And, just so the workforce remained committed to the new owners, the payout included a bonus for continued performance that could double the original payment.

Summary

Employee involvement is the first supporting competency practiced by the Best Plants. They know that the full participation of their workforces is needed in the quest for world-class status. Best Plants begin to build an active and involved workforce by opening as many lines of communication as possible between workers, their trade unions, and management. These lines enable the initiation of a two-way exchange of information that builds trust and knowledge in the workplace.

Best Plants know that their workforces are an appreciating asset and they invest in their employees accordingly. The plants use a wide variety of methods to train employees and call upon both internal and external resources to provide workforce education. Training is a continuous process at the plants and learning is viewed as a lifelong pursuit.

Employee participation requires many opportunities to become involved in workplace improvement activities. Accordingly, Best Plants create innumerable ways for employees to participate in activities beyond their immediate jobs. Traditional suggestion systems, more modern idea-implementation systems, and team-based improvement activities are all offered to Best Plants' employees. Further, the plants do not expect workers to participate in workplace improvement efforts at the risk of their jobs.

Finally, Best Plants reward and recognize the achievements of their employees. They celebrate employee successes and share the financial rewards that their employees have earned.

> High-technology equipment and facilities can easily be bought, but the development of world-class competitiveness requires the development of world-class *human* resources.
>
> —*Rockwell International, Autonetics Electrical Systems*

CHAPTER 6

SUPPLY MANAGEMENT

Supply-chain management is a discipline that has undergone a tremendous reappraisal in recent years. As more and more companies pursue the competitive advantages inherent in an intense focus on customers, total quality in product and process, and organizational agility, they often find that their journey is leading them outside the walls of their own plants and into areas of their suppliers' operations that they had never before seriously considered.

Supply management reappraisal has been driven by a number of realizations. For instance, many of the Best Plants discover that as they begin eliminating their internal organizational barriers to world-class levels of performance, the quality and timeliness of the raw materials and parts they purchase become the source of some of their largest problems. They are also sometimes surprised to find that their own supplier practices and policies are the highest obstacles in their path to improvement. Even manufacturers with well-established supplier strategies can suddenly be confronted with the hard fact that they cannot reach the breakthrough levels of improvement that the marketplace is demanding without completely redesigning their working relationships with their suppliers. All these circumstances lead to one important conclusion: A Best Plant's supplier is just as important a supporting factor of its performance as any internal component of success.

The truth of this conclusion is immediately apparent when Best Plants examine the goods they produce from their customers' perspective. In a customer's eyes, a plant's products are not some mix of components and materials from different suppliers that have been assembled or processed by yet another company. They see a *single* product purchased from a *single* source, and when a defect or failure occurs, they look to the company who sold the product for satisfaction. Accordingly, knowing that the ultimate responsibility for product quality, cost, and service is theirs, Best Plants extend their corporate view to include the entire supply chain. This extended perspective suggests that suppliers be treated in the same manner as a plant's employees, with respect and honesty and as an appreciating asset. It also requires that suppliers be held accountable for the same levels of service, quality, and agility as the Best Plants establish for their own performance. In other words, a world-class supply management strategy and process is a supporting competency of Best Plants.

Best Plants start creating excellence in the supply chain by analyzing and evaluating the process through which they purchase the materials they consume in their business. They seek to align their supply strategies with their major business objectives—customer service, quality, and agility. In many cases, this means a complete overhaul not only of the supply process, but of the underlying assumptions and rationale that drive it.

One goal of this new supply strategy is the establishment of close and extended connections between suppliers and Best Plants. Best Plants open new avenues of communication, become involved in the development of their suppliers' operations, and invite suppliers to participate in the development of their own products, processes, and cycles. There are various ways to establish this closeness to suppliers, but all are aimed at the same target: a seamless relationship between plant and supplier that fosters service, quality, and agility.

Finally, the Best Plants aggressively pursue the same philosophy of continuous improvement externally with their suppliers that they do internally in their own facilities. The plants help suppliers build their capabilities because they know that the operational improvements their suppliers realize will also benefit the Best Plants themselves.

Creating a Supply Strategy

The cost per unit on a supplier's invoice has traditionally been the driving force of manufacturing supply strategy and, if the off-the-record complaints of primary suppliers to many of the nation's major industries are true, there are plenty of companies where unit cost is still the be-all and end-all of supply management. In this thinking, the lowest-priced material or part is the best choice, and the company that provides it is the supplier of choice. It is a simple system that is, of course, based on fallacious thinking.

Cost per unit is only one element of the supply decision making utilized by Best Plants, and many times it is not the most important element. Instead, Best Plants consider the total cost of procurement, a figure that includes a much wider range of supplier characteristics. For example, total cost will consider the timeliness of shipment, which is a key support of JIT production systems, and the quality of parts, since defective materials and parts can add tremendous expense in returned products and service.

Best Plants find that when they tally the total cost of working with a supplier, the final figure usually bears little resemblance to the invoiced cost per unit. John Michaelos, John Crane Belfab's quality manager, clearly makes this point when he says, "Long-term partnering eliminates traditional administrative cost, such as individual purchase orders, and promotes continuous process improvement, long-term investments, and value engineering. Certification ensures product quality and eliminates non-value-added expenses of incoming inspection."

As a direct result of this understanding, Best Plants restructure their approach to supply management. Sony's San Diego Manufacturing Center undertakes what it describes as "a very lengthy courting stage" when considering a new supplier. Explains vice president Stephen Burke:

> *The basic process and what we look for can be broken into the following four steps. The supplier:*
>
> 1. *Must have the capability to make our parts or provide services*
> 2. *Must be able to meet or exceed our quality standards at current production levels*
> 3. *Must be able to maintain our quality standards when volume increases*
> 4. *Must be committed to continue to work with us to reduce the total cost to provide us goods and services.*

Once a supplier passes muster at the San Diego Manufacturing Center, it becomes "a member of the Sony family" and is given greater access to company data. The plant shares its business direction, provides monthly forecasts, and establishes strong communication links to its suppliers.

Gilbarco's Greensboro, North Carolina, plant redesigned its supply management strategy using a supplier qualification and a part certification process known simply as the Supplier Partnership Program. "Gilbarco's strategy," says manufacturing strategy manager Jerry Smith, "is to optimize our supplier base by upgrading existing suppliers or moving to world-class suppliers. We are establishing partnering relationships with single-source suppliers while making the transition to 'best value' instead of 'lowest price' as our supplier selection criteria."

As might be expected, improved quality is a major objective in Gilbarco's supplier strategy. Suppliers take full responsibility for the quality, reliability, and workmanship of all components, assemblies, or materials they provide to the company. Gilbarco's suppliers are required to use SPC to reduce variation in their processes and prove their goods meet the plant's requirements. Once accepted, suppliers can usually expect a single-source contract.

The plant selects suppliers using a team-based process that includes a formal Supplier Quality Systems Survey and, once approved, suppliers are eligible for training in areas such as quality, JIT, SPC, and other support at no charge. Gilbarco also created a part-certification program that allows purchased materials to pass directly into the manufacturing processes in the plant. This program is applied on a part-by-part basis and requires that suppliers provide process documentation that includes a Cpk of 1.33 or better. Once certified, incoming inspection is waived and parts are delivered directly to the point of use.

In 1989, when Dallas, Texas–based Marlow Industries realized that reaching world-class status meant revamping its entire supply structure, it analyzed its supplier base with the complementary goals of reducing its total cost of procurement and increasing the quality and timeliness of supplied materials. The company targeted the top 80 percent of its purchases from suppliers for improvement and established a goal of reducing the number of its critical production suppliers by half within a three-year period. There are many benefits to a reduction in the supplier base, not the least of which is the fact that it allows a plant to devote more attention to individual suppliers. At the same time, the company created a Supplier Performance Index (SPI) that measures the individual and combined quality and delivery performance of their suppliers against a world-class benchmark. Marlow communicates SPI results to its primary suppliers on a quarterly basis, and, internally, the plant's TQM council and senior management review the results with an eye toward continuous improvement.

Reflecting the intensity of Marlow's customer focus, its supplier process also includes a feature asking suppliers to rate the company as a customer. Marlow gathers this feedback on its own responsiveness, professionalism, and conformance to requirements from its suppliers' perspective as a way to improve its own supply management performance. On winning the Best Plants award in 1993, Marlow was able to report a reduction in its supply base from 204 to 128 suppliers, with 80 percent of the company's top suppliers maintaining quality levels that enable the plant to accept materials and place them directly into production inventory without inspection.

As one studies the supply strategies of Best Plants, it soon becomes obvious that suppliers are viewed as long-term business partners. At Fisher-Rosemount's Chanhassen, Minnesota, plant, for instance, the supplier development program is an extended process that first approves purchased materials and processes and then forms key alliances for continual improvement. To work with the Chanhassen plant, all suppliers must be listed on its Approved Suppliers List. Addition to this list requires a supplier to undergo an evaluation of its total business capability, parts and material evaluations on both first-run and production quantities, a system audit, and to establish 12 months of performance history.

For suppliers who provide critical parts and services or provide a high-value input to the plant, a formal Supplier Alliance is established. In this case, a cross-functional supplier team is formed at Chanhassen to work with the supplier to identify target areas for improvement at both the supplier's operation and at the plant. In 1993, Rosemount had 13 Supplier Alliances in effect. Among the results of this approach: an improvement in supplier on-time delivery from 90 percent to above 95 percent and an increase in supplier quality from 70 percent to above 90 percent.

From a supplier's perspective, the certification and partnership programs in place at Best Plants represent a major investment in time, commitment, and effort. Yet the business results are usually worth it. Suppliers can often expect single-source contracts for the entire life cycle of a product and, perhaps of much greater value, the opportunity to grow along with their customer. At Miami, Florida–based Symbiosis, for instance, the corporate-intent statement known as "The Symbiosis Way" specifies the company's objective "to create mutually beneficial partnerships with suppliers." One measure of that commitment is illustrated by the fact that 20 suppliers accounted for 80 percent of the Best Plant's purchased material in 1992, and the same number of suppliers accounted for same amount in 1995 . . . after Symbiosis had registered a 35 percent cumulative growth in sales and a 50 percent increase in product mix.

In return for being given first opportunity at new business, Symbiosis' Key Suppliers are expected to work to jointly establish performance goals. Continuous-improvement meetings are held monthly and attended by cross-functional representatives from both the supplier and Symbiosis. These meetings address current problems, ongoing cost reduction, and quality improvement programs, design changes, etc.

During the start-up phase of new projects and during problem investigations, the company's buyers, engineering, and quality employees travel to their supplier's plants to provide technical support, quality-planning assistance, and to develop procurement strategy. For the suppliers who are working on more mature products, the plant's buyers turn over Symbiosis' MRP spreadsheets and, when needed, train suppliers to do their own material resource planning. Thus, suppliers have the plant's entire production forecast and are enabled to more knowledgeably plan their own production schedules.

Supplier Integration at Varian Associates' Oncology Systems

The objective of supply management strategy at Varian Associates' 1992 award-winning Oncology Systems business unit is to develop world-class supplier partners capable of providing 100 percent defect-free parts delivered on a just-in-time basis directly to the point of use on the Best Plant's manufacturing floor. Toward this end, a supplier certification process and a partnership program dubbed Value Managed Relations (VMR) are being pursued with the plant's suppliers.

The Oncology Systems unit turns suppliers into world-class manufacturing partners using a three-stage supplier integration process. The first stage is a qualification process that earns a prospective partner a spot on the unit's Qualified Supplier List. The second stage results in a certified parts agreement. The third results in a supplier partnership formalized with a Value Managed Relations (VMR) agreement.

Only Oncology Systems' top-tier suppliers are awarded a VMR contract. This contract elevates a supplier to the rank of "virtual business partner with plant," according to manufacturing manager Jim Younkin. It features supplier benefits such as long-term (usually three- to five-year) exclusive contracts, access to training, tooling, and technical assistance, streamlined administrative procedures including billing and invoicing, and improved scheduling visi-

Managing the often complex supply base of a Best Plant requires a prioritization scheme. Starting with over 1,500 production-material suppliers (a figure that has since been reduced to 500), Rockwell International's Autonetics Electronic Systems Division uses a four-tier classification system that allows the plant to concentrate its supply management efforts on those suppliers who have the greatest impact on product design, quality, cycle time, manufacturability, cost, and delivery. Project administration manager Don Teague describes the system as follows:

> *Tier 1 suppliers are those aligned with El Paso's current and future technology direction. They provide the high-technology products that are critical to the success of the product. These suppliers become members of El Paso's Integrated*

bility. The goal: to move the plant closer to a small, manageable supplier base comprised of world-class organizations who are willing to work in close partnership to gain competitive advantages for both parties.

The task of managing the supplier improvement and integration process at the Best Plant has been delegated to a series of empowered cross-functional teams, as follows:

Supplier Quality Team. Monitors supplier quality performance. Takes corrective action on repetitive problems. Forms Supplier Corrective Action Teams including supplier representatives to resolve specific supplier problems.

Supplier-Base Management Committee. Approves and removes suppliers from the supplier base. Selects and develops VMR relationships with top-tier suppliers.

Supplier Certification Committee. Recommends and develops certified suppliers. Monitors performance.

Supplier JIT Committee. Coordinates JIT delivery of components. Develops JIT arrangements with key suppliers.

Additional supplier partnership activities include the participation of key suppliers on product development teams. Oncology Systems also conducts periodic supplier surveys to determine needed improvements in supplier relations and an annual supplier award program to provide recognition and incentives. The company also established a supplier technical support center.

Product Teams (IPTs) during the initial design of a product and work as partners for the mutual success of the identified project/product.

Tier 2 suppliers provide commodities that have already reached the mature phase of their technology life cycle. In many cases, these suppliers have already attained Rockwell's certified supplier status and receive a small degree of oversight based upon the dollar value of their commodity. Supplier performance trends are tracked for quality, cost, cycle time, and delivery. Because of their mature performance history, these suppliers are already members of Rockwell IPTs and support new product teams as they arise.

Tier 3 and Tier 4 suppliers support IPTs with low-cost and high-quantity material common to the electronics industry. Their material is not design-

critical and is normally available as needed. Their quality and delivery performance levels are well established within the industry and tracked. Rockwell has worked to reduce the total number of these common suppliers and has achieved 66 percent reductions in the past five years.

The El Paso plant also maintains a Supplier Rating Incentive Program that is used to determine "best value" suppliers. All the plant's suppliers are included in the rating program, which is based primarily on quality and delivery metrics.

As noted above, sometimes it is a plant's own supply processes and practices that stand in the way of performance improvement. Exxon's Baytown, Texas, site manager Ray Floyd calls its efforts to streamline the way it purchases Maintenance, Repair, and Operations (MRO) supplies "our most important supplier partnership strategy." The two-stage improvement project began with a fundamental redesign of the internal process that the site uses to plan, acquire, store, and use MRO materials. The new process abandoned the functional silos of the old system, creating in its place a multifunctional Service Delivery Unit (SDU) which performs all MRO purchasing functions. By 1993, one-quarter of the site's MRO activities had been converted to SDUs at a 30 percent savings in labor and a 100 percent elimination of paperwork processing errors.

The second stage in Exxon's streamlining of MRO purchasing was the adoption of an Integrated Supplier strategy, which is based on a benchmark program created by General Motors' Saturn Corporation. By using single-source suppliers for MRO commodity classes, Exxon reduced its supplier base by 25 percent in 1992, and again in 1993. "We currently have a single integrated supplier for the commodity class of office products, furniture, and equipment," explains Floyd. "We are negotiating integrated supply for three more commodity classes—pipe, valves, and fittings; electrical supplies; and industrial supplies—which will be in place by year-end [1993]. Within three years, we will have only 12 suppliers for all of the commodity MRO needs of the business."

Getting Close to Suppliers

Once the Best Plants move beyond a simple unit-cost relationship with suppliers, they commit themselves to the same kind of relationship they maintain with their own customers and employees. Partnerships of this kind require a closeness with suppliers that is often described by the word *seamless.* In a seamless environment, a Best Plant and its suppliers can communicate without undue restraint, and goods can flow back and forth between their operations without restrictive barriers.

Copeland Corporation's Sidney Scroll plant offers a good illustration of how a seamless supply partnership works in practice. Operations director Randy Rose describes the relationship with supplier, Osco Foundries:

> *Typical of our strong philosophy of supplier partnerships is that with Osco Foundries, the shell casting facility that manufactures all Copeland Scroll castings. Osco has been involved in the Scroll program since 1985 and has been instrumental in working with Copeland on our shop floor to eliminate quality problems and to improve the machineability of the Scroll castings. Osco has signed a Confidentiality Agreement with Copeland which covers both the formulation of the metal and the configuration of the Scroll castings, as well as how the castings are machined. The only way we could reach the current state of manufacturing efficiency is through sharing and understanding both companies' processes. Copeland has also provided technical support to Osco in process development, process control, and metallurgy, and has used techniques, such as Taguchi experiments, to identify the root cause of problems and find satisfactory corrections.*
>
> *Osco is a member of the New Product Team and is intimately involved in all aspects of the casting design and machining process. The best way to achieve the lowest-cost raw material and finished component is to leverage the design process by utilizing the supplier's expertise and achieving the lowest true cost for the component. By being an integral part of the new product development team and attending meetings with our design engineers, Osco has been able to assist us in achieving the best quality and cost combination for this critical part of the compressor.*

Product development is one area in which Best Plants commonly turn to the suppliers for help in cutting cycle times and costs and improving quality. (See Chapter 8 for more detail.) At Super Sack, for instance, vice president Dave Kellenberger says, "We recognize that one of the keys to our business success is to select the right partners. Our products and services develop from our partnerships with creative suppliers and demanding customers."

A good example is a newly developed formed-liner machine that was designed and built by this Best Plant and installed in a supplier's plant. The company sent its technicians and engineers to the supplier's plant for the installation process. Super Sack is also playing a major role in helping one of its suppliers identify and develop a supply source of its own for a new product. Super Sack obtained sample materials and set up and participated in meetings and trial runs in their customer's plant with the subcontractor in attendance. All three companies worked together to determine how the highest-quality, lowest-cost end product would be developed.

Grand Rapids, Michigan–based Delphi Energy & Engine Management Systems' supplier strategy is based on General Motors' Supplier Quality Improvement Process. This process provides guidelines for new supplier assessments, sourcing decisions, production-part certification, and the initial production of supplied components. It mandates continuous-improvement activities and performance monitoring, and details a process for assisting suppliers with the resolution of quality problems. The Delphi plant follows this model and developed four main supply strategies, as manager Peter Gallavin explains:

1. The first is to establish a high expectation for the supplier. In order to produce a quality product, the suppliers must meet or exceed the plant's quality standards. This expectation is communicated to the supplier at every opportunity.
2. The second strategy is to have manufacturing personnel of the plant work closely with the manufacturing personnel of the supplier. The supplier's manufacturing employees need to understand how the plant uses the component: method of assembly, component function, final customer requirements, etc.
3. The third strategy is continuous improvement. By working to improve the supplier's quality, delivery, and cost, the plant benefits in improvements to the final product. In this regard, the plant tracks the quality and delivery performance of all direct-material and most indirect-material suppliers on a weekly basis.
4. The last strategy is constant communications between the plant and the supplier. This communication allows the plant to be proactive in dealing with possible supplier concerns and the supplier understands that he is an integral part of the plant's manufacturing process.

The Delphi supply management process includes key supplier participation in the plant's new-product development cycle. For example, as a supply contract requirement, one of Delphi's major suppliers is a member of the product-focused team that is charged with the development of future customer programs. Suppliers are also asked to review the initial concepts and make recommendations regarding the expected performance, manufacturability, packaging, alternative materials and design, and target cost. "This approach," says Gallavin, "leverages our engineering resources and ensures continuity of effort between our customer, our supplier, and ourselves."

Timken's Faircrest Steel Plant Builds a Supplier City

Getting close to suppliers is the big idea behind The Timken Company's Supplier City, an industrial development designed to house companies that supply critical goods and services to the company's facilities in northeastern Ohio. Timken, which owns the 10-acre site located across the street from the company's 1994 award-winning Faircrest Steel Plant in Canton, Ohio, hired a developer to build, manage, and maintain the property.

Based on the "vendor city" concept that was successfully pioneered by PPG Industries in Lake Charles, Louisiana, Timken decided to create the supplier city instead of constructing a new building to warehouse MRO stock. Besides slashing inventory-related costs and eliminating the expense of building and staffing a new warehouse, having suppliers located so close to the Faircrest plant helps the company in its efforts to reduce its supplier base and substantially improves delivery times.

Linked by an EDI system, the plant's employees can order products and services from Supplier City residents and, in return, get deliveries within 15 *minutes* of the order's receipt. Qualified employees of the plant are provided access to supplier's warehouses, so if a part or product is needed on an emergency basis or after regular business hours, they can literally walk across the street and pick it up. The Supplier City distributors are further linked to the company's facilities by a computerized inventory system known as PERMAC.

Timken focused on MRO suppliers as the first residents of Supplier City and rewarded each of those that took the leap with a 20-year Master Purchase Agreement to supply specific product groups. In order to discourage an unhealthy dependence on the steelmaker, the agreements contain no guarantees, nor are dollar amounts specified, and suppliers are encouraged to limit their sale volume from Timken purchases to 25 percent.

With a half-dozen suppliers on location when it won its Best Plants award in 1993, Timken had already slashed inventory by $3 million. Another half-dozen suppliers were building facilities in Supplier City, which should double the savings.

The Best Plants use a wide variety of methods to increase contact with their suppliers. At Wilson Sporting Goods' Humboldt, Tennessee, plant, where most raw materials, manufacturing supplies, and MROs are single-sourced, suppliers are permitted full access to manufacturing facilities and processes, testing equipment and labs, and Wilson employees at all levels of the organization. To support the plant's JIT strategy and to allow suppliers to smooth their own production flows, first- and second-tier suppliers are provided with six-month rolling production forecasts and 30-day product-specific release schedules.

Formal cross-functional supplier partnership meetings are held twice a year, with supply-focused technical, quality, and sourcing meetings held on a more frequent basis. Supplier/customer teams and task forces are established on a regular basis to identify new business and improvement opportunities and issues. (See the plant's Individual Profile for more information.)

Meetings and conferences are another common communication avenue between Best Plants and their suppliers. Varian Associates' NMRI unit hosts "Supplier Days" which provide the plant with an opportunity to present its suppliers with information on the state of the business and reiterate the importance of the role suppliers play in the plant's success. It also provides an open forum for the exchange of ideas on new technology and process trends.

General Motors' award-winning Cadillac Motor Car Division holds regular "Partners in Excellence" conferences with its major suppliers to share current and future information about the business. The focus of these gatherings is on developing strong partnerships and mutually beneficial approaches toward new programs. And Kennametal's Solon, Ohio, plant invites key suppliers to an annual "Vendor Day" at the plant, which is used to foster improved working relationships, communicate goals, and discuss recommendations for improvement.

Continuously Improving Supplier Performance

A major element in the supporting competency of supply management is its focus on continuous improvement. Continuous improvement in supplier performance and the quality of supplied materials help the Best Plants create and maintain their competitive edge. As we have already seen, continuous improvement in the supply chain is a prominent feature of Best Plants' supply management strategies and it is also an ongoing emphasis in the plants' supplier communication programs.

Continuous improvement is a key element in Allen-Bradley's (A-B) corporate supply strategy, which is in effect at its award-winning EMS1 unit. The corporate program approaches supply management strategy from a total-cost-of-procurement perspective, where price is only one of several measurement criteria. At this plant, parts and materials quality, supplier responsiveness, schedule compliance, cost of communication, and cost of inspection are also key factors of total cost. But just making the grade isn't enough at this Best Plant.

Each of A-B's major suppliers is required to submit an annual long-range improvement plan. This plan specifically addresses component quality, part certification, cost containment, responsiveness, cost of communication, and cost of schedule noncompliance. The progress toward the plan's goals is continuously reviewed throughout the year by the supplier and its corresponding A-B buyer.

A-B's major suppliers also participate in the plant's Supplier Self-Certification Program. Through this program, suppliers review processes and procedures related to the overall delivered product quality. Purchased components must successfully advance through a series of certification steps until incoming inspections at the plant are no longer necessary.

Feedback is an important part of the A-B's supplier focus. Every month, two reports are provided to vendors. The first is the Parts Per Million Quality report which summarizes all of a supplier's parts defects over the past 13 months. The second is the Total Cost of Procurement report. This report captures the costs associated with six measurable areas of total cost and converts them into a single figure per $1,000 of business. Detailed breakdowns of major cost contributors are also provided, so the supplier and the company can easily identify the areas in which the most significant levels of improvement are possible.

The results of this ongoing focus on improvement are substantial. By 1992, A-B had recorded a 93.6 percent reduction in defective ppm over a six-year period. The total cost of procurement was reduced by 54.8 percent over the prior three-year period, with inspection costs reduced by 77.8 percent. Purchased parts inventories were also significantly reduced using JIT techniques which were enabled by the increases in quality and supplier responsiveness. Total costs have been continuously reduced with the help of jointly conducted value analysis.

Similar to A-B's monthly reports are the supplier "report cards" issued by Lockheed Martin's Government Electronic Systems plant in Moorestown, New Jersey. These report cards are produced on a quarterly basis and include a 12-month summary of supplier performance, including the supplier's "market share" for the commodities provided, a quality rating, a delivery rating, and a responsiveness rating. The measures are individually weighted and combined to create an overall supplier rating. Supply performance excellence is celebrated with a recognition program called the AEGIS Excellence Award (after the plant's major Navy contract). This award is given to contractor personnel, including suppliers, for significant achievements on the AEGIS Program.

In its pursuit of world-class status, the Moorestown plant also utilizes two other supplier improvement programs: the Supplier Process Baseline and Proven Quality Supplier programs. The Supplier Process Baseline program focuses on the review and improvement of program-critical supplier processes, and the Proven Quality Supplier program is the plant's part certification system.

The Best Plants' own continuous improvement programs often provide the impetus for the improvement of supplier quality. At Edy's Grand Ice Cream, for example, when employee problem-solving teams were assembled to review customer contact letters and prioritize the quality issues they represented, the teams discovered that 90 percent of all customer complaints were caused by two conditions: *insufficient ingredients,* that is, complaints about not receiving the amount of ingredient that is expected, and *foreign objects,* complaints stemming from finding something (such as an almond shell) that is not supposed to be in the product. The teams further found that seven flavors of ice cream accounted for 75 percent of all customer complaints.

Since the ingredients in ice cream are introduced directly into the finished product, the teams then turned their attention to the suppliers of the ingredients in the seven flavors. In the process, they developed a vendor certification program that emphasizes both product safety and consistency. In this program, a team member, the plant's quality manager, and employees from the supplier examine and improve the supplier's processes so that the ingredients used to make the plant's ice cream are of the highest quality. In 1993, the plant reported that already low complaint rates were being reduced by an additional 4 percent per year.

Lansdale, Pennsylvania–based Ford Electronics & Refrigeration Corporation established a Total Supplier Improvement Group to serve as a focal point for all of its supplier-related continuous-improvement efforts. Formed in October, 1992, the group's mission is to significantly reduce purchased-part defect levels and encourage suppliers to adopt a zero-defect mind-set. The components of this strategy include: mutually agreed upon expectations and measurable goals, regularly scheduled on-site meetings for all suppliers, real-time process performance data, clearly defined roles and responsibilities for all supplier representatives, and a negotiated supplier charge-back procedure. In the eight months prior to winning its award, the Improvement Group recorded the reduction of supplier defects per million parts from 7,255 to 1,030.

As with the Best Plants' internal quality efforts, teams are often the vehicle of choice in continuous improvement with suppliers. MEMC Electronic Materi-

Air Products & Chemicals and Matlack Continuously Improve Service

Air Products & Chemicals' South Brunswick, New Jersey, plant perfectly illustrates the results of a supply relationship based on continuous improvement with trucking company Matlack, Incorporated. Starting in 1989, teams from the trucking company and the plant began meeting each month to identify permanent solutions to the transport problems.

The agenda for these meetings is fixed. First, a review of the previous month's actions; then, an analysis of newly discovered service failures, including what the company calls a "near miss" (a service flaw that doesn't directly involve the customer); and finally some good news, a sharing of instances where the trucker performed above the customer's expectations.

The joint improvement team also takes a long-term view of supply performance by establishing objectives for a full year. Prior to 1992's Best Plants award, these goals included a zero-defects promotion, a campaign aimed at raising the plant's profile among the trucking company's employees, and a program to involve Matlack's drivers in the service improvement process.

As for the results of the effort, Matlack's defect-free service levels rose to 98.44 percent by 1991, the supply partnership efforts spread to seven other Air Products' plants, and the partnership between the South Brunswick plant and Matlack received the American Society of Transportation and Logistics 1991 Shipper/Carrier Partnership of the Year Award.

als' Saint Peters, Missouri, plant adopted the team structure for its work. Explains manufacturing manager Perry Lee:

> *The supplier partnership strategies and practices are managed through our Supplier Improvement Program. Currently, we have 13 teams working, in some cases, for over five years, with 17 suppliers. Team members include Manufacturing, Process Engineering, Research & Development, Purchasing, IQA, Materials Management, and others as needed. Goals are developed which coincide with the objectives of the manufacturing department using the product in order that they meet our continual improvement efforts.*

Feedback is continually given to the supplier through interactive meetings between the supplier, supplier teams, and the user areas. Part of the feedback is a certification process which provides a formal framework for efforts to improve quality and lower "cost of ownership." Certification is a four-step process, including a yearly audit, used by MEMC to bring suppliers in line with our performance standards. A quarterly report is sent to the supplier rating them on quality, delivery, defect rates (ppm), paperwork correctness, cost, and services.

Summary

Supply management, the second supporting competency of the Best Plants, is a logical extension of a plant's internal emphasis on customer focus, quality, and agility. Suppliers, like employees, represent a tremendous resource in the attainment of the Best Plants three core strategies, and to ignore their capabilities by focusing solely on the invoiced price per unit of supplied materials is to miss an important competitive advantage.

Best Plants take a much broader vision of supply management by adopting strategies that evaluate purchased materials and goods in terms of their total cost of procurement. A total cost perspective takes into account unit cost as well as other important measures, such as the timeliness and accuracy of deliveries, the responsiveness of the supplier, and the quality of the materials themselves. In support of a total cost perspective, Best Plants limit their supplier base as a way of also limiting the time and cost of managing the purchasing function. This reduction to fewer supplier partners allows the plants to devote greater efforts in assisting key suppliers in the improvement of the quality of their materials and in the reduction of their total cost.

Best Plants' supply strategies also foster long-term supply partnerships that are characterized by a willingness to invest in continuous-improvement, single-source, life-of-product purchasing agreements, and a lowering of the barriers between supplier and plant. The resulting closeness that is fostered supports JIT production systems, faster new-product development cycles, and continuous reduction in costs and defects.

Begin supplier development and partnering early in the improvement process.

—Gilbarco, Incorporated

CHAPTER 7

TECHNOLOGY

Two hundred years ago, it was the ability to understand and utilize newly discovered advances in technology that ushered in the industrial era and heralded the rise of manufacturing as a major economic force. One hundred years ago, it was the ability to adapt and organize advances in technology that gave birth to the first great manufacturing empires. Today, it is the ability to integrate the latest technological advances that drives Best Plants toward their strategic goals of world-class customer service, quality, and agility.

Technology has changed the manufacturing industries so radically and so quickly that manufacturing's image in the minds of the general public can't keep up with its reality. Ask someone on the street to describe a steel mill or a machine shop and you will likely get a description of an operation that was out-of-date 50 years ago. The insides of the Best Plants would astonish most of the people who use their products.

In place of the sweatshops of the popular imagination, they would find Best Plants operating in sterile conditions that surpass the requirements of a hospital operating room, computerized control systems that regulate the flow of products and processes, machine shops run by computer programmers, and communications networks that make it possible to hold a team meeting even when its members are scattered across the globe.

There appears to be no end to current technological possibilities available to Best Plants, and given the present conditions, there seems sure to be more, not less, technological choice in the future. Those plants that can best understand, adapt, and utilize technology will gain a substantial advantage in the marketplace. For this reason, technological adeptness is the third supporting competency of the Best Plants.

The applications of technology in use at the Best Plants are as varied as the plants themselves. Each plant supports its production and information needs in a different way; each resides at a different point on the spectrum, from low-tech to high-tech. As you may already suspect, however, no matter where they are on the technology curve, the Best Plants share four common approaches, "lessons" if you will, to the application of manufacturing technology.

The first lesson of the Best Plants is that adoption of new production technologies must support the business objectives of quality, agility, and productivity. This may sound simplistic, but there are many notable examples of companies that installed production lines featuring every imaginable bell and whistle—only to create more problems than product. Best Plants build quality, agility, and productivity into their manufacturing equipment.

The second lesson of the Best Plants is to use information technologies to create networks that put the right knowledge in the right place at the right time. The great advantage of information technology lies in its ability to create an unrestricted flow of real-time data. The Best Plants, however, do not use information networks to simply speed the flow of data, they also use them to create new and more effective avenues of communication between people.

The third lesson of the Best Plants is to leverage investments in technology through the integration of production and information systems. The results of this integration are factorywide, even organization-wide, networks that allow real-time measurement, communication, and response. The integration of information and production systems represents the latest technological thrust in many areas, creating previously undreamed-of effectiveness among the Best Plants.

The fourth lesson of the Best Plants is to seize a competitive advantage in the marketplace by using technological advances to create innovative new products and services. In some instances it is the operational improvements realized by technological applications that allow the plants to create these products and services. In others, it is the study of new advances in technology for the specific purpose of creating products that dominate their markets.

Building In Quality and Productivity

The first demand that Best Plants make of new production technology is that their expenditures support the strategic objectives of their companies. An examination of the plants' production equipment and machinery applications reveals that quality, both in process and product, and increased productivity are the two principal reasons for the Best Plants' investments in production technology.

At TRW's Vehicle Safety Systems' Cookeville, Tennessee, plant, for example, the defect-free operation of the roughly 6 million automotive air-bag modules the plant annually produces is a critical issue. Accordingly, many process mistake-proofing devices are designed into the plant's traceability system, which utilizes radio frequency identification technology to reduce material handling costs and maintain inventory accuracy.

In this real-time system, bar-coded serial numbers are scanned at each operation in the production process. Contained in these numbers is information that ensures that all the proper components are present in each unit. At every step, the cell controller analyzes each unique module's serial number before allowing the individual workstations to begin their part of the manufacturing cycle. If a product is not defect-free, or if the cycle is manually broken before the process has been completed, the unit is automatically rejected.

Industrial Engineering section head Walter Marcum explains some of the considerations that went into the plant's production technology decision-making process:

> *Automation is not an across-the-board decision. While implementing 25 assembly lines in four years [14 of them in a single year], TRW Cookeville had to make some key decisions regarding capital investment in automation versus manual operations. TRW learned that to optimize resource and capital investments during such an aggressive ramp we needed to select an equipment supplier who could get involved with our product designers as well as plant engineering and manufacturing technicians. As a result of close plant involvement with our key equipment suppliers, we feel that the appropriate amount of automation has been successfully integrated with manual operations. This success is due to cooperative efforts between product design, equipment suppliers, and operations.*
>
> *Rather than focus on excess automation, we opted to direct much of our efforts toward the implementation of comprehensive mistake-proofing technology into our equipment. With the rapidly changing technology associated with inflatable restraints, too much automation can result in an inflexible system and a gross waste of capital. Working as project partners with equipment suppliers gave us the opportunity to get leading-edge technologies which meet customer needs in low-ppm defects, employee needs in safety and ergonomics, TRW's need for a reasonable capital investment, and yet remain flexible to accommodate changing technology.*

Building process and product quality was also a major focus at both of the award-winning Timken Company plants when they implemented new production technologies. The 1992 award-winning bearings plant in Bucyrus, Ohio, installed electronic SPC (ESPC) on all of its production lines to provide automated process control. The plant's ESPC system is built around internally developed software and a personal computer with the capability for automatic or keyboard data entry. "The real power of this system lies in the capturing of real-time data for several parameters per process," says general manager James Benson. "Currently, the Bucyrus Bearing Plant has 99 ESPC computers attached to 123 gauging stations which are used to control approximately 1,500 parameters on over 400 processes."

The plant also utilizes real-time, nondestructive testing in its heat-treat process, an addition to its quality tools that drove out-of-tolerance defects down 85 percent. Its computerized vibration analysis equipment detects and calculates the vibration pattern inherent in bearing assemblies. The Bearing Signature Analysis equipment generates a pass/fail quality evaluation and includes diagnostic tools for troubleshooting and improvement efforts.

The same equipment is used in the plant's TPM efforts to monitor and evaluate cutting and grinding machine performance. Vibration probes are attached to the plant's machinery to provide a view of the vibration pattern created by the movement of parts during production cycles. The findings are compared to the past maintenance history of the machine and used to make predictions of future failure and establish schedules for preventive maintenance.

Timken's second award-winning facility, the Faircrest Steel Plant, is a technology benchmark, the details of which are included in the plant's Individual Profile. To mention just one of its notable quality-based technologies, its Elkem inspection unit, which has helped the plant virtually eliminate surface defects in its steel, is only the fourth unit ever built and one of only two units in the United States (the other is also in a Timken plant). The machine produces a consistently accurate surface inspection at a rate that is 2.5 times faster than previous inspection methods.

Like Timken, Copeland's Sidney Scroll plant also uses computerized SPC to help control process variation, but warns against blanket implementations. Says operations director Randy Rose:

> *SPC technology had been introduced, but is not used extensively. A total of 12 computerized SPC stations are utilized in the Scroll Plant. The machining, the scroll set, main bearing, lower bearing, and muffler plate lines all have computerized SPC at workstations with critical feature segments. In assembly, computerized SPC is utilized to control bearing assembly operations and automated bolt torqueing machines. Computerized SPC is used where it is advantageous for process control, but for many operations, because of extremely high capability, computerized SPC is not warranted.*

Honeywell's Industrial Automation and Control plant in Phoenix, Arizona, doesn't always sell all the high-tech production control solutions it makes. In 1988, for example, the plant made a significant departure from the traditional hot/cold testing of electronic assemblies with the introduction of environmental stress screening (ESS). By rapidly changing temperatures from –20 to +80°C, this process triggers early-life failures in electronic components through accelerated aging. With the introduction of ESS, the plant was able to reduce its historical 4 percent field failure rate to less than 1 percent. Honeywell IAC also uses ESS as a tolerance analysis test in new designs. All new product developments now require ESS testing prior to release.

By closely monitoring and controlling critical parameters with proprietary equipment, the plant has been able to produce surface-mount assemblies that exceed military performance specifications. Similar control technology is used on the wave solder process, where 43 parameters are monitored and controlled on-line to ensure optimum quality throughout the process.

Lord Corporation created a computerized TPM system to support process quality in its Dayton, Ohio, plant. This system maintains daily, weekly, and monthly maintenance schedules for all of the machining production teams. The maintenance team is also using the software to schedule and prioritize work orders.

The Dayton plant, which features computerized control of all its processes, also invests in technology as a way to build its operational agility. Its production equipment includes a flexible manufacturing line consisting of three Toyoda horizontal machining centers that are serviced by a remote guided vehicle (RGV). The RGV is capable of accessing a storage rack containing up to 18 different pallets and delivering them to each specific machining center in the line. This system allows the plant to leave numerous jobs set up at the same time, permitting faster turnover on small lot sizes.

As detailed in its Individual Profile, the General Motors/Cadillac Hamtramck Assembly Center in Detroit, Michigan, was originally envisioned as a world-class example of cutting-edge manufacturing technologies. Unhappily, the initial results were far less than stellar. Yet this plant went on to win the Baldrige National Quality Award five years later (in 1990) and stands today as a fine example of how the proper applications of technology can build a plant's production capabilities. Manufacturing manager Roy Roberts explains the plant's transformation as follows:

> *The Detroit-Hamtramck facility was originally designed to be a working laboratory for advanced technologies. However, some of the initial technologies were unproved and not statistically capable. Over the past five years*

Ford's North Penn Electronics Is a Manufacturing Technology Flagship

Dedicated in 1990, Ford Motor Company's North Penn Electronics Facility (NPEF) in Lansdale, Pennsylvania, produces electronic power-train and vehicle-control products for braking, suspension, and speed control. In all, the facility manufactures over 200 different products, shipping 100,000 units per day to more than 50 Ford and affiliate customers worldwide. It is also a showplace of manufacturing technology.

The key manufacturing processes utilized in the plant include: automated leaded component; insertion and surface-mounted device placement; manual and automated odd component insertion; wave and reflow soldering; in-circuit testing; ceramic substrate fabrication; laser scribing and resistor trimming; and environmental, vision, and computerized testing. In addition, fully integrated computer systems are used throughout the plant for materials management, warehousing, component delivery, production performance, and financial tracking.

The Electronic Engine Control (EEC) product lines were completely revamped between 1987 and 1993. Stand-alone manual insertion operations were replaced by in-line robots along the Odd Component insertion lines. At a topside surface-mount assembly operation, changeovers between different model types are performed automatically in less than two minutes.

[1986–1991], the noncapable technologies have been modified or eliminated. In addition, new processes have been instituted. The following are a sampling of advanced technologies that are in place:

- *Detroit-Hamtramck (D/H) is the only U.S. manufacturer with the capability to process "tri-coats" using two-component urethane paints.*
- *D/H utilizes numerous state-of-the-art robotics in the body and paint shops.*
- *D/H employs up-to-date manufacturing information systems, including maintenance-monitoring and performance-reporting systems.*
- *D/H uses a broadband coaxial communication network with video capability to transmit valuable information about the business to TV screens throughout the plant.*

Vision systems are utilized to guarantee proper solder printing and correct component placement on the boards. Process verification includes X-ray analysis of circuit board assemblies. The plant's CIM system ensures machines run the correct program with the correct parts and provides continuous monitoring of machine operations, quality, and performance.

The EEC product lines run 79 hardware-family types and 300 different calibrations. Throughout the production process, bar-code readers are utilized to accurately track components and products. A serial number is assigned to every unit to enable accurate data collection, traceability, and inventory tracking. Advanced bar-code-reading technology used at the plant ranges from handheld wands to fixed, high-speed scanners that utilize a vibrating mirror raster and are capable of scanning 1,000 times per second, reading bar-code labels on the fly, and accommodating multilocation bar-code labels. Among the results of this technological wizardry: EEC's 900-unit-per-hour production rate previously required 252 operators. Now, the same production rate requires 147 operators, a savings to the plant of $20,000 per day.

Nor is that the end of the story. NPEF applied the same thinking to its Electronic Distributorless Ignition System product lines, installing a fully automated Weldon integrated manufacturing system. Productivity levels before the switch stood at 17 units per operator and a total of 850 units per day. After the changeover, the line produced 8,000 units per day at a rate of 54 units per operator. This new technology also produced a 77 percent reduction in scrap and an 85 percent reduction in cycle time.

- *A vision system is employed in the body shop to ensure the dimensional integrity of the body.*
- *Computer simulations are applied to improve bottleneck operations and predict the effect of model and facilities changes before they occur.*
- *In-process electrical checks are made to ensure the quality of electrical components and systems before they are assembled into the vehicle.*

Robotics, once a subject relegated to science fiction writers, is a common technology in the Best Plants. Unisys' Pueblo, Colorado, Government Systems plant uses robotics throughout its operation. In addition to its fully automated, 143,000-square-foot Material Management Center, a robot named Huebert is "employed" to paint parts-identification color codes on printed circuit card assemblies. Huebert replaces the traditional manual method that utilizes a toothpick. Another robot conducts automated conformal coating operations that result in enhanced process yields and an economical replacement for the manual coating method previously in use.

Robotics aren't Pueblo Operations' only advanced use of technology. The plant also acts as a Surface Mount Technology Center of Excellence, a core capability that helped reduce the manufacturing and test costs for the U.S. Navy's AN/UKY-44 computer by 50 percent. The surface-mount technology operation uses autoinsertion equipment and, in 1993, was being upgraded to support medium-volume, fine-pitch (0.015-in) automated assembly.

Robotics is an expanding technology at Nippondenso Manufacturing's Battle Creek, Michigan, plant, where in 1993 the company was in the middle of an expansion from 49 to over 100 robotics units. The plant uses robotics to free employees from hazardous tasks, to reduce repetitive operations, and to improve product quality. In order to maximize the company's investment, over 200 employees participated in an internally developed in-house training program designed to teach safe operation, repair, and improvement techniques for robotics equipment.

Nippondenso also developed a labor-saving Automatic Guided Vehicle (AGV) System to transport molded parts between the plant's Molding Department and the Fan Shroud assembly line. The AGV is directed by radio signals from a computer that executes commands entered by employees on the Fan Shroud line.

Jumping the Barriers to Communication

The second lesson derived from the Best Plants' technology applications is the extensive use of electronic information and communication networks to remove the physical barriers between people and to put all the data that employees need to make business decisions at their fingertips. Not convinced? Sam Byrd, marketing manager at Siemens Automotive's Newport News, Virginia, plant, clearly reiterates the point:

> *Information technology has been the key to rapid productivity growth. It provides real-time information on all aspects of manufacturing performance at every level of plant operation from the hourly workforce to general management. Information technology provides the ability to respond immediately to variation in manufacturing operations, to respond to customer issues, and to support rapid changes in customer requirements.*

Varian Associate's Nuclear Magnetic Resonance (NMR) business unit perfectly captures the potential of information technology with the assembly instruction system it installed for employees on the factory floor. The information system uses computers, digitized images, parts lists, and detailed assembly instructions to prepare and present a single, comprehensive assembly documentation package for each of its orders.

The computerized documentation files are created at a stand-alone workstation which is equipped with a camera, VCR, and a digitizing computer. The captured images are transferred to an engineering station for editing, highlighting of critical tasks, and the addition of a parts list. Once complete, the files are stored in a server from which they can be distributed and readily updated.

On the plant floor, assemblers are able to call up the instructions, parts lists, and images on terminals located in their work areas. They can scroll forward and backward through the file at a keystroke and, in some cases, can obtain several detailed views of a critical step. NMRI's Ed Sehrt describes the benefits of the system:

- *Assembly instruction can be created as the product is developed, allowing for a greatly reduced start-up and learning phase. Product changes can be documented and brought on-line and obsolete documentation is quickly and completely replaced at the same time.*
- *Since all images are full color with highlights, annotations, and explanations where needed, they serve as a perfect tool to facilitate cross training. It has been Varian's experience that this system reduces training time by a measurable amount with a minimum of quality problems.*
- *For portability, the system provides for high-resolution, full-color hard copies of selected screens which can be used off-line. Assembly instructions for customer-site retrofits can be created and distributed worldwide to service engineers and customers. And mistakes and errors can be quickly documented and provided to internal and external suppliers for corrective actions and as a reference.*

The 1994 award-winning NMRI business unit adopted another productivity-building information technology application with its CIM Bridge network. CIM Bridge connects the plant's Palo Alto, California, CAD-based design systems and engineers to the production equipment at TEMPE Electronics, an Arizona-based circuit board manufacturing plant. The software, from which the name CIM Bridge derives, links and translates the various CAD software formats used at the unit for use in the manufacturing equipment.

The data files that contain the engineering schematics are manipulated in a UNIX-based system through the Network File System. While on-line, CIM Bridge translates the data for toolpaths, assembly sequences, and parts orientation. The files include board type and dimensions, bored-hole dimensions, point-of-reference coordinates, and the type and location of all circuitry.

The CIM Bridge application also allows Palo Alto's engineers to make real-time design changes. Because files can be simultaneously accessed, the system also allows on-line, real-time problem solving by engineers at both NMRI and TEMPE. As soon as a problem is identified, it can be solved. Further, the need for hard-copy documentation is almost completely eliminated by the system's electronic storage capabilities and universal access.

Allen-Bradley's (A-B) EMS1 unit developed a similar application which allows it to translate engineering transfer file (ETF) data from its CAD design stations into machine programs in five to seven minutes. In-circuit test fixture and test program development has also been automated using the same ETF data. The result: Fixtures can be designed and manufactured in a few weeks, and test programs can be debugged and implemented in a few hours. All benefits aside, A-B manager John Field warns:

> *Advanced technologies and CIM are not implemented as ends in themselves. When utilized, they must serve a business purpose. They are a means to the achievement of the objectives of the strategic business plan. We determined that CIM and technology-based manufacturing were the tools necessary for us to survive in the manufacturing of world-class solid-state products.*

Connecting people to foster greater communication and productivity is also the objective of a pilot communication program at Ford's North Penn plant. Named "Person to Person" (P2P), the pilot project is adding interactive video technology to the plant's personal computer–based workstation network. P2P brings video-conferencing capabilities to employees' desks. The program emulates a two-way picture as well as transmitting video clips, documents, sketches, and photographs. Real-time "chalkboard" capability allows participants to modify these drawings on-line for everyone present to view simultaneously. P2P also enables multisite conference calls at all nine of the Electronics Division plants around the world.

The need for information technology at the operator level is another focus at NPEF. Explains plant manager Dudley Wass:

> *Operators used to receive inadequate information because our information systems were not developed around their needs. This was a very important lesson*

for NPEF since it impeded further improvements on the production floor. Now, NPEF has a completely reengineered production information system, the Flexibility Enabling Manufacturing System (FEMS), that provides operators with critical information at their own workstations on a real-time basis. Operators and engineers don't have to wait for start-up meetings anymore to learn about the top-five rejects on the line.

Rockwell's Autonetics plant in El Paso, Texas, developed a factorywide tracking system (FWTS) to provide its Integrated Product/Process Teams (IPTs) with real-time feedback on key levels of performance. SPC quality and test data distributed through FWTS allows manufacturing, quality, and reliability personnel to analyze problems and develop corrective actions. Production work-in-process is tracked via bar-coded input to FWTS terminals located throughout the factory floor. Cycle-time measures and reports are provided, along with IPT productivity performance, on demand.

The information network also proves useful for the plant's Opportunities for Improvement program. Manager Don Teague explains:

El Paso has learned that user-friendly systems must be in place to facilitate the easy input of employee ideas, reporting problems, and the documentation of actions taken. El Paso has implemented Opportunity for Improvement (OFI) and Technical Issues Communication (TIC) system to answer this need. The OFI and TIC systems utilize user-friendly computer terminals located throughout the factory floor to record employee ideas or problems, assign responsibility, ensure timely responses, and maintain a permanent record of the actions taken. These systems ensure that El Paso employees receive timely and documented responses to their continuous-improvement ideas.

Unisys' Pueblo Operation devotes as much of its resources on its information network as it does on production technology. The plant's Optical Storage System maintains 50,000 engineering drawings using laser disk technology. Employees are able to view and print the drawings on a laser printer from their workstations via the plant's local area network (LAN). A wide-screen, rear-projection system was installed to streamline the time and effort spent preparing presentations. It includes a computer interface and is able to display virtually any chart, data, video, or drawing in the plant's various information systems. A Career Development Software system running on the plant's LAN allows employees to use training programs at their workstations rather than going to the Career Development Center and finally, a videoconferencing center rounds out the package.

Integrate Production and Information Technology

The third lesson of the Best Plants is to capture the maximum leverage from investments in production and information technology by integrating them into plantwide control and communication systems. Computer-integrated manufacturing (CIM) technologies and enterprisewide integration strategies are among the advanced results of this effort.

Not surprisingly, since it is their business to provide the most advanced manufacturing system control solutions, the Best Plants in the industrial control industry feature some of most highly evolved systems. For example, Foxboro's process control systems plant in Foxboro, Massachusetts, extensively utilizes CIM technology. All the plant's business systems, including financial, material, quality, and design functions, are integrated and function as a single, seamless entity. This system, when combined with a plantwide computer network, has created the much vaunted "paperless organization."

Shop floor data collection is integrated as well. Bar-code scanners with handheld terminals, touch screen monitors, diskettes, and auto uploads from networked PCs are some of the technologies employed to collect data; manual and postprocess data entry has been eliminated. This results in the reduction of non-value-added operations and improvements in database accuracy. All the business unit's transactions are automated and, in most cases, one transaction in one business system triggers the appropriate changes in all other systems.

The thinking that has guided the Foxboro plant's integration efforts is crisply described by production manager V. S. Venkataraman:

> *The philosophy is the same that guides our journey to continual improvement; management, ownership, and participation in all plant process designs; detailed involvement and employee participation in everything; a team-oriented approach. No technology or automation project is implemented that adds overhead cost to the operation or reduces the flexibility of the operation. Simplicity is legislated. No customization or creeping elegance is allowed.*

Of course, the Best Plants in other industries are not far behind in their integration efforts. MEMC Electronic Materials' Saint Peters, Missouri, plant is designed to remain competitive as a supplier of silicon wafers to the semiconductor industry, where technological change is constant and the parametric controls required by customers are ever more narrowly defined. Since each semiconductor device requires a tailor-made wafer, and customer wafer specifications are rarely identical, manufacturing agility is critical. In 1994, for example, MEMC manufactured over 2,000 different wafer types; 90 percent of them did not exist three years earlier.

The Saint Peters plant remains on top of this rapid-change environment with interactive control and data acquisition/information management systems. The plant uses COMPASS, a finite-capacity scheduling program that is capable of detailing the variations between orders, to schedule production and communicate the various specifications of each order to production employees.

During the production process, computer terminals in each of the plant's operations track an order's progress and add manufacturing data to an electronic file that moves with the product through each fabrication step. Employees update the file as each order passes to the next operation. Finally, this production documentation package passes into the Customer Information System (CIS), where it becomes immediately available to the plant's customers. (See Chapter 2 for more on CIS.)

Baxter Healthcare's North Cove IV Systems plant is breaking new technological ground as a member of Consortium III. This process industry partnership is funding the development of POMS, a process operations management systems software application tool that can integrate all manufacturing operations using a module-based system. This system will replace the cumbersome batch record documentation and specifications with electronic data collection. North Cove already uses POMS applications for SPC, scheduling, and package printing. It will adopt future POMS applications as they become available.

The technological features of Ford's North Penn facility have been discussed at length in this chapter, so it should come as no surprise that the company has leveraged its investment by integrating the various systems at work in the plant. In June 1993, the plant launched a next-generation CIM system called FEMS (Flexibility Enabling Manufacturing System). This system utilizes object-oriented technology to create software applications that provide production operators with real-time data. In addition, FEMS provides the ability to perform rapid changeovers and supports the manufacture of multiple products on the same line. Like Baxter's POMS application, FEMS is comprised of flexible software modules that allow programmers to rework a single segment of the plant's systems without having to change the other modules. A module-based system shortens development time, reduces costs, and provides an integrated approach to linking manufacturing information systems.

Exxon's Baytown Site Adopts New Technologies

Exxon's Baytown site controls its processes using a system developed with the help of another Best Plant award winner, Honeywell. (Under the terms of a joint development partnership, Baytown received its system before it became commercially available.) Based on Honeywell's TDC 3000 process manager hardware, the Baytown system features:

- Alarms and reporting for deviation from operating directives
- Integral SPC analysis and adjustment for identified process biases
- Closed-loop rheological property control on finishing extruders
- Real-time availability of process data distributed through fiber-optic LAN to desktop computers throughout the plant
- Historical process data, which is available through a LAN to all desktop computers
- Shared LAN data acquisition and data analysis tools
- Process simulation computers to train operators on unusual operations and upset recovery without actual unit involvement

In another technological first, Exxon's Baytown facility was the first industrial site to install fence-line monitoring equipment using Radian Corporation's FTIR long-path technology. This technology features continuous atmospheric testing for 170 different chemicals and is sensitive to the parts-per-billion level. Explains site manager Ray Floyd, "As a result . . . we are able to say with certainty that nothing we do is hazardous to the environment or the community."

Creating an Advantage in the Marketplace

The fourth lesson of the Best Plants is to utilize new technologies to create a competitive advantage in the marketplace. They accomplish this feat in two ways: (1) by adopting new technologies in the development of new products and services and (2) by using new technologies to build process capabilities that will help them to make superior products.

Siemens Automotive's Induction, Fuel, and Emission Components plant in Newport News, Virginia, provides a good example of the latter strategy. It raises the performance levels of its fuel injectors and simultaneously increases its production volume capabilities by developing highly refined grinding, finishing, and measuring methods. In the needle-grinding operation, for instance, the plant utilizes two high-precision cylindrical grinders that feed a two-spindle superfinishing machine. Seat grinding is done with high-precision internal grinders. All operations are highly automated. High-precision gauging for dimensions, finish, and roundness is transferred directly to the Shop Floor Information System for automated SPC analysis and to collect historical data on process capability.

The results of this investment are twofold. The plant has recorded productivity improvements, such as a 61 percent improvement in production rate per machine for seats and over 118 percent improvement for the needle-grinding operation. And, in response to continuing customer demands for improvement in vehicle performance, fuel economy, and emissions, product performance specifications have been continuously improved.

One revealing measure of this plant's technological commitment lies in the fact that 30 percent of 1995's annual operating budget was earmarked for information systems and production technology. Approximately 43 percent of that figure was invested in flexible manufacturing information systems, specifically aimed at shop-floor performance and quality measurement in the fuel injector final assembly clean rooms. Another 13 percent was spent refining measurement systems and on automated gauging for the high-precision machining and grinding, and super-micro-finishing of critical injector component parts.

John Crane Belfab also invested in new production technology to increase its ability to satisfy customer needs. In 1993, it created the Machined Products Group (MPG), a new minibusiness. This production cell is composed primarily of 10 CNC mills and lathes. It utilizes CIM technology to provide hardware to the company's other business units. Components designed at various company locations on CAD stations are passed via data link to a CADRA CAM system which is used to write programs for MPG's machines. The programs are directly fed to each machine via the company LAN. The results: short-run products, custom-designed for Belfab's customers.

To add to its capacity to develop guidance systems for customers such as the Department of Defense and NASA, Rockwell's El Paso Autonetics plant purchased an electronic Theodolite (T-3000). This product is used to determine exact measures of latitude, longitude, and azimuth. The technology integrates the electronic output of the T-3000, a Global Positioning Satellite receiver for an accurate time signal, and MICA—the Multiyear Interactive Computer Almanac from the U.S. Naval Observatory—to produce the precise position and motion of celestial objects. The resulting bearings are used to calibrate a mirror located on an Azimuth Reference Station, which is then used as a reference point for testing guidance, navigation, and control systems.

What that all means, according to manager Don Teague is that "this device and procedure will allow the facility to provide a unique technological capability and considerably enhances the plant's posture to assemble, test, and align inertial guidance systems. Very few sites such as this are currently available in the Free World."

Some of the Best Plants create their own technological breakthroughs. In 1985, Tennessee Eastman's Chemicals from Coal program won the Kirkpatrick Chemical Engineering Achievement Award for its pioneering efforts in developing new technology aimed at manufacturing chemicals from a coal base. The new process, which produces acetyl chemicals, replaces a petroleum-based process and results in a decreased dependence upon petroleum feedstocks. It also establishes the Kingsport, Tennessee, site as the world's largest producer of acetic anhydride from coal—at a billion pounds per year.

Super Sack is another Best Plant that creates its own technological innovations in order to build market share. In addition to building most of its own manufacturing equipment, the company develops proprietary bag-filling systems for its customers. This equipment incorporates programmable logic controllers and touch-screen controls for operator ease.

"PLC technology has allowed us to offer a new product to the industry, which can be made by programming in a number of dimensions for the product, resulting in a finished customized product," says vice president David Kellenberger. "This has allowed us to produce a line of unique bags which satisfy a large customer need and are not able to be offered by our competition because the equipment is not available in the marketplace."

Fisher-Rosemount's Chanhassen, Minnesota, plant is the largest pressure-transmitter manufacturing facility in the world. The company combines production technology with product innovation to stay on top of its market. Examples include the following:

- Seven patents for its application-specific integrated circuit (ASIC) technology make the company the only supplier of low-power microprocessor-based transmitters in the world.
- Solid-state sensor technology uses less power, creates a more stable process, and includes silicon micromachining, thick- and thin-film

processing, silicon fusion bonding, and ceramic processing for sensors that are inherently more stable.

- Finite element analysis and custom thermodynamic models help the plant produce mechanical packages that minimize stress on the sensor, increasing the stability and robustness of the final device.
- A patented laser-welding process for exotic construction materials prevents the mixing of dissimilar metals and improves corrosion resistance.
- Precision grinding and metal cutting operations are capable of processing a high volume of parts with extreme precision while maintaining the flexibility to make multiple product-mix changes every hour.

The leading result of these innovations is that Fisher-Rosemount has maintained its worldwide market leadership position in pressure transmitters since the early 1970s.

Summary

The third supporting competency of Best Plants is technological adeptness. The plants understand and utilize technological advances to achieve their core strategies of customer focus, quality, and agility. Although the way in which technology is applied among the Best Plants varies according to the unique needs of their organizations, the plants do follow four common application paths.

First, the Best Plants ensure that their investments in manufacturing technology support improved product and process quality, enhance their productivity levels, and support their ability to respond quickly to customer needs. Automation and robotics are commonly used, but applied carefully, to eliminate repetitive and dangerous tasks. The plants also invest in testing and maintenance technology that gives frontline employees the information they need to maintain and improve process and product quality.

Second, the Best Plants create electronic information and communication networks that support employee empowerment and eliminate the physical barriers between people and business functions. Production information systems provide workers with all the information they need to make defect-free products. Communication networks enhance the transfer of knowledge.

Third, the Best Plants leverage their investments in technology by joining information, communication, and production systems to build real-time systems capable of managing and operating an entire facility. Often, these systems take the form of CIM networks and enterprise-integration strategies. These systems must be responsive to fast-changing environments, so they are often module-based for easy adjustment.

Finally, the Best Plants apply technological innovation to build market share. They are adept at utilizing new technology in the creation of innovative products and services. They also apply technology to build process capabilities to levels that allow new advances in product performance.

> Don't overcomplicate processes; keep them simple.
>
> —*Copeland Corporation, Sidney Scroll Plant*

CHAPTER 8

PRODUCT DEVELOPMENT

New products and the refinement of existing products are the life's blood of Best Plants. Without a product to build, there just isn't much point to a manufacturing plant, and the Best Plants know the importance of a robust product development strategy and process to their long-term welfare. Therefore, whenever possible, the plants become experts in the creation and development of products. That ability, which is a primary requirement in the journey to quality, agility, and customer satisfaction, is the fourth supporting competency of Best Plants.

The level of a plant's product design and development responsibility differs widely among the Best Plants. Some of the plants have only a small degree of input into the process; corporate-level staff consults them regarding the manufacturability of a design, but little else. Others among the plants are active partners in large portions of the design process, but do not control the process itself or its outcome. Still others, however, are the sole captains of their product fleets. No matter what level of product development responsibility is invested in a plant, one thing is certain: Best Plants live or die by the success or failure of their products.

Accordingly, there are few Best Plants willing to completely pass the product development buck. They know that, at the very least, their input regarding the manufacturability, quality, assembly, and procurement of components, parts, and materials of a design can have an immense impact on the speed with which a product appears in the marketplace, its final cost, and the level of acceptance it enjoys with customers.

The product development and design efforts at the Best Plants are characterized by two major themes, both of which relate to integration. First, these processes are usually run by cross-functional teams that integrate the demands, needs, and limitations of a wide selection of product "constituents." Best Plants involve their extended enterprise in their product development cycles. Often, these teams include representation from all the major functions in their business, their suppliers, and their customers in the process of creating new goods.

Second, the product development process embraces a wide variety of tools and techniques. The integration of product development methodologies and technologies enables the Best Plants to reduce a product's time to market, ensures that the characteristics of the product match the needs of its intended customers, and guarantees the ability of a plant to profitably manufacture the product.

Further, these two trends—an integrated team-based structure and an integrated approach to the tools and techniques of product development—are intimately related. The combined talents of a team and its ability to act concurrently, that is, to conduct multiple activities at the same time, are a necessary support of an integrated process that utilizes a wide-ranging collection of development techniques. Conversely, a strategy that interweaves many different product development techniques cannot reach timely completion without a structure that allows for a concurrent effort.

Teaming Up for Product Development

The Best Plants approach to product development is overwhelmingly team-based. Internal teams, with memberships spanning the full range of plant and business functions, are at work constantly in these facilities. In several cases, the plants have created development labs, a sort of common ground where the various functional experts can model the manufacture of new products to gain greater insight into the ramifications of their designs. In others, the entire workforce has been included in an enterprisewide effort to develop new product ideas and concepts. Even the product development process itself is subject to team-based improvement approaches.

In keeping with their extended-enterprise perspective, the Best Plants build extended-product teams. They invite their suppliers into the development process to reduce cycle times and ensure the procurability of their designs. Most important, they include their customers in the development process as often as they are able. Customer input directly feeds the idea-creation process, shapes the initial prototypes, and tests and approves the final product.

As one part in a five-year program that fundamentally changed the entire company's structure and culture, teams were established as the operating structure of choice in the product development process at Super Sack's Fanin County, Texas, facilities. Starting at the top of the organizational chart, the plant's Strategic Planning Team, which is composed of employees representing sales, marketing, engineering, manufacturing, quality, and the company's leadership, meets monthly to oversee product development. The team approves product development capital, sets project goals and deadlines, and reviews the progress of current projects.

"Our manufacturing processes have been improved by involving our teams in the design process from the start," says vice president David Kellenberger. "Sometimes, team involvement may increase the design time for new products. Reaching consensus can take longer for a diverse team than a single manager. The important difference is that our teams are also more positive about changing to a new product because they have been integral to the design. This results in higher quality and productivity."

Marlow Industries introduces an average of six new thermoelectric cooler designs each month, most of which are custom-developed to perform to its customers' specifications. Heavily reliant on concurrent and process engineering, Marlow's development system accomplishes its new product goals using a cross-functional design effort lead by Release to Production (RTP) teams. RTP teams meet weekly to oversee the design and technical details of developing products.

Marlow has established an RTP team for each market segment served by the company. The teams are composed of operators and engineers and they are, according to vice president Ed Burke, "the most effective source for our reductions in the product development cycle." The team concept coupled with a detailed development procedure helps the plant ensure its devices are being built "right the first time" and also helps the company maintain a continuous focus on the all-important flow of new products.

One measure of the success of Marlow's design and development process is clearly evidenced by the fact that the company was able to shift its primary business away from military customers, who accounted for 75 percent of sales in 1989, and build a new base of commercial customers, who by 1993 accounted for 55 percent of the company's annual sales. Further, this shift was accomplished, without a loss in sales volume or profits, with a team-based process that has recorded 100 percent first-article success and a 50 percent reduction in product cost.

MEMC Electronic Materials of Saint Peters, Missouri, employs cross-functional product development teams in its search for continuous reductions in design cycle times. MEMC's teams include participants from quality assurance, manufacturing, and manufacturing engineering, as well as R&D. Interestingly, the company says it is purposely moving away from lab-based product design and prefers to develop new product enhancements on the production floor, where the more practical considerations of manufacturability are better understood.

The Saint Peters silicon wafer manufacturing plant uses a formal seven-step process for designing products and services that incorporates a "mechanistic" approach to design phases and clearly specifies team responsibilities and tools. To create an interdisciplinary perspective among team members, the company often gives its R&D engineers a few production responsibilities. Conversely, manufacturing process engineers are assigned some R&D project responsibility.

"Both manufacturing and development engineers have significant statistical training and use statistical design of experiments to shorten development cycle time," explains manufacturing manager Perry Lee. "The broad perspectives and interaction obtained as a result contribute significantly to mastering the cycle time of specific new enhancements."

In the past, many manufacturing plants have had little input to the product design process. But with increased use of team-based functions and greater understanding of the contribution that manufacturing employees can make to the process, their involvement is increasing. Diane Breedlove, operations director of Coherent's Auburn, California, plant explains manufacturing's role as follows:

> *The Electronic Products Division is the manufacturing arm for Coherent Laser Division in Santa Clara. Products are designed and developed in Santa Clara and sent to Auburn for manufacturing. In order to shorten the product development cycle between the two divisions, a very close relationship in the early stages of the design is coordinated. Auburn's purchasing, engineering, management, and direct-labor employees are represented during the prototype and first-article releases.*

Milwaukee Electric Tool's Jackson Plant is another plant that has little direct product development responsibility. Nevertheless, it plays an active role in the design process. Manufacturing-cell team members participate in new-product development from the outset of the cycle under its design for manufacturability focus. The result, according to plant manager Harry Peterson, is the elimination of prototype and manufacturing start-up problems, and a 35 percent reduction in the cycle's time to market.

Northrop Grumman's Naval Systems' plant in Cleveland, Ohio, involved its entire workforce in the development of new products. While owned by Westinghouse Electric, the plant reduced its dependence on military contracts by actively soliciting and developing ideas for commercial products from its workforce (for details, see Individual Profiles). As soon as these ideas were ready for the drawing board, the plant used a concurrent engineering approach, which included marketing, design, purchasing, and manufacturing staff, to fast-track the development process. In this way, the plant ensured the products were developed with minimum cost, maximum operator efficiency, and the highest possible yield. This plant collocates all key development personnel to promote teamwork and communication.

Another way in which the Best Plants bring together all the different functions involved in product development process is through the establishment of development labs. Cadillac's Hamtramck Plant created such a facility in its Assembly Line Effectiveness Center (ALEC), where hourly production-line employees come to actually assemble the prototypes of future products. The 20,000-square-foot facility acts as a simulated assembly environment, and the design-for-manufacturing input it generates from production workers is known as the "voice of the assembler."

ALEC is an important element in the plant's team-based simultaneous engineering (SE) approach that involves all disciplines early in the design and engineering phases of new-product development. These teams are also responsible for the continuous improvement of current products. All of the SE teams include plant representatives, and nearly all include hourly workers. Up-front employee involvement in the development of new car models provides "hundreds of improvement suggestions" and allows Cadillac to bring the voice of the assembler into the product design process over 18 months sooner than in its earlier development methodologies.

General Motors' Delphi Energy & Engine Management Systems plant created its development lab, in this case a short-run, highly flexible manufacturing cell, to evaluate the new manufacturing-process concepts the plant required in its product development efforts and to reduce the high cost and extended lead times of its existing product development process.

Established in 1992, Delphi's Development Lab's mission is to develop processes, prototypes, and low-volume production parts that are competitive in price and delivery and that meet or exceed customer expectations. These objectives, according to the plant, are accomplished through team structure, manufacturing flexibility, and the creative application of new technologies. Goals that support the development lab's mission are as follows:

- Reduce lead time in bringing products to market.
- Continue to improve product quality.
- Reduce cost and delivery time of prototypes.
- Reestablish a higher level of trust, pride, and enthusiasm by:
    - Recognizing the full worth of all people
    - Providing employees a voice in the decision-making process

Treating each other with respect
Open and regular communication
Enhancing involvement and self-motivation

- Train employees in new technology and manufacturing methods.
- Shorten manufacturing learning curve for new products.
- Develop and implement new process technology that can be applied to low- and high-volume production.
- Incorporate standardized documented processes.

Team-based development strategies are not relegated to speeding new products to market. In some cases, the Best Plants use teams to reengineer the product development process itself. For example, in 1991, Ford's Electronics Division established a cross-functional team of engineers and analysts from around the world to streamline its engineering change process. This process improvement team created a new methodology that categorizes engineering changes according to their complexity (complex, intermediate, or express) and defined specific process steps for the changes in each category.

The new development process was fully implemented at Ford's North American and European facilities in late 1992, and within two months had reduced the processing time for engineering changes from 235 days to 82 days. "The key," says plant manager Dudley Wass, "was simply defining a better process, communicating it to everyone involved, measuring the time of each change, setting specific goals, and regularly reviewing status."

At Johnson & Johnson Medical's El Paso, Texas, Artcraft plant, time savings and other improvements of the product development cycle are championed by a Value Analysis Team. Established in June 1992, the team is "committed to develop best-cost pack and gown products with fast and flexible processes, utilizing both our customers and an internal team effort to become market leaders." Led by Artcraft's IE/Technical service manager, the value analysis team is composed of employees from marketing, sales, R&D, manufacturing, technical services, and market research. Customer input is also regularly included as part of the analysis process. The team's initial effort eliminated 60 days from the 320-day development cycle, a reduction of 18.8 percent.

As should be obvious by now, a team-based product development cycle need not, indeed should not, be restricted to company employees. Extending the teams to include customers adds substantial value to the development process. For instance, Fisher-Rosemount's cross-functional team-based development process is enhanced by a design partnership program that actively solicits customer input early in the design phase of new products. The company identifies potential customers for these partnerships based on industry affiliation, image as an industry leader, and worldwide business impact.

The pressure-transmitter manufacturer and its selected customer-partners enter into proprietary information agreements to facilitate the unrestricted sharing of design and process data. Multiple meetings with each partner are conducted to identify product features, design approach, and project status. The customer-partners also provide early field testing of beta-design hardware. Says operations manager Amy Johnson:

> *The advantage of this program is that information is given directly from the customer to the design team. Team members are allowed to ask what-if questions that help them focus their efforts on the most feasible solutions early on in the process.*
>
> *These two approaches [cross-functional teams and design partnerships] proved to be critical in defining the performance and functional requirements of the Model 3051C pressure transmitter. DuPont, ARCO Chemical, Mead Paper, and ICI participated in the design partnership program. The cross-functional design team was taken to a facility off-site to focus their development efforts. It is estimated that the design time of the 3051C was reduced by 18 months through the use of these programs. Today, the Model 3051C is recognized as being a world-class high-performance pressure transmitter, selling 10 times the annual volume of the product it replaced.*

Unisys' Pueblo Operations facility also depends heavily on customer participation in the development process. "Twenty-five percent of our employees from top management down have frequent customer contact as work is performed on a product," states staff consultant Rudolph Krasovec.

At the beginning of all product and service programs, this plant establishes a program team and designates an "Executive of Interest" to support the customer. Customers are encouraged to contact these people for manufacturing and material-related issues. Although these customer contacts are clearly identified in the beginning of the program, the plant also encourages frequent, informal contact between customers and employees. According to Krasovec, customers are formally brought in to the product development and manufacturing process at the following points:

- Production readiness review. *These reviews by the customer evaluate our ability to perform contract requirements.*
- Testing cycles. *These include all qualification and production tests. Two important test phases exist: the first phase is to ensure the test equipment com-*

bined with the acceptance software is exercising the product as required, and second is to perform the First Article Test, which verifies that the hardware meets the design performance requirements.

- Physical configuration audit. *This is a physical audit of one of the first articles built. It confirms that the hardware and materials meet the requirements of the product drawing definition and contract.*
- Program review meetings. *These are periodic reviews of the contract, schedule, quality, engineering, and new business or future requirements.*
- Formal acceptance *of our products by the customer's quality assurance personnel.*

Pueblo Operations chooses concurrent product development (CPD) as its primary technique for reducing its products' time to market. Over 100 engineers involved in new-product development attended CPD training classes, and collocation of both development and manufacturing personnel is practiced to improve communication among team members.

At Cadillac's Hamtramck assembly plant the voice of the customer plays as important a role in the development process as the voice of the assembler. Cadillac uses its Market Assurance Process to identify current and potential customers' needs and wants, and then matches those requirements with engineering and manufacturing capabilities. As with assembler participation, the company is bringing customers into the development cycle earlier than ever before, thereby cutting process time and more efficiently building its customers' needs into the final design.

Using a Full Product Development Toolbox

The Best Plants product development team efforts are supported by a variety of design tools and techniques. Techniques, such as design for manufacturability and systems engineering, provide standardized methodologies that the plants can follow again and again in their quest for competitive advantage in the marketplace. Design tools, such as stereolithography, CAD/CAM, and finite element analysis software, reduce their cycle times and development costs while enhancing the quality and accuracy of the products.

It should come as no surprise that most of the Best Plants are adept at applying design-for-manufacturing principles to the product development process. Siemens Automotive's Newport News plant reports that their principal product, the Deka multipoint fuel injector, was specifically designed with manufacturability and ease of assembly in mind. Thanks to the design technique, this injec-

Gilbarco's Teams Cut Time to Market

Between 1990 and 1994, Gilbarco cut the time to market for its new products by 70 percent on virtually all new products. How? "Through the use of cross-functional teams," answers manufacturing strategy manager Jerry Smith. "These teams *always* include the hourly production employees who will be assembling the product."

During the development of the company's Legacy product, integrated point-of-sale and dispensing-equipment systems for gas stations, a team of hourly and salaried workers reviewed the line's conceptual drawings and manufacturing diagrams for ease of production. During the prototype phase, every affected production worker reviewed the new design and provided more input on its manufacturability. Hourly workers, who also meet with the suppliers involved in the product line, were instrumental in all aspects of production start-up, including process definition and production-line layout. "It is Gilbarco's philosophy that no one knows better how to design or assemble a part than the hourly production worker who does the work," declares Smith.

The company's product development process is guided by the principles of concurrent engineering, a fixed part of its overall manufacturing strategy. Concurrent engineering processes seek to reduce cycle time by performing a variety of development tasks simultaneously instead of sequentially. By its very definition, concurrent engineering depends on teamwork.

At Gilbarco, the development process starts with the formation of a series of cross-functional teams that, at one time or another, include customers, suppliers, design engineers, manufacturing engineers, and materials, quality, marketing, and production employees. The duties of these teams include defining customer requirements, product design, process design, supplier selection, the purchase of new equipment, and production-line layout. Further, the plant empowers its concurrent engineering teams to make decisions, allocate resources, and spend funds with minimal senior management review, thereby saving more time.

The team-based product development process has been a resounding success for Gilbarco. The plant's design-to-market cycle time was reduced by 70 percent and the Legacy product line's reliability is 65 percent greater than previous models.

tor requires only four high-precision grinding operations versus the thirteen grinding operations required by its principal competitor.

The process for building the Deka injector was designed to allow for automated calibration during assembly as opposed to calibration by part-matching as required for previous designs. The assembly and test operations for the product are also fully automated. "The focus on simplification of the manufacturing processes and automated assembly, calibration, and testing has enabled the division to establish itself as the worldwide low-cost producer of multipoint fuel injectors," declares marketing manager Sam Byrd.

The Newport News facility additionally supports its new-product development efforts with the structural, magnetic, and fluid dynamic analysis tools, the adoption of solid modeling, and state-of-the-art technology for measurement and visualization of fuel-spray preparation. The result is a more efficient product development cycle, clearly evidenced by the fact that the technical development staff at the plant has remained relatively constant "while sales have increased by a factor of 2.9, and 75 new customer programs have been obtained."

In 1983, GE Fanuc established a Producibility Engineering group to master the principles of design for manufacturability and to incorporate those practices into the plant's product development. This group remains an active entity within the product development function. Some of the results of this initiative: Almost all the plant's circuit boards use a standard 8.7- by 11-inch form; setup times have been reduced by up to 90 percent; and the number of different components used at the plant has been significantly reduced.

Among the other "design for . . ." practices, General Motors' Delphi plant reports that design for procurement practices, which it utilized in conjunction with a supplier of a component used in the plant's Direct Hydraulic Valve Lifter design, resulted in significantly reduced manufacturing costs and shortened time to market. Marlow Industries reports that its reliance on design for quality has resulted in 100 percent first-article acceptance.

Ford's North Penn Electronics Facility (NPEF) serves as a divisional proving ground for new-product development. Routinely, new products are prototyped, debugged, and launched at the facility before their production is transferred to other plants. In its role as the organization's product development expert, NPEF maintains a strategic technology plan that reaches eight years into the future to identify emerging manufacturing technologies and link them to specific product programs. The goal is to ensure that the plant remains on the leading edge of the electronics industry. The plant's Advanced Process Technology Group uses this plan as a road map for the development of manufacturing processes and as a primary influence on product development efforts. NPEF maintains facilities for all phases of process and product development. Prototype labs within the plant are equipped to duplicate existing processes, as well as to provide for new process developments.

The plant follows a product development process known as PDP II, which is utilized by Ford Electronics Division in its drive toward shorter development cycle times and high-quality standards. PDP II is a flexible, standardized process that utilizes cross-functional product teams and program management concepts to meet the different product development requirements of its customers. Value engineering is applied to new product programs in a joint process with product engineering and key suppliers. In addition, PDP II encompasses all of the necessary disciplines to achieve first-pass design success.

The plant also features a unique product development tool called the Manufacturing Design Rule Checker (MDRC). MDRC is a new technology that allows drafters and product designers to review the manufacturing process requirements of a design, including such details as component pad and hole dimensions, required part-to-part spacing, and equipment and tooling clearances, prior to releasing it to production. MDRC alerts the designer to potential conflicts and problems in the design and suggests changes to the design to improve manufacturability.

According to the plant, the use of MDRC has reduced the average number of feasibility issues in new designs by 97 percent and reduced the time to conduct a feasibility analysis by half. The early detection and correction of design issues helps NPEF achieve its high rate of first-pass design success. Engineering changes have also been dramatically reduced during the development and launch phase, leading to shorter overall product development cycles.

The Best Plants utilize a world-class toolbox of product development aids. At Symbiosis, for example, the bells and whistles on their product development process include Pro/Engineer (Pro/E) software that enables the company to create feature-based solid models for use in product tooling and mold design. The company's Pro/E software is also used to create stereo lithography–type prototypes and to conduct finite element and failure analyses.

Symbiosis uses Autodesk 3D Studio software for product visualization and marketing presentations. The company can import the Pro/E-developed model into its Autodesk package and create a photo-realistic image and a fully articulated three-dimensional animation of its designs. The 3D animation is used for motion studies and functional simulations.

Like Symbiosis, Varian Associates' Oncology Systems pursues rapid prototyping using techniques such as stereo lithography to generate a plastic model of its products from computer-generated designs. The models aid in the design review process and can often be used directly to create production tooling. "These tools allow visualizing, developing, reviewing, and iterating a design early in the design cycle," explains manufacturing manager Jim Younkin, "in addition to automating the process of generating final production drawings."

Marlow Industries also cut its development time by utilizing electronic networks. The company realized a significant reduction in design and layout time by linking its photo artwork supplier and the company's CAD systems using EDI. "This sped up the entire artwork cycle, resulted in better first-pass yield, and reduced our costs by 60 percent," says manufacturing vice president Ed Burke.

At Unisys' Pueblo Operations plant, simulation techniques and stress screen testing are used extensively during the design phase in an effort to reduce the time needed to introduce new products in full production. Design reviews are conducted via videoconferencing, and board layout and design parameters are electronically transferred from the plant's design suppliers.

Testing throughout the development cycle is also a regular occurrence at Lockheed Martin's Government Electronics Systems division. Explains quality manager Bill Adams:

> *Another critical feature of the development process is testing of the product. As the multilevel specifications are developed, test requirements are also generated. These test requirements cover each integration and test step and are translated into measurement test plans and test procedures. We use the concept of "build a little, test a little" to allow phasing of the design execution in a logical order, where the critical pieces can be tested. Failures can be replaced, thus reducing the complexity in downstream efforts. As digital modules are assembled, they are tested, then nests are tested, and finally, the entire system is tested. This multiple-step process has contributed significantly to the success of the [U.S. Navy's] AEGIS Program.*

Summary

Historically, manufacturing plants have had little responsibility in the corporate product development process. Often, they were simply handed a finished product design and a corresponding process plan and were told to get to work. Today, plants are integral partners in the development process. Best Plants take an active role in the design, planning, and creation of the products they are called upon to manufacture; they know that their future depends on their abil-

Integrated Product Development at Rockwell's Autonetics Plant

Rockwell International's Autonetics Electronic Systems plant in El Paso, Texas, has linked a wide range of product design and development tools and techniques together to create their Integrated Product Development (IPD) strategy. IPD combines the design for assembly, manufacturability, quality, and procurement techniques with a concurrent engineering approach that is, of course, team-driven.

An IPD team, whose members possess all the expertise and different perspectives required to plan and design for the product's total life cycle, is formed at the beginning of each project. This team supports initial customer communication on product requirements and remains in place through all phases of product design, production, and delivery.

The design teams utilize a systems engineering approach to product and process development. In systems engineering, IPDs attempt to achieve an optimum balance of all product elements, including design, manufacturing, quality, reliability, logistics, procurement, and customer requirements.

The Autonetics plant then adds a liberal dash of development tools, including manufacturing simulation models for flow-time optimization; parametric and life-cycle cost models; and CAD/CAM design tools. What does it all add up to? A 96 percent improvement in quality yields, a 44 percent improvement in productivity, and 100 percent on-time contract deliveries.

ity to quickly develop cost-effective and customer-focused goods. Accordingly, product development is the fourth supporting competency of the Best Plants.

As with so many activities, the Best Plants use a team structure for their product and process development work. Cross-functional teams ensure that their new products will be infused with the needs of all the parties involved in the production of the new goods. They also allow concurrent approaches to the development process, thus speeding time to market. The plants create a common meeting ground for their product teams through collocation and the creation of development labs. Best Plants also reengineer their design and development processes using cross-functional teams.

Best Plants use a variety of tools and techniques to wring the best efforts from their development processes. Design for manufacturability, quality, and procurement methodologies are commonly used to build the best possible products and processes. High-technology tools and supporting concepts such as finite element analysis, solid modeling, and rapid prototyping cut time and cost from the development process and build in product quality and development agility.

> Assume that your competition is capable of advancing and applying technology *at least* as fast as you are and adjust new product goals accordingly.
>
> —*Fisher-Rosemount, Incorporated*

CHAPTER 9

ENVIRONMENTAL RESPONSIBILITY AND EMPLOYEE SAFETY

The fifth supporting competency of Best Plants is the ability to create a manufacturing process that is environmentally sound and a workplace in which people are safe. This responsibility is a part of the fundamental viability of any business. Without both conditions in place, a company cannot effectively pursue the core strategies of Best Plants. In fact, without a safe and environmentally sound plant, the ability of a business to continue to operate profitably at all must be questioned.

This has not always been the case. The results of the industrial excesses and lack of knowledge of yesterday's manufacturers are well documented, and the consequences of their actions have become some of today's most pressing environmental and health problems. However, among today's Best Plants, there is a determined and concerted effort to operate in a way that reflects the best interests of the community and employees.

Environmental Responsibility

The Best Plants take their responsibility for the environmentally sound operation of their facilities very seriously. They make large investments in employee time and production equipment in order to substantially reduce any negative impact their work might have on the environment. The Best Plants set ambitious goals for themselves, often adopting new regulations long before their required implementation dates and, in some case, substantially outstripping their goals.

Setting Responsible Goals

In the fast-changing world of environmental science, it often seems that new methods, technologies, and regulations are announced every day. The state of the art of responsible environmental practice is constantly evolving. Best Plants respond to this continuous change in environmental programs and guidelines quickly and by setting ambitious goals for their own environmental objectives.

For instance, Air Products & Chemicals' South Brunswick, New Jersey, plant "strives to reduce all emissions and waste generation through capital and process improvements as well as procedural modifications and high standards of housekeeping," according to superintendent Patrick Loughlin. The plant is guided in its efforts by the codes of management practice outlined in the Chemical Manufacturers Association's Responsible Care Initiative and has taken a lead role in making the public a partner in its environmental activities (as described in the plant's Individual Profile).

Community outreach is not the only environmental accomplishment of the Air Products South Brunswick plant. It also takes a proactive approach to its compliance with all federal, state, and local regulations, and it sets continuous-improvement goals of its own to increase the effectiveness and efficiency of the plant's environmental controls. For example, the company set SARA Airborne's cumulative reduction goals of 85 percent by year-end 1992 and of 87 percent by 1995. Cumulative hazardous waste reductions were targeted at 87 percent in 1992 and 93 percent in 1995. The plant meets its goals using a project-based approach that, in 1992, had already resulted in a cumulative solids-per-pound-of-product discharge reduction of 24 percent, emissions reductions of 83 percent since base year 1987, and hazardous waste reductions of 70 percent in the same time period.

MEMC Electronic Materials set the ambitious goal of totally eliminating the generation of hazardous wastes at its Saint Peters, Missouri, plant. Guided by a corporate initiative that targeted several of the plant's air emissions processes, hazardous air pollutants were reduced from the 350,000 pounds generated in 1991 to 2,330 pounds by the end of 1994—a 97 percent reduction rate.

Ozone-depleting chemical (ODC) emissions, which exceeded 620,000 pounds annually in 1991, were targeted for elimination in another MEMC-initiated program. This goal was achieved by the end of 1994, a year ahead of schedule. During the same time period, solid waste generation was reduced by 51 percent.

Baxter Healthcare's North Cove, North Carolina, plant set the highly laudable goal of zero waste to landfill by year-end 1996. In this effort, the IV Systems plant focuses its quality improvement teams on waste management efforts and actively benchmarks the successes of other companies. In the pursuit of its overall goals in environmental stewardship, the plant removed all underground storage tanks, eliminated the use of Freon as a cleaning agent, began recycling all lubricants and hydraulic oils, eliminated SO_2 emissions by burning waste wood for steam generation, and started a recycling program for paper, cardboard, plastic, metal, wood, and Styrofoam. The plant attained Small Quantity Generator status for its efforts and has won numerous environmental awards, including North Carolina's Best Recycling Program award in 1991.

Best Plants don't simply comply with governmental regulations. Often, they become active community and corporate advocates for environmental stewardship and responsibility. The Foxboro Company's Environmental Group has been actively working with local, state, and national environmental agencies for over 20 years. It is also active in national organizations, such as Associated Industries of Massachusetts, Environmental Manager's Steering Committee, and the Board of Certification of Wastewater Treatment Operators, to promote environmental stewardship among other businesses. The group actively participates in seminars, workshops, and local radio and TV programs. Its goal in these activities is to promote and share its environmental knowledge.

Taking a Proactive Stand

Prevention is the name of the game when it comes to environmental initiatives. Building processes and products that do not require the use of hazardous materials or generate pollution is the goal of the Best Plants. In 1989, for example, Varian Associates and all of its business units adopted a pollution prevention program to permanently replace the "end of the pipe" solutions approach it had previously used. Toward this end, the company implemented purchasing controls that restrict the order of certain materials, established maximum levels for chemical inventories, and identified less hazardous alternatives to the chemicals the company uses in its products and processes. Varian, by the way, is one more of the many Best Plants that committed itself to the 33/50 Toxic Emission Reduction Program. It has met all of the requirements of that program long before the scheduled target dates.

Siemen's Automotive addresses environmental considerations early in its design process. It adopted a formalized Environmental Responsibility Policy that contains a mechanism that requires a full assessment of the environmental impact of all new products and processes starting at the concept stage. Mandatory design reviews for environmental impact are also required in the company's Product Program Management Process. Among the results of these required tasks is a 40 percent reduction of VOC emissions per injector, for which the company credits process and test equipment modifications, the elimination of all Class I ODC solvents by replacing the old process with an aqueous cleaning process, and a 70 percent reduction in solid waste through use of returnable dunnage and off-site recycling. The plant banned the use of cadmium and cadmium-containing materials in all new products and also eliminated its use in existing products, subject to customer approval.

Rockwell International's El Paso, Texas, plant uses its Integrated Product Teams (IPTs) to design environmental integrity into its products and processes. The plant's engineers, as members of the ITPs, identify and eliminate hazardous chemical requirements in new product designs and create hazardous waste elimination plans for all new production programs. This proactive approach begins during the proposal stage. Prospective designs are evaluated by design, manufacturing, and process engineers, and hazardous substances are replaced with "friendlier" chemistries and processes.

The environmental scrutiny continues after the contract award and through the engineering model, operational model, and production phases of the plant's work. Standardization of adhesives, revision of machine maintenance schedules, extension of shelf life through testing, and the modification of product drawings are among the techniques the plant uses in its efforts to create a greener business.

Results show that the effort is paying off. The El Paso manufacturing facility reduced its generation of hazardous wastes by 50 percent in a two-year period, an improvement that qualified the El Paso facility as a Small Quantity Generator. Additionally, the facility eliminated the use of Ozone-Depleting Substances (ODS) by 85 percent, beating the Montreal Protocol 1999 target for complete elimination by four years.

Other techniques the El Paso plant uses to reduce waste include:

- Recycling, distilling, or filtering of chemicals to reduce waste disposal. This program saves the plant $68,000 in new chemical orders each year.
- Involving chemical users in the monitoring and adjustment of chemical orders for optimum usage quantities, thus eliminating excess chemicals in the manufacturing process.
- Replacing outdated batch-cleaning equipment with new, near-zero-waste, environmentally friendly equipment.

The Rockwell plant also established a Hazards, Elimination, Loss, Prevention (HELP) Supervisor Program to improve worker safety. Under this program, the facility is segregated into four hazard-elimination areas. Four members of management are appointed for a 12-month period, and each is assigned responsibility for one hazard-elimination area. Their duties include environmental health and safety inspections, plant equipment and fire inspection, and the auditing of team meetings to ensure that environmental safety remains a constant focus.

Gilbarco, Incorporated, implemented a series of process and operational solutions to improve its environmental status. Manufacturing strategy manager Jerry Smith describes them as follows:

- Elimination of chlorinated cleaning solvents for degreasing and replacement with aqueous-based systems.
- Implementation of a system to review all chemicals for health and environmental impact before they enter the plant either as samples or production supplies. This review is conducted by an Industrial Hygienist and each item accepted is given unique ID.
- Elimination of all lead-based paints from operations and the elimination of all suspected carcinogens from the manufacturing process.
- Installation of powder paint system.
- Installation of two self-contained, fire-protected chemical vaults for the storage of hazardous waste prior to off-site transfers.

The Timken Company's Faircrest Steel Plant is a major user of recycled materials; it melts down the equivalent of 2,000 cars per day to feed its steelmaking operations. The plant reuses municipal scrap and all of the scrap produced in the operation, thereby reducing the volume in landfills. The plant also features state-of-the-art pollution control equipment, an investment representing 10 percent of the construction costs for the facility.

In 1994, Timken received a certificate of achievement from the Environmental Protection Agency for exceeding the goals of the agency's 33/50 Program three years ahead of schedule. Of the 17 chemicals listed by the EPA in the initiative, company's emissions and off-site releases consisted primarily of several heavy metals in electric-arc-furnace (EAF) dust, created when scrap metal and alloys are melted in electric-arc furnaces, and a degreasing agent, III-Trichloroethane, used in its bearing operations.

In 1993, the company had reduced its use of these chemicals by 92 percent, well ahead of schedule. William Fladung, Timken environmental control manager, explains how the reductions were achieved:

> *We worked with purchasing to switch the disposition of our Faircrest Steel Plant's EAF dust from solidification and disposal in a hazardous waste landfill to high-temperature metals recovery. This move actually resulted in a savings*

for the company. EAF dust, which is high in zinc, from our Harrison Steel Plant was already going to high-temperature metals recovery.

The EAF dust collected in the bag houses, a facility containing numerous bags to filter out dust particles, is no longer disposed of in a landfill. Instead, it is recycled so that zinc and other metals are recovered. As an extra benefit, the company saves money with this process.

Ford Motor's North Penn Electronics Facility is a leader in the development and application of the new inert-gas wave-soldering process. At one time, 350,000 pounds of Freon per year were consumed in the plant's wave-soldering processes. Today, the inert-gas wave-soldering process, also known as a *no-clean* process, has eliminated all Freon cleaning agents from production. This new process is also projected to reduce the plant's operating expenses by $1 million per year.

To further aid in reducing atmospheric emissions, the plant converted its solvent-based conformal coating operation for printed wiring boards into a solventless coating process. This reduced the usage of Volatile Organic Chemicals (VOC) by 70,000 pounds per year, or 25 percent. The plant is researching the use of non-VOC-based fluxes to further reduce emissions, and the use of methyl chloroform in the production of thick-film hybrid products was totally eliminated.

North Penn also places a major emphasis on recycling by returning, reusing, and reselling packaging materials. The returnable packaging programs have eliminated more than 1 million pounds per year of waste in the form of cardboard, pallets, and plastic, which has saved $300,000 per year.

Unisys' Pueblo Operation created its own extensive recycling program starting in 1992. It is based on three simple strategies: Reduce consumption, reuse if possible, and recycle what is left. To achieve these goals, this plant also began reusing all incoming packaging and packing material and implemented a policy of refusing to accept packages that contain loose packing material. Chip carriers are now returned to their manufacturers for reuse. The plant also recycles all paper and aluminum. In 1992, it recycled 27 tons of paper, more than any other organization in its Colorado community. In addition, over 1,000 pounds of aluminum was recycled in one six-month period.

Employee Safety

Employee safety is the second half of the Best Plants' fifth supporting competency. The Best Plants believe that people are their most important assets.

Super Sack Manufacturing's Environmentally Friendly Designs

"Not only do we not use or create any hazardous materials, but we avoid putting hazardous materials into the environment as well," declares Super Sack Manufacturing vice president Dave Kellenberger. At this Best Plant, products have environmental responsibility built into their designs.

Among its design solutions to environmental concerns is Super Sack's all-polypropylene bag, which is reconditionable and ultimately recyclable into polypropylene resin. The company has also developed bags that are rated for multiple trips rather than for a single use. This design created a whole new business in the cleaning and repair of the reusable bags. Recently spun off from the company, the new business is housed in a 20,000-square-foot plant located next to the Bonham plant.

Super Sack pioneered the process of printing on woven plastic with water-based inks. To do so, it built its own printing presses and worked with ink suppliers to develop a new ink. The company even built its own treatment system to remove ink solids from the plant's wastewater. All of its inks are CONEG-approved, and the in-plant cleanup removes even the color.

Further, all of the materials used in the company's standard products are food-grade and composed of 100 percent polypropylene to allow for complete reprocessing. All scrap material is sent to recycling facilities. Says Kellenberger: "We are in the process of setting up to reprocess and reuse our own scrap internally with the 1994 acquisition of a company that provides our liners and will be recycling wastes and providing fabrics."

The plant's recycling team captures and recycles all polypropylene, polyethylene, corrugated, and paper core waste. The money generated from this activity is used for plant beautification projects. "This sends the message that recycling has direct environmental benefits as well as direct economic benefits," adds Kellenberger.

Accordingly, they protect their employees. The plants also know that the trust, responsibility, and accountability that they have worked so hard to develop within their workforces is dependent on their employees' belief that the plants are sincerely concerned with their employees' safety and well-being.

Employee safety is an integral component of the operating systems at Best Plants. Manager Peter Gallavin of General Motor's Delphi Energy & Engine Management Systems begins to suggest why this is so when he says, "Our focus on safety is directly related to our emphasis on total customer satisfaction. It is our belief that any employee who is preoccupied at any level with their personal safety and well-being cannot focus totally on product quality."

Stop a moment and replace the words "total customer satisfaction" and "product quality" in Gallivan's quote with any other objective of a business that requires the participation of its employees. It makes no difference what the words are, the statement remains accurate—people who are not safe cannot perform to their best ability. The resulting loss is twofold: Neither the employers nor the employees have achieved their highest potential.

Planning a Safe Workplace

Best Plants approach the task of creating a safe workplace in much the same way that they plan and implement any other organizational improvement effort. This process usually includes the establishment of an operating structure, the ongoing commitment of people at all levels of the organization, and clear objectives, strategies, and tactics.

A good sense of the long-term commitment required to create effective safety programs in the workplace is evident in site manager Ray Floyd's description of the ongoing development of Exxon's Baytown Chemical Plant's (BCTP) safety strategies:

> *Because this is such a serious business, we are continuously improving our safety systems. BTCP is now embarking on our fifth cycle of safety improvements. Originally, we operated a traditional safety practice. In the second cycle, we adopted the Du Pont layered safety audits and related practices. In 1988, we adopted as a third cycle an Exxon Chemical system called Safety Management Practices (SMP).*
>
> *In 1992, we consolidated our SMP standard with other standards for health, environmental, and community awareness to create the fourth cycle of progress: Operating Integrity Management Practices (OIMP). We believe that through deployment of OIMP, our policies, standards, and practices are among*

the best in industry, are well documented, and are thoroughly well known to our people. With that in place, we are now focusing our fifth cycle of improvement on human factors, including personal responsibility for safe behavior and team responsibility for the safety and behavior of team members, including safety support networks to deal with personal issues which may adversely impact safe behavior.

In 1993, Floyd was able to report that the results of the plant's efforts included a 1992 National Petroleum Refiner's Association Safety Achievement Award for more than 2 million hours of work without a lost-time injury. Before a single lost-time injury occurred in 1990, the 3,500-acre facility, which is staffed with several thousand people, had run for more than 15 years with no such injuries. Many of the facility's operations continue to have long records of perfect safety. For example: in the Laboratory, 31 years without a lost-time accident; among the Riggers, 27 years; and in the Mobile Equipment Garage, 27 years.

The Best Plants' safety and wellness strategies are usually deployed through a formal operating structure, such as the one in place at South Brunswick's Air Products & Chemicals. Air Products' safety system, which has been in place since the plant opened in 1974, is led from the top of the organizational chart. Its Central Safety Committee is chaired by the plant manager and meets monthly to oversee the plant's safety programs. Part of its job is the review of the progress of the plant's team-based safety initiatives. These improvement efforts are undertaken by the plant's Corrective Action Teams.

To involve the rest of the workforce in safety awareness and keep everyone alert to safe practices, monthly safety meetings are held for all employees, and each shift begins with brief Spotlights on Safety. Additionally, a Safety Suggestion Committee, which is composed of hourly employees and production supervisors, reviews safety suggestions and is empowered to commit capital funds for improvements. An annual safety program audit is conducted by the company's safety personnel to assess performance and plan for continual improvement.

"The focus is on safety performance and includes the plant employees and contractors," explains production superintendent Patrick Loughlin. "The years of training and commitment to safety have resulted in nearly four years without a lost-time injury and over one and a half years without a recordable injury."

Many of the Best Plants spread the responsibility for safety through the ranks of their employees. Like Air Products, Advanced Filtration System's Champaign, Illinois, plant also uses a safety committee and safety team structure. Its Safety Committee, which includes two elected representatives from each shift and all of the plant's supervisory personnel, also serves as a kind of organizational clearinghouse for safety information. The committee hears reports on safety performance, suggestions, and problems. They also arrange for outside professionals, who present training in areas such as regulations and practices specific to bulk chemicals and ergonomics and ergonomically designed equipment.

Shift-specific teams report to the committee and meet at least once per month to conduct safety training, collect written safety suggestion cards from employees, and to prioritize and implement their suggestions. The plant's kaizen teams are also empowered to investigate and implement safety improvements.

Team-based safety strategies are common among the Best Plants. The Timken Company's Bucyrus Bearing Plant employs a full-time Safety and Security department that is responsible for maintaining a safe work environment, accident investigation/prevention, environmental compliance, and first-aid administration, but the plant's self-directed work teams also maintain an active interest in improving workplace safety. General manager James Benson says:

> *Safety is seen as an important issue at this plant, to the point where it has been established as a Key Role position on our self-directed work teams. One of the most significant accomplishments coming from this has been identification of potential safety hazards when dealing with short tubing from our steel supplier. Through this Key Role position, minimum acceptable tubing lengths have been increased to a size that does not create a safety hazard when placed in the material-handling equipment.*

Super Sack Manufacturing has deployed all of its safety responsibilities to the work-team level. Each of the plant's production teams now has a safety coordinator who is responsible for the safety of his or her work area. The safety coordinators also form a team of their own and are responsible for managing the plant's entire safety program. The coordinators conduct safety inspections, fire and tornado drills, accident investigations, plant safety meetings, and training sessions. The safety coordinator role, like that of team leader, rotates. Every three months, the job shifts to another member of the production work team, thus allowing everyone in the plant the opportunity to be actively involved in the plant's safety program.

Walking the Safety Talk

With a strategy and structure in place, the Best Plants turn their attention to employee education and the continuous improvement of workplace safety. At Honeywell's Industrial Automation & Control (IAC) plant in Phoenix, Arizona, for example, employee work teams conduct safety training, review safety issues and accidents, as well as create an awareness of the importance of employee safety.

The plant also maintains an in-house graduate-level course in Ergonomics that teaches the plant's employees how to reduce repetitive-motion injuries. This course promotes and teaches the use of fail-safing concepts to eliminate safety hazards.

Safety, a key element in the quality of work life at Lockheed Martin's Moorestown, New Jersey, Government Electronic Systems plant, enjoys a high priority in the plant's highly successful partnership between management and the labor union. This resulted in an alliance called the GES/IUE Competitive Initiative, which established a comprehensive work-safety program that in 1993 included 13,000 employee-hours of safety training and certifications.

Another component of the program is an ergonomic task force that addresses human-factor issues and has initiated a three-year plan designed to increase employee awareness, identify trends, implement changes, and provide funding to improve employee well-being. The ergonomic task force has been instrumental in identifying areas that need redesign and has worked successfully with tool designers to implement permanent improvements.

In March 1991, when safety awareness was designated as a primary goal at John Crane Belfab, the plant began an aggressive safety awareness program. One element of that plan is a safety suggestion award program in which employees who identify safety hazards or safety ideas are paid $10 in cash on the spot. The company also adopted incentives, such as barbecues cooked by management, team luncheons, and extra time off on holidays, to reward accident-free performance.

Among the results Belfab achieved: a 257-day stretch without a lost-time accident, an 83 percent reduction over the previous year; and an 80 percent reduction in cuts and laceration injuries over a two-year period (historically, these injuries account for 60 percent of all the plant's injuries). The Volusia Safety Council recognized Belfab for the most outstanding active safety program in 1991.

Johnson & Johnson Medical's El Paso, Texas, plant received the highest rating on its annual Corporate/NATLSCO Environmental and Safety Audit. Contributing to that ranking are zero-access guards that are monitored in quarterly audits on all the plant's machines, annual employee training in conformance to federal and corporate health and safety guidelines, and an ongoing safety incentive program.

The plant also employs a full-time industrial engineer devoted solely to ergonomics projects. In 1993, over $500,000 was dedicated to those projects, and 26 improvements were completed in the United States and 340 in Mexico. An ergonomics exercise program, adopted after employee benchmarking projects, was established in March 1993 and enjoys 100 percent participation.

Milwaukee Electric Tool reduced its OSHA incidence rate from 16.42 in 1991 to 5.77 in 1994, an improvement of 65 percent and a rate only slightly more than half the national average for plants in its industry. The plant accomplished this by eliminating difficult manual and repetitive operations, selecting automated processes, and reviewing all assembly methods for ergonomic improvement. A large portion of the improvement is attributed by the plant to the elimination of cumulative trauma disorders.

General Motor's Delphi plant in Grand Rapids, Michigan, boasts of one of the lowest workmen's compensation rates of any GM manufacturing facility. Among the elements of its safety program are continuous monitoring and reportage of air quality data, lockout training for the entire workforce, and the use of side shields for every employee. As a result of the latter initiative, eye injuries at the plant have been reduced 60 percent. Ergonomics training and a falling-hazard protection program whose goal it is to eliminate as many falling-hazard situations as possible (rather than simply training employees in the use of protective equipment), round out the picture.

Ford's North Penn Electronics Facility credits its 56 percent reduction in lost-time injuries, an achievement that was recognized by the National Safety Council with an Award of Honor in 1993, to several hours of safety awareness education and other training courses aimed at improving employee health and quality of life. The courses include Guidelines Responsibilities and Safe Practices awareness, Energy Control and Power Lookout, and various seminars on diabetes, cancer, hypertension, weight reduction, smoking cessation, nutrition, and grief counseling.

Nippondenso Manufacturing's vigilance in safety issues and an investment in fitness facilities has paid off with a worker's compensation rate that is one-fifth the national average for companies that share the plant's SIC (Standard Industrial Code). Says employee Karen Cooper-Boyer:

> *We run numerous programs to encourage safe working, track and prevent lost-time accidents, address repetitive strain injuries, and promote health and fitness for associates and their families. Our new on-site Recreation Center for all associates features tennis, volleyball and basketball courts, a softball diamond, a putting green, and top-of-the-line workout equipment. Professional fitness trainers are on hand to provide individual physical assessments and workout advice, and to coordinate sports and health activities.*

Gilbarco reports a sterling record in its Best Plants application. Between 1989 and 1994, the company won four Awards of Honor and one Award of Merit from the National Safety Council. One award cited the plant's achievement of over 3 million safe work hours, a record for this plant's SIC.

The Greensboro, North Carolina–based company assigned a safety coordinator to address safety concerns, training, and help conduct safety audits in each of its self-directed production work teams and an active Corporate Safety Committee to oversee the company's safety strategy. Ergonomics teams composed of medical and engineering personnel attack and eliminate deficiencies in Gilbarco's production system. For example, performing a certain hose-connection operation was so difficult that assemblers were experiencing hand and wrist injuries, and an engineering design change eliminated the problem.

Summary

The fifth supporting competency of the Best Plants is their ability to create environmentally responsible operations and a safe workplace for their employees. This responsibility is a fundamental one that all companies must undertake. Best Plants know that they cannot successfully pursue the three core strategies of quality, agility, and customer focus until these conditions are established.

The Best Plants accept their environmental responsibilities with gusto. They respond quickly to new environmental developments and set an example by their rigorous compliance with governmental regulations. Further, they set ambitious environmental goals of their own and invest the time and capital necessary to attain those goals.

Best Plants also know that employee safety is a fundamental responsibility of manufacturers, as well as every other business. They plan and implement their safety programs with the same attention to detail that they focus on quality and productivity. They walk their safety talk by ensuring that their employees have a safe and healthy environment in which to work.

> We recognize that a shared world means a shared responsibility for the environment and accept this responsibility.
>
> —*Chesebrough-Pond's USA Company*

CHAPTER 10

CORPORATE CITIZENSHIP

The sixth and final supporting competency of Best Plants is corporate citizenship. Unlike the previous five supporting competencies, corporate citizenship's impact on the core strategies of quality, agility, and customer focus is not always a direct one. Yet it is just as important a support in the attainment of their objectives as the competencies with more obvious connections, such as employee involvement or product development.

The skills levels and learning potential of the people that compose the Best Plants' workforces, for instance, is a direct result of the ability of their communities to provide them with a strong educational foundation. In fact, the learning baseline created and maintained by local public schools establishes ground zero for the Best Plants' educational programs. The economic well-being of the people who live in the Best Plant's hometowns, states, and nation has a tremendous impact on the condition of the markets for their goods. And, as we have already seen, Best Plants define "customers" to include employees, their families, and their neighbors. For these reasons and more, Best Plants must act as community-builders.

Community-building is a very familiar role to manufacturing plants. After all, manufacturers are builders—they transform raw materials and parts into a finished product that is worth more than the sum of its elements. Best Plants produce wealth, and the wealth they produce is an important measure of the health of our national, regional, and local communities.

In their roles of community-builders, Best Plants create internal structures and processes to govern and manage their citizenship efforts. They encourage and support volunteerism among their employees. In fact, much of their work as good corporate citizens is employee-initiated and -driven.

Best Plants are especially active in their support of public and private educational- and career-development initiatives. The plants and their employees maintain high profiles in local schools and, as we shall see, their work with disadvantaged segments of the population is especially notable. Finally, Best Plants put their money where their mouth is: They support charitable causes, both local and national, with direct financial support in often impressive amounts.

Creating a Structure for Citizenship

As with the development of any other Best Plants' supporting competency, corporate citizenship needs a structure and a process that will allow it to take root and grow inside an organization. Some of the Best Plants create this structure in the form of a dedicated community relations function; others establish special teams and work groups. In either case, all are charged with developing the plants' citizenship efforts and responding to special needs of the community as they appear.

Honeywell's Industrial Automation and Control plant coordinates its efforts with other Honeywell operations in the Phoenix, Arizona, area through a divisional-level group called HERO, an acronym for Honeywell Employees Reaching Out. HERO oversees the financial and volunteer efforts of the entire division. It is chaired by a board of directors that is elected from and by the local employees, and its administrative process is handled entirely by the elected board members and additional volunteers from the general workforce. Approximately half of the HERO board members are representatives from its manufacturing employees.

HERO's main job is to select community organizations offering a broad range of services to receive both financial and volunteer support from the company. This is not an insignificant task: In 1993, Honeywell employees participated on a total of 48 community boards. And, in addition to individual involvements and commitments, the Phoenix-based Honeywell divisions made annual charitable donations through cash outlays, equipment, grants, and scholarships in excess of $2.5 million.

At the plant itself, a majority of the work teams chose to skip their traditional holiday celebration and used the savings to "adopt" less-fortunate families. Plant employees also participate regularly in community blood drives, telethons, Special Olympics, and community cleanup activities. In an interesting twist on customer focus, the plant partnered with a customer in a cosponsorship of the Great Salt River Cleanup. Some 90 employees helped collect debris along a stretch of the river and the plant contributed $10,000 to defray additional costs associated with the cleanup.

Dallas, Texas–based Marlow Industries' citizenship efforts are guided by the Marlow Industries Community Reach Out, or MICRO, Team. Composed of a mix of hourly and salaried employees, MICRO selects and organizes the charitable activities of this small business. In 1992, company employees participated in programs including Adopt-a-Grandparent, Children's Medical Center Christmas Parade, Scottish Rite Hospital Christmas Gifts, and Adopt-a-Family. In addition, the company makes financial contributions to the American Heart Association, Ronald McDonald House, and Children's Medical Center.

Sony's San Diego Manufacturing Center (SDMC) governs its citizenship efforts through the SDMC Community Involvement Council. This seven-member group meets monthly to consider all donation and product requests. In fiscal year 1993, the council fielded 208 requests and granted just over half of them. The plants' contributions totaled $275,979. Vice president Stephen Burke details some of the charitable programs included in that figure:

- *United Way/CHAD Campaign.* A committee of employees volunteered to run a companywide campaign to educate all employees about United Way/CHAD and encourage them to contribute. Results for fiscal year 1993: 50 percent employee participation; employee contributions of $118,652; company-matched contributions $59,326; extra campaign budget contribution $6,000; total United Way contribution $183,978.
- *March of Dimes WalkAmerica.* A group of 125 Sony employees and their families participated in a 25-kilometer walk to support March of Dimes. The walkers raised over $5,400 in pledges and the company matched their pledges to $2,500 limit for a total contribution of $7,900.
- *New Alternatives, Inc. & Casa de Amparo.* Sony employees "adopted" these homes for abused children. Employees volunteer their time to organize plantwide food, clothing, and holiday gift drives for the children. They also assist the house counselors by reading, feeding, and playing with the children. For this support, the employees received the 1993 Outstanding Leadership Award from the Leadership Institute.

SDMC provides technological support to six area high schools and three community colleges. Its partnership with Orange Glen High School was started in 1989 to mutually enhance the quality of life at the plant and the quality of education at school. Activities conducted on behalf of the school include scholarship, recognition, and career programs for the students; technological support for the school's CAD Lab and ROP Electronic Assembly and Electronic Technician courses; team-building management training for teachers and school district administrators; plant tours; and cash and product donations. In return, Orange Glen High School provides the company with classroom space large enough to train 150 employees at a time on an as-needed basis; the school chorus sings at the SDMC's annual holiday party; and its Agriculture classes supply poinsettia plants to decorate the plant during the holidays.

Nippondenso Manufacturing (NDUS) coordinates its citizenship effort through the Community Relations Committee, which is empowered to ensure that corporate donations reflect the company's commitment to the needs and interests of the plant's employees and the community at large. "NDUS has established itself as a leading corporate citizen in the Battle Creek [Michigan] area with outstanding support of community education, health programs, civic development, and the arts," says plant employee Karen Cooper-Boyer. "We practice positive corporate citizenship on local, state, and national levels to preserve and enhance the vigor and quality of life and the environment in which we live, work, and conduct our business."

The plant splits its support among a variety of causes. Educational efforts include support of the Battle Creek Math/Science Center, Kellogg Community College, Western Michigan University, and the Michigan Colleges Foundation. Community causes include participation in the local food bank, Toys for Tots collections, and annual record-breaking in-house campaigns for United Way and the March of Dimes.

NDUS also supports and recognizes the volunteer efforts of individual employees with its Warm Hearts Award. The program is named for the favorite phrase of a former company executive: "One must always have a warm heart for the people." The employees who are nominated for the award receive certificates of appreciation, and the two top winners receive $100 checks for their favorite charities.

Giving Time for Citizenship

Being a good corporate citizen means more than a financial commitment to the needs of society. It also means becoming an active participant in the work of community-building. The Best Plants' employees give of themselves—their time, brains, and brawn—in their citizenship efforts, and the Best Plants support that commitment.

A fine example of this point, and a unique support of citizenship efforts among the Best Plants, is in effect at 1995 award-winning Super Sack Manufacturing. Super Sack encourages community involvement by adding financial

Johnson & Johnson Medical on the Border

Johnson & Johnson Medical (JJMI) is well known for its emphasis on corporate citizenship. Its credo, reproduced in Chapter 1, makes social responsibility an integral part of its business goals. Nowhere is that focus more evident than in the plants located along the border of United States and Mexico, including the Artcraft facility in El Paso, Texas, and its twin plant in Juarez, Mexico.

"Most critics of the twin-plant industry are probably not aware of the extent to which companies like JJMI are involved in their local communities," explains Artcraft plant manager John Morrison. Well, here's a short rundown: JJMI emphasis on health care is evident in its support of the Juarez General Hospital, which serves the poorest segment of the population in Juarez, Mexico. The company's involvement is based mainly on a multiyear grant that is funding the establishment of fire safety systems and programs in the hospital, equipment repairs, the creation of a hospital-wide infection prevention program, and the refurnishing of the emergency room.

Additional grants, in the amount of $50,000 each, made under the auspices of JJMI's Community Health Care and Access to Care programs, have been given to Centro Medico del Valley, Incorporated, which provides medical services in areas of El Paso County's Lower Valley that have limited health-care facilities, and to the Brownsville Community Health Center for the development of programs for the care of pregnant teenagers. Professional training for nursing executives and Head Start office directors, both of which which have been attended by local residents, is supported under the company's cosponsored educational programs.

In addition to corporate programs, there are many plant-level contributions. In the 10 years between 1983 and 1993, JJMI collected over $1 million in employee gifts and matched contributions for the United Way. Other programs include: U.S. Savings Bonds, Adopt-a-School, Keep El Paso Beautiful, Project Del Rio, Adopt-a-Clinic, Junior Achievement, Bridge to Employment, School Nurse Fellowship Program, MDA Bowl-A-Thon, Christmas in April, and Children's Miracle Network. The giving, as well as the list, goes on.

incentives to their employees' compensation. Vice president David Kellenberger explains:

> *We support our members' involvement in all civic activities. Our pay system rewards members with extra pay for each hour spent in any civic or nonprofit community activity. Our members documented over 3,000 hours of community service in 1994.*

In addition, since 1989, Super Sack has sponsored Summer Reading Programs for the Bells/Savoy, Texas, elementary school students, and in 1993 the program was expanded to include sponsorship of the Super Reader program in Bonham, Texas. In support of these programs, the company provides financial rewards for reading and an end-of-program party, as well as volunteers to work with the children. The program, according to Kellenberger, is a reflection of the company's commitment to lifelong learning. Further, since the vast majority of the plant's employees come from these towns, it is also "a mutually beneficial investment in our future and the future of these students."

One of Exxon's Baytown, Texas, chemical facility's four plant goals is to "maintain our license to operate." That sounds quite businesslike, but when Baytown translates that goal into two values (the first is internal operating integrity and the second is external involvement in the community), the result is corporate citizenship. "Since doing that, we have fielded more than 250 regular volunteers in various areas," says site manager Ray Floyd. "They constitute our most important community outreach. We support them through our volunteer involvement fund and through various larger monetary grants."

With Exxon's help, employee volunteers have organized themselves into a number of in-house citizenship groups. The Science Education Workgroup, for example, has at least two volunteers in each of the 23 schools of the plant's local school district. The Black Resources Network, Hispanic Access Council, and American Association of University Women groups all act as tutors, mentors, and role models for at-risk children in the community while simultaneously supporting the site's emphasis on diversity.

Given the resources and the size of the employee base of the company, it can field large teams of volunteers. On January 26, 1993, for example, 152 employees from the area's Exxon plants, along with the appropriate equipment and tools, created the first wheelchair-accessible playground in the city . . . in a single, 11-hour day. During the park-in-a-day effort, more than 17,000 pounds of prefabricated playground equipment, sand volleyball courts, horseshoe pits, benches, and covered picnic tables were installed on a

20-acre site adjacent to the Baytown site. The company supported the effort with a $50,000 grant.

Community emergencies are another instance where employee volunteerism can mean as much as, if not more than, money. In the fourth quarter of 1992, when tornadoes destroyed the residential community of Channelview, the Exxon Heavy Rescue Team was the first group on the scene that was properly equipped to deal with the disaster. In addition, within 24 hours, employees had contributed, collected, and delivered more than 10,000 pounds of relief supplies for the affected families.

In 1993, when the flooding of the Mississippi River turned communities all along its banks into disaster areas, several of the Best Plants responded to the emergency. Forty percent of the county in which MEMC Electronic Materials is located was flooded, and the company jumped into the relief efforts with volunteers and cash contributions.

Milwaukee Electric Tool (METCO) responded to the massive flooding with newly formed Disaster Relief Teams. Six employees, all on company time, were dispatched in a van towing a trailer containing $25,000 of cleanup tools. The team joined with local volunteers in the Quad Cities region (near Saint Louis, Missouri) and distributed free tools, training, and labor as needed.

METCO, which likes to put its tool-building skill to work in its citizenship efforts, is also the largest corporate supporter of Habitat for Humanity International. "As a professional power-tool manufacturer serving the building trades, we have the same goals as Habitat building homes for people all over the world," explained CEO Dick Grove on the occasion of a donation of 950 power tools valued at $143,000. "We recognized that Habitat's volunteers, which include some of our employees, need heavy-duty tools to do professional work, and decided to increase our contribution to Habitat projects."

As part of a company-sponsored program, an hourly Project Build team at General Motor's Delphi plant in Grand Rapids, Michigan, designed, built, and installed a number of special-purpose machines for Goodwill Industries. This particular project resulted in the full-time employment of 20 individuals with varying disabilities who could not secure employment in a nonsheltered environment. Other community involvement and charitable activities include the United Way, where two of the plant's employees work full-time for three months on the annual campaign and to which employees make a significant contribution. Plant employees have also worked with community agencies to build swing sets in school playgrounds and exercise stations in city parks.

Employee volunteerism, bolstered by corporate financial support, is the citizenship model at Greensboro, North Carolina's Gilbarco, Incorporated.

Employees are active in a variety of educational partnerships, including tutoring activities at Guilford Primary School, Junior Achievement, and Guilford Technical Community College, where employees provide training, leadership, and support to a TQM program. Company executives, managers, and workers also take leadership roles on a long list of organizations ranging from the local United Way to the YWCA.

It is ironic that Unisys' Pueblo, Colorado, Government Systems operation was sold and closed in 1995, given the time it had invested in promoting its community to other businesses. In 1993, on applying for its Best Plants award, 85 percent of the company's senior managers were serving on the boards of community groups. Heavily involved in the Pueblo Economic Development Corporation, the plant played a direct role in persuading 19 companies, representing 3,600 new jobs, to relocate to the area. The plant also assisted the city government in its strategic planning process.

Supporting Educational and Career Development Initiatives

Many of the Best Plants' citizenship efforts focus on educational and career development services for the community at large. The plants and their employees are active in the public schools at all levels and among several segments of the disabled population. There is a sort of enlightened self-interest present in these activities; Best Plants know that their efforts to raise the educational levels and capabilities of the local population will be well repaid when they need to hire new employees.

Employees at Rockwell International's Autonetics Electronics Systems plant developed a 10-week, 60-hour career development and employee expectation program for El Paso's Lighthouse for the Blind. The program assists people with visual disabilities in developing the interpersonal and relationship skills they will need in the workplace. The classes teach career skills such as résumé preparation, interviewing techniques, employment networking, and introducing participants to common business practices.

The El Paso facility's employees also actively support El Paso Community College's Women in Technology program. Through this program, employees provide advice and financial assistance to women who are single parents, heads of households, and/or widows who wish to enter training programs for previously nontraditional occupations such as drafting and welding. Other of the plant's employees increase student awareness of engineering, space, and the sciences by making presentations to local schools.

Continental General Tire's Mount Vernon plant is southern Illinois' largest employer. Its 2,000 employees represent over 100 area communities. So it is little wonder, according to Floyd Brookman, that "youth and education are dear to the hearts of plant personnel, and many hours and dollars are spent each year toward their support." Among those activities are employee visits to local schools to educate students about General Tire itself, management techniques, and employment skills. Plant employees also teach classes in management, quality assurance, mathematics, computer science, and engineering at the college level.

Ford's North Penn Electronics Facility (NPEF) has an active relationship with the Methacton School District, where plant manager Dudley Wass serves as a member of the Executive Advisory Committee. The plant used its technological process expertise to help the district develop its own long-term technology plan and donated computer equipment to support it. NPEF hosts tours for the school district that have included the school board and math and science teachers. The plant also made a $25,000 grant to develop a high school course intended to introduce students to career opportunities in manufacturing.

Siemens Automotive plant in Newport News, Virginia, is another plant that focuses a good portion of its citizenship efforts on education. Says Sam Byrd:

> *Key educational activities include: financial and management support of a regional effort to increase minority student participation in engineering and science, including summer intern positions; support of a middle school science awareness program through plant visits and classroom presentation; summer intern positions for teachers; and senior management participation on the industry advisory boards of regional universities and community colleges.*

Industrial engineering manager John Martino of Baxter Healthcare's IV Systems plant in North Cove, North Carolina, can rattle off a host of educational initiatives supported by the plant and adds, "The plant's future is driven by these activities." Among these activities are Adopt-a-School volunteers, academic achievement awards sponsorships, plant tours for students, sponsorship of a mobile classroom for preschoolers, career shadowing for students, teacher shadowing, and career fair participation. Individual workers at the plant, which is the largest employer in the area, serve on the local school board and at the local community college as a trustee, instructors, a business adviser, and a maintenance adviser. For these, as well as its many other citizenship activities, the North Cove plant received the McDowell County Volunteer Industry award in 1993.

Symbiosis Corporation's Commitment to the Hearing-Impaired

An active corporate citizen in the Miami, Florida, area, Symbiosis Corporation has developed a particularly comprehensive program designed to assist members of the deaf and hard-of-hearing community. Explains vice president Scott Jahrmarkt:

> Even before the passage of ADA [Americans with Disabilities Act] legislation, Symbiosis has had a special commitment to hiring the hearing-impaired. Special relationships have been established with appropriate referral agencies, and outreach efforts have been made not only to employ, but to offer encouragement to those with impaired hearing.

The multifaceted approach that the company takes to this commitment is a model of corporate citizenship. The company works with the Dade County Public Schools in an annual Job Fair & Career program designed specifically for hearing-impaired students. At the fair, Symbiosis employees who are deaf discuss, in sign language, their careers at the company. Symbiosis also produced an orientation film about itself and its products that is closed-captioned for the hearing-impaired. This film is used for new employees, as well as to recruit people who are deaf.

Inside the company, a TDD (Telecommunications Device for the Deaf) is available for use by employees to communicate between their homes and work. The company also hired a qualified sign language interpreter as a full-time member of the Human Resources staff to assist in hiring and serving Symbiosis' deaf employees.

For these efforts and more, Symbiosis has been repeatedly honored as Employer of the Year by the Deaf Services Bureau of Dade County. More rewarding, however, may be the knowledge that, as Jahrmarkt says, "Through this program, Symbiosis is making a meaningful contribution to a disabled segment of our society."

Supporting Charitable Organizations

The last, but certainly not the least important, way in which Best Plants demonstrate their corporate citizenship is through the outright financial support of charities. At Lockheed Martin's Moorestown, New Jersey, plant, for instance, $320,000 was donated to United Way of Burlington County in 1993. Per capita giving at the plant has increased steadily since 1991, despite a shrinking employee base. In addition, each year a key executive is loaned on a full-time basis to United Way Fund staff for the fall campaign.

Copeland's Sidney Scroll plant is giving $45,000 over a three-year period to the Alpha Community Center toward the purchase and renovation of an existing building in downtown Sidney (Ohio). Some of the ongoing programs at the Center include the tutoring of children, day care and overnight camping for disadvantaged children, and meals for the homeless.

The plant is also highly supportive of Shelby County's United Way program, which funds approximately 20 local charitable organizations. In 1993, a corporate contribution of $81,000 and employee contributions totaling $113,439 were made to the program. The company is also contributing $100,000 over five years to The Salvation Army's capital-building project.

The charitable donations of Baxter Healthcare's IV Systems plant rival its educational involvement. Over a three-year period they included: walkathons that generated a $30,500 donation, United Way campaigns that generated 4,403 individual donations and a total donation just shy of $200,000, and an Angel Tree, which provided $90,000 worth of holiday gifts for 600 children.

Summary

The final supporting competency, and the last piece in the Best Plants' puzzle, is corporate citizenship. Best Plants are community-builders. They are active participants in the growth and well-being of their communities. They lend their financial support and encourage their employees to participate in a wide range of educational and charitable organizations.

The Best Plants establish a structure and process for their citizenship efforts. This usually takes the form of work groups or dedicated community-relations personnel. A formal organization gives focus to and encourages support for citizenship activities.

The citizenship efforts of Best Plants include a strong focus on educational and career-development programs aimed at (1) helping the people of their communities attain a greater measure of security and (2) participating in charitable activities designed to help the less-fortunate. This support, which often includes cash donations, is accompanied by a full measure of volunteer activities by Best Plants' employees.

> We are responsible to the communities in which we live and work and to the world community as well.
>
> —*The Johnson & Johnson Credo*

SECTION 2

AMERICA'S BEST PLANTS

INDIVIDUAL PLANT PROFILES

Section 1 presented a holistic vision of the world-class manufacturer. It was based on the commonalities between the 62 Best Plants. Collective excellence is not, however, a replacement for individual accomplishment. Rather, to achieve the best possible results, they must coexist. The following profiles, one for each of the *IndustryWeek*'s America's Best Plants award winners, celebrate individual excellence.

There is much to learn from these portraits. First, it is impossible to read them without beginning to comprehend, and enjoy, their sheer diversity. The profiles conclusively prove that there is no single path to success—every one of these plants constructed its own road. The next time a business guru promises off-the-shelf success, keep that lesson in mind. Second, approach the profiles prepared to learn, adapt, and implement. Who knows which idea you borrow will contain the seed of your own company's success?

The profiles are presented in a standard format for ease of use. Each includes:

- *A narrative description* of each plant. These narratives are authoritative—each represents days, often weeks, of work on applications on the part the employees of each plant and an on-site visit and formal report by an *IndustryWeek* editor, whose name follows the text and whose work places the reader inside each plant.
- *Best Plants working tips,* each of which identifies an instance where the plant hits the best practice target square on center. These tips, which you will find marked with topical identification icons and set in boldface type throughout the profiles, offer a quick visual reference to the areas in which each plant excels and, more particularly, act as a guide to the many unique programs, tools, and techniques used within the plants. Keep the tips in mind as you read the text; they will lead you to some of the most valuable lessons of the individual Best Plants.

- *Contact information* for each plant. The profiles end with a listing of the plant's address, telephone number, a benchmarking contact name at the facility, main product listings, and number of employees. There are any number of uses for this information. Benchmarkers, job hunters, salespeople, and potential customers, vendors, and partners will find the best manufacturing prospects in the United States of America listed here.

The Best Plants profiles are presented in alphabetical order for easy reference. Readers who would like to contact the *IndustryWeek* editor who conducted the original site tour can find telephone, mail, and E-mail addresses in the Editor Directory on page 415.

ADVANCED FILTRATION SYSTEMS, INCORPORATED

Champaign, Illinois

EMPLOYEE INVOLVEMENT
AGILE MANUFACTURING
ENVIRONMENT AND SAFETY

When management requires workers to fill out paperwork, they tend to concentrate on the documentation to prove they are making improvements. We want our employees to concentrate on making improvements.

—STAN ZWICK, PLANT MANAGER

You cannot help but be impressed with the smooth and orderly production flow when you walk through the filter-making plant of Advanced Filtration Systems Incorporated (AFSI) in Champaign. But the star performer in this plant is not the equipment—it's the workers. That is why when Stan Zwick walks you through the plant he lets the workers tell you about the hundreds—if not thousands—of changes they have made to turn the AFSI plant into such a highly efficient operation for making liquid filters for construction equipment.

And don't ask Zwick to tell you which changes are responsible for which improvements—or to what degree. "We don't put a lot of effort into documentation," says Zwick. "We will check the improvement, sometimes in scrap rate or maybe reduced setup time or downtime, but we downplay the record keeping. We can see [the improvements] daily. We don't have to be told about them in writing."

That nonstatistical bent may horrify number crunchers who always look for the impact to the bottom line. But AFSI's worker empowerment has helped create a highly efficient plant that reduced waste 90 percent between 1990 and 1992. The number of product defects, as determined by a customer quality audit conducted in 1991 and the first quarter of 1992, was zero. (It had been 26.5 per 1,000 units in 1988.) That improvement occurred during a period when monthly unit production tripled and the number of workers declined by 20 percent. A 1992 Best Plant award winner, AFSI also eliminated methylene chloride and polyurethane solvent flush from its processes and reduced drastically the residual drum wastes of three substances: isocyanate, polyol, and adhesive. It also recycles all of its paperboard and polyethylene wrap.

Even more remarkable, these improvements have been produced at a plant considered state of the art when it began operations in 1988 as a joint venture between Caterpillar Incorporated and Donaldson Company Incorporated. "When the plant began production," recalls Zwick, "we had a vision of what we thought was right. But we found out that the highly automated aspects were too automated, too complex, and that we had to go to a more practical approach of simplifying operations." AFSI had invested millions of dollars in equipment, but neglected to train its workers in how to use the equipment. In order to maximize the capabilities of the equipment, it began in-house training to teach workers how to apply statistical process control, team building, and kaizen techniques to their jobs.

◎ **"When we got to kaizen, we taught it out on the floor—not the classroom," recalls Zwick, "and we asked workers to solve specific problems on our machinery.** We started with three basic teams in three different areas—packaging, roving, and can forming. We let them pick their own problem, choose how to approach it, and decide what was important." Often, workers would choose to solve a quality problem and wind up with a secondary benefit such as cost savings, reduced setup time, or a reduction in waste. "When they see they are making changes to their benefit, things start feeding on themselves," he says.

◎ **One team videotaped how it changed rolls of polyethylene film and found that everyone was working, but that "no one was doing anything."** As a result, tools needed to change dies and fixtures were put nearer to the operation, and different handles and gauges were installed to make the changeover easier. The net result was an eight-hour process cut to less than 90 minutes.

Another example: A worker who felt that machinery breakdowns were hindering production efficiency suggested regular equipment shutdowns for maintenance. "[Before] the routine shutdowns, we were behind in the production schedule," says Zwick. "After we started regular shutdowns, we began to increase our production efficiency." As a result, all operations are now shut down one shift per week for preventive maintenance or "to do something to improve the process."

In packaging, workers have improved efficiency and productivity, while cutting setup time from 12 hours to 2 hours through a variety of changes, very few of which would be considered major or breakthrough ideas, says Tom Hardesty, associate technician in the packaging area. "Everyone has different ideas. And when we get everyone together, you get a little better idea."

That is why Zwick is reluctant to pinpoint specific changes that have led to the plant's productivity improvement. "It's not one or two things, but the combination of a thousand little things that you do in operations, design, and maintenance," he says.

◎ **Similarly, the waste reduction and recycling was accomplished through an employee analysis of every waste stream in the plant and the implementation of many simple changes.** The company worked with suppliers to change the packaging of incoming materials. And now it purchases much of its raw materials on the same-size pallets used for storing the products in the plant. In addition, the team decided to eliminate, through substitution, a substantial number of chemicals or oils that it used minimally each month. "Just by looking at our operations, we eliminated 90 percent of our waste," says Zwick. Besides, "giving away a lot of things [for recycling] saved us a lot of money" because of the reduction in disposal costs.

Worker empowerment has been successful for three reasons, maintains Zwick. First, workers can make changes on their own—including expenditures up to $500. Second, "we never eliminate a full-time worker" when changes increase efficiency. Third, "we don't require workers to spell out, or track, how much improvement occurs because of a change."

—MICHAEL A. VERESPEJ

Advanced Filtration Systems, Inc.
Contact: Charlie McMurray, General Manager
3206 Farber Dr., Champaign, IL 61821
Telephone: (217) 351-3073
Employees: 134
Major products: Liquid filters

AIR PRODUCTS & CHEMICALS, INCORPORATED

South Brunswick, New Jersey

TECHNOLOGY
ENVIRONMENT AND SAFETY

I know chemistry and engineering pretty well, but as far as running that plant [the workers] know a lot more than I know.

—PATRICK LOUGHLIN,
PRODUCTION SUPERINTENDENT

With all due respect to plant manager Michael Berus and the rest of the brass at the Air Products & Chemicals Incorporated plant, it is abundantly clear that the foot soldiers run the show. And even though there is still technically an "us" (the workers) and a "them" (management), the barbed wire that distinguishes the two in so many factories is utterly lacking in this plant.

This 18-acre facility in South Brunswick manufactures polymer emulsions that are used in the production of nonwovens, adhesives, paints, paper coatings, and building products. These compounds are integral components of consumer products such as latex paint, acrylic caulk, wallpaper coatings, paper products such as Baby-Wipes and paper towels, and specialty adhesives used for the seals in microwave packaging.

A 1992 Best Plants award winner, this operation's 22 products are generated via a state-of-the-art distributed process control system. The technology runs the process, from raw material charging, batch sequencing, run steps, and process control to data acquisition. ◎ **A PC-based integrated production-quality and distribution-management system not only operates on the local area network (within the plant), but also allows data to be transmitted weekly to corporate headquarters in Allentown, Pennsylvania.** There it is used by business, product development, and technology personnel.

It would be easy to rely on the system's capability to generate on-line statistical quality control (SQC) charts in addressing quality problems, acknowledges Patrick Loughlin, production superintendent. And indeed, there is a tremendous historical value in such data gathering. Nevertheless, production operators still create SQC charts for every product and property by hand. "If something is out of control on the computer, it will give you a little beep," says Loughlin, "but it doesn't show you exactly where it fits in with everything else you've been doing." Generating charts the old-fashioned way, on the other hand, "gives you immediate feedback."

The reality of management's role here is unmistakable—to serve the workforce. And while Berus is indeed located at the bottom of the organizational chart, the upside-down structure was hardly inspired by altruism. Rather, the workers' pride in themselves and their work demands that they essentially run the show. "I don't think anyone," confides Thomas Orabona, a 16-year-veteran operator, "even the Japanese, can make this product better than we can."

The trust and accountability are so high here that operators work more hours unsupervised than they do with supervision. Joseph Ruszczyk, the production supervisor, works an eight-hour shift Monday through Friday. The rest of the time, nights and weekends, four operators run the plant. "They each have a specific job which they rotate every shift," notes Loughlin. "They make the decisions, and they do what they need to do."

Members of the South Brunswick team are hard pressed to keep ideas to themselves. Although suggestions often come about spontaneously, quality and safety meetings are held on a monthly basis to encourage workers to share ideas for improvement. In addition, several corrective action teams (CATs) exist throughout the plant and address a variety of issues. One of the most notable was the Yield Improvement CAT, established in 1989, which sought to end a five-year period of stagnation during which yield improvement stalled at 96.9 percent. Members generated 86 ideas aimed at improving yields at the first meeting alone. Since then, all 86 have been implemented, along with subsequent suggestions. Although not every idea worked, the net overall effect is a one-percentage-point increase, which translates to an annual reduction in raw material costs of $350,000.

◎ **The great irony of waste has not been lost on this workforce. As Berus notes, first "you lose the product," and then "you pay to get rid of it."** That problem was solved by installing a recycling system in the loading areas designed to vacuum product back into the plant. The result is an annual savings of roughly $150,000 in raw material and effluent costs (the price the plant pays to deposit waste into the sewage system).

Loughlin notes that most of the "good stuff" in terms of ideas comes from the operators. Though some old-school managers find this level of exchange threatening, it suits him fine. "From a practical point of view, I'm thrilled because you can see the results, the cost savings, the improved quality, and the commitment you get from [the workers] to do a good job and really feel they're part of the success of the plant."

The high level of competence that permeates every part of the South Brunswick operation enables Berus to turn his focus outward. ◎ **His campaign is based on a solid platform of safety and environmental waste reduction, and is neatly wrapped in Responsible Care, an initiative developed by members of the Chemical Manufacturers Association.** Responsible Care is a guiding philosophy based on six codes of management practice: community awareness and emergency response, pollution prevention, process safety, employee health and safety, the distribution code of management practice, and product stewardship.

Berus views the role of the plant he oversees in the community very seriously. Responsible Care is, in his eyes, "a way of encouraging chemical companies to get out into the community, to tell people what you're doing, what you make and how you make it, what chemicals you have, and the things you're doing, so that they don't go by your plant and feel like there's some sort of barrier there, like you don't want them to know what's in there." Toward that end, Berus enlisted the support of other area plants and, in October 1991, formed a community advisory panel made up of concerned citizens from the area. The group meets monthly to discuss activities within the facilities.

—Tracy Benson-Kirker

Air Products and Chemicals, Inc.
Contact: Mike Gibbons, Plant Manager
11 Corn Rd., Dayton, NJ 08810
Telephone: (908) 329-4086
Employees: 55
Major products: Vinyl acetate/ethylene copolymer emulsion

ALLEN-BRADLEY COMPANY

Milwaukee, Wisconsin

. . . you really have to get down to improving the process, improving the process, improving the process.

—J. DANIEL CHALLGREN, VICE PRESIDENT OF OPERATIONS

TECHNOLOGY
ENVIRONMENT AND SAFETY
EMPLOYEE INVOLVEMENT

Only five months after making its razzle-dazzle public debut, Allen-Bradley Company's EMS1 facility took center stage again—this time as one of 1992's Best Plants.

Taking its name from Allen-Bradley's carefully nurtured Electronic Manufacturing Strategy, EMS1 produces printed circuit board assemblies (PCBs) for the company's Operations Group's industrial automation products. Among those products are small logic controllers, solid-state motor starters, and machine-vision systems. Its objectives are to achieve unsurpassed product quality, reduce time to market, provide required manufacturing capacity, implement capacity to handle new products, significantly improve customer satisfaction, and achieve the lowest possible total life-cycle cost for each product.

EMS1 takes a seamless approach to PCB manufacturing in a low-volume, high-product-mix environment. A variety of circuit boards can be produced on the continuous-flow assembly line. Lot sizes are as small as one. In addition, as many as 10 different types of circuit boards may be in production at any one time. ◎ **The facility's conveyor system can carry each panel—in effect, a "pallet" carrying different products—through all the process options, including robotic assembly, wave soldering, and surface mount. Or, by using an overhead loop, it can bypass a process that is not appropriate.**

When Allen-Bradley officially unveiled EMS1 in May 1992 (fully automated production as an integrated facility had begun in 1991), this manufacturing powerhouse was producing more than 1,000 panels in a 40-hour workweek. On each of those panels' boards may be a total of more than 500,000 integrated circuits, resistors, capacitors, and other electronic components. And don't look for anything but unrelenting quality production from EMS1, declares J. Daniel

Challgren, vice president of operations in the Operations Group. During the facility implementation process, for instance, the improvement in process yield, or the percentage of "good" output, was a reflection of the lofty "no defect is acceptable" policy. Top-surface-mount first-pass yield improvement was 3.17 percent; post-wave-solder first-pass yield improvement jumped 12.89 percent. The total combined facility yield registered a 15.23 percent improvement.

Vincent R. Shiely Jr., former EMS1 manager, says that when it comes to quality in this kind of manufacturing, the facility is state of the art. Those unfamiliar with PCB manufacturing, says Shiely, will point to six sigma and say that is the only acceptable level of production quality. The fact is, however, a first-pass yield in the upper 80th percentiles is standard in PCB manufacturing. EMS1, though not six sigma yet, boasts a much better percentage.

The EMS1 implementation teams searched out the best in class in electronic manufacturing. "We knew what was out there in consumer electronics," explains Challgren, who himself came to Allen-Bradley from that side of manufacturing. "We knew we had to benchmark. We started with Digital Equipment Corporation, DELCO, Northern Telecom, and Allen-Bradley's own World Contactor Plant. We also visited plants in Japan—Nippondenso and Sony. What we learned was that the development of our strategy was right on target."

Starting in 1989, a three-phase implementation of EMS1 began. Since then, notes Challgren, the goal has been "to achieve a competitive advantage in the manufacturing of low-volume, high-mix PCBs." And the results speak for themselves. Consider, for instance, EMS1's capability to automatically transform engineering transfer file (ETF) data, generated by computer-aided design stations, into numerical control machine programs for all applicable equipment in five to seven minutes per machine. "In-circuit test fixture and text program development," reports the company, "have also been automated using the same EFT data. We can now have fixtures manufactured in a matter of a few weeks, and test programs can be debugged and implemented in a matter of hours."

In the area of environmental stewardship, as well, EMS1 is setting new standards for performance. ◎ **It implemented the use of no-clean wave-solder process, which eliminates the use of chlorofluorocarbons (CFCs) or solvents to clean assemblies after that process.** That helped the Operations Group to cut its CFC use 62 percent last year.

Overall, says president Don Davis, Allen-Bradley has taken all it has learned about manufacturing and management in the last 10 years "and placed it in EMS1." Certainly, nowhere is that any clearer than in what the company proudly refers to as the "human side of CIM." When it came to staffing EMS1, Allen-Bradley decided to hire from within the ranks of its loyal, highly motivated workforce. Of course, this cutting-edge venture meant that training would be not only important, but critical. ◎ **Allen-Bradley, therefore, enlisted the help of Milwaukee School of Engineering (MSOE) to develop a novel educational partnership.** MSOE started by presenting basic math skills and helping employees use those skills to build a familiarity with surface-mount technology.

Such schooling has helped to ensure the success of EMS1's workforce. Personal responsibility, says Shiely, is the hallmark of EMS1. At the start, "we did unusual things out on the floor. The first thing was to set up work teams. What did we make the people responsible for? Basically, everything."

"I've had a case where I've had an operational failure, and the operator will say, I'll run third shift. That way, the back of the line will have work when they come in at 7 A.M.," reports James Feiertag, who has succeeded Shiely as EMS1 manager. "This guy has just finished a shift at 4:30, saying, 'Let me go home and get some sleep. I'll run work from midnight to 7.' "

—Joseph F. McKenna

Allen-Bradley Co., Industrial Control Group
Contact: Alicea Fernandez-Campfield, Manufacturing Manager
1201 S. Second St., Milwaukee, WI 53204
Telephone: (414) 382-2000
Employees: 76
Major products: Circuit board assemblies

BAXTER HEALTHCARE CORPORATION

North Cove, North Carolina

QUALITY
EMPLOYEE INVOLVEMENT
ENVIRONMENT AND SAFETY

Quality is for everyone, not just the quality organization.

—JOHN MARTINO,
INDUSTRIAL ENGINEERING MANAGER

After visiting a customer, one employee of the 1994 Best Plant award-winning IV Systems Division of Baxter Healthcare Corporation expressed her surprise. She had noticed that the hospital operating room, where open-heart surgery was in progress, appeared to have less stringent conditions regarding cleanliness than her production cell back in the North Cove plant. Surprising . . . until you realize that lives also depend on the quality of North Cove products.

Baxter's 1.4-million-square-foot plant at the foot of the Blue Ridge Mountains has been a benchmark for the production of a variety of large-volume intravenous solutions and related products for more than 20 years. Indeed, the Federal Drug Administration asked the plant for a training video illustrating world-class practices within the pharmaceutical industry. And, of 44,000 Department of Defense suppliers, the Baxter plant was one of only 15 recognized for outstanding customer service during Operation Desert Storm. It is also a facility where customer reject rates are nonexistent. Instead, complaints representing every documented customer concern are added together to reach a reject rate of 1.5 parts per million (ppm).

The IV Systems Division at North Cove thrives in a heavily regulated, mature industry where perfect quality is merely an entry fee. Cost and service are the market's driving forces today, and Baxter Healthcare continues to set the pace, as it has since 1973. What's the secret? The Quality Leadership Process—QLP.

◎ **"QLP is an umbrella that can drive the proper business alignment of strategies for overall operational excellence in concert with customer-driven goals," explains industrial-engineering manager John Martino.** QLP is North Cove's guiding light. It is a quality-driven business plan that the facility depicts as a Parthenon-like structure. The QLP structure is supported by five foundation stones: continuous improvement, total employee involvement, prevention, defect-free work, and the fulfillment of customer requirements. On this foundation rest 11 manufacturing strategies—pillars, so to speak—that represent a gamut of world-class practices. The pillars, which change with the state of manufacturing, include strategies such as computer-integrated manufacturing, statistical process control, benchmarking, and simultaneous engineering. Together, they support QLP's roof, which is North Cove's world-class goal statement and its Manufacturing 2000 vision: "To be a world-class organization through continuous improvement, a focus on the customer, and commitment to operational excellence which far exceeds our competition."

In and of itself, QLP is not highly unusual; many companies spend time and resources formulating impressive quality strategies. Much less common, however, is North Cove's comprehensive deployment of this vision. Take the pillar of flexible manufacturing systems and its primary measure, time to release, for example. *Time to release* is the amount of time that elapses between the packaging of the last piece of a batch and its formal release for shipment. Releasing product requires a complete documentation package, which grows larger at each stage of the manufacturing process and must be in perfect order before North Cove ships finished goods.

◎ **Walk into any of North Cove's three dedicated manufacturing cells, the labs, the sterilization department, or any other area in the plant that is part of the manufacturing process and ask for their time to release.** The answer always comes without hesitation. That intensity of focus has driven time to release from 190 hours in 1988 down to 50 hours in 1993.

Similar accomplishments are recorded for each pillar in North Cove's QLP strategy. Over the years, the pillar of total preventive maintenance has evolved into total productive maintenance (TPM) and now has come to be called reliability-centered maintenance. ◎ **The plant, a charter member of the Maintenance Roundtable, has trained all 2,200 employees in the philosophy of TPM.** It also shares data and practices with other leaders in the area, such as DuPont and Ford Motor, and its TPM apprenticeship program serves as a benchmark in industry. Among the results: Manufacturing delays caused by facility downtime reduced from 36 hours in 1991 to 25 minutes in 1994.

Maintenance teams, with more time to concentrate on major tasks, have returned the investment in TPM many times over. One such team is pleased to show a visitor compressor bearings that have been redesigned to increase service life from 6 months to 19 months. Annual savings: $350,000. Another team records only one failure in 40 cumulative months of run time for five trucks.

The same trucks sport redesigned beds that feature quantum reductions in repair times and parts costs below those of the original manufacturer.

QLP's pillars also extend well beyond operational definitions of world class. One such pillar is labeled "public responsibility," and it receives as much attention as any other. ◎ **This is a plant that vowed to completely eliminate waste to landfills by 1996 and that has been regularly recognized for its environmental initiatives.** Between 1991 and 1994, North Cove reduced nonhazardous waste to landfills by 68 percent, eliminated the use of Freon as a cleansing agent, and removed all underground storage tanks. All lubricants and hydraulic oils are now recycled, and steam for the plant is generated by burning waste wood purchased from the nearby furniture industry.

North Cove employees give more than 30 percent of the blood donated in the county, donated almost $200,000 to the United Way, and, via the employee-originated Angel Tree program, distribute Christmas gifts to 600 children each year. The plant, which ranks as the area's largest employer, is a close partner with local educational institutions, offering shadowing opportunities for both students and teachers as well as career fairs. It also serves as an adviser to the area's community college.

—THEODORE B. KINNI

Baxter Healthcare Corp., IV Systems Division
Contact: John Martino, Industrial Engineering Manager
P.O. Box 1390, Marion, NC 28752
Telephone: (704) 756-4151
Employees: 2,399
Major products: Pharmaceutical solutions

CHESEBROUGH-POND'S USA COMPANY

Jefferson City, Missouri

AGILE MANUFACTURING
QUALITY
SUPPLIER RELATIONS

We realize that the process of how work actually gets done is more important than how we manage the function.

—JOSEPH ROY, PLANT MANAGER

It is obvious when you drive onto the property of the Chesebrough-Pond's USA Company plant in Jefferson City that they have spiffed up the place. Freshly planted shrubs, trees, and flower beds adorn spacious, well-tended grounds. A tasteful new sign announces the plant's name. And once inside the building, your eye falls on pleasing new furniture, wallpaper, and artwork in the spotlessly clean reception area, offices, and corridors.

Although the facility is 26 years old, it looks spanking new. But the newness extends beyond outward appearances. Far more important are improvements on the production floor—particularly in processes, management practices, and culture. These changes are not so visual. They are so sweeping and successful, however, that they have earned the 475,000-square-foot facility near the Missouri River a 1992 Best Plants award.

Products manufactured on the plant's three-shift operation are everyday ones: a variety of skin-care lotions (Vaseline Intensive Care lotion is its biggest seller), Q-Tips cotton swabs, Cutex nail polish and nail-polish remover, and the recently introduced Mentadent toothpaste. But the plant's multifaceted activities under its no-nonsense theme, "Quest to be the Best," are anything but everyday. This explains those visual improvements. "We practice total quality in whatever we do," says plant manager Joseph Roy. "We can't be credible in talking about quality unless our plant has a quality appearance."

The plant's Quest to be the Best revolves around the simple operating philosophy that Roy put into place when he arrived as plant manager in 1989: Provide defect-free products to customers on time. "Everything flows from that philosophy," reflects Roy, who came to Jefferson City after managing Chesebrough-Pond's Chicago plant. "Once we adopted it, everything we had to do became very apparent."

◎ **The most far-reaching move has been to replace traditional function management techniques with process management. Although the plant retains several key functions—manufacturing, personnel, and accounting, for example—it now is organized around work flow.** As a result, cross-functional, multiskilled, self-directed process teams have replaced managers in each of the plant's four product areas. First-line supervision, which once numbered 13 people, is down to 7. Hourly associates perform former supervisory functions, such as making production-line assignments, scheduling overtime, determining the length of production runs and changeovers, handling cost control, requisitioning parts, and placing work orders.

◎ **In the key area of quality, responsibility for quality control now tests 100 percent in the hands of associates, who follow the plant's 13-step total quality process.** In fact, the plant now has only two quality inspectors, and their roles have changed from on-line inspection to end-goods auditing. Associates themselves initiate and implement quality improvements. Through the plant's C-QUIC (Continuous Quality Improvement Challenge) process, associates suggest ideas or "challenges." Cross-functional Quality Action Teams then implement the challenge. A listing of current C-QUIC projects, which by 1992 numbered more than 300, and their status is posted throughout the plant.

To back up its empowerment effort, the plant has invested heavily in training. Each associate will receive five days of training this year and six days next year—up from 4.6 days in 1991. Training includes personal computer skills, which are essential, because the plant has become high-tech. For example, computers flash up-to-the-minute sales data that make possible the scheduling of production on the basis of customer demand rather than sales forecasts. "We used to run a shade for two days on our nail-polish line," recalls Roy. "Now we change 8 to 12 times a day." Also, an interactive, on-line network of computer workstations throughout the plant gives all associates access to scheduling information.

This system will be installed in suppliers' plants as well. "That'll make it possible," comments production manager Ray Bruen, "for them to resupply us without any communication between us." The computer linkup with suppliers is just the latest in the plant's program of strengthening partnerships with more than 300 suppliers. ◎ **Each month, for instance, delegations of associates visit suppliers' plants; supplier employees also visit Jefferson City for four days of TQM training.** Too, the plant conducts quarterly supplier reviews and a supplier award program, and involves suppliers in product development. Nearly half of key suppliers operate on either a kanban or just-in-time basis.

The plant's sharp improvement in quality, productivity, cost, and other key measures gratifies corporate management at Chesebrough-Pond's USA, a Greenwich, Connecticut–based subsidiary of Unilever United States Incorporated. Senior vice president of Operations Paul Dolan, who was the first manager at Jefferson City, points out that the total quality "framework" exists at each of the company's six U.S. plants. "But Jefferson City really grasped the concept and ran with it," he says. "The things they've done in the area of empowerment are especially dramatic."

Plant Manager Roy concedes that implementing the change wasn't easy. "People were comfortable with the way we'd been operating for 22 years," he says. "Their natural reaction was, 'Why change when we're good at what we do?' " It took nine months, Roy reports, "just to get people to realize we were serious" and to get managers converted to their new roles as coaches. Still, he "continues to be amazed" at the way the new concepts are taking hold. "The challenge now," he says, "is to avoid complacency. We have to continue to instill the realization that no matter how good we are, the competition is still there—and getting better too."

—William H. Miller

Chesebrough-Pond's USA Co.
Contact: Robb Vrbicek, Plant Manager
2900 West Truman Blvd., Jefferson City, MO 65109
Telephone: (573) 893-3040
Employees: 600
Major products: Skin-care lotions, cotton swabs, nail polish

CINCINNATI MILACRON, INCORPORATED

Batavia, Ohio

AGILE MANUFACTURING
EMPLOYEE INVOLVEMENT

. . . it was demonstrated that a relentless honing of the company's manufacturing prowess is key to a successful competitive strategy.

—MARTIN LAKES, MANUFACTURING MANAGER

"Find a better way, a better tool, a better method" is not a quote from someone's recently contrived mission statement. It is an operating maxim at Cincinnati Milacron, which originated in 1884 with founder Frederick A. Geier, at a time when the company was known as the Cincinnati Milling Machine Company. Today, the people at "the Mill" still heed the founder's exhortation to improve the process.

At the company's Plastics Injection Machinery Business and its 1994 Best Plants award-winning Batavia operation, the historic mandates for manufacturing process improvement and continuous product improvement efforts have become the fundamental underpinning of the company's audacious business strategy called "Wolfpack." Conceived in 1985 as a response to offshore competition in Cincinnati Milacron's plastics injection machinery market, the strategy has expanded until it embraced the entire company. Wolfpack, so named for the aggressive, cooperative hunting habits of wolves, permeates the entire corporate culture, but some of the strategy's most profound implications involve the production floor.

"For example, the number of hours required by Wolfpack-designed machines to produce the same dollar volume of sales is generally half those required by the products they replace," says Ray Ross, president and chief operating officer. Cincinnati Milacron needs half the time, half the floor space, and half as many other assets to produce the same amount of product.

"Key to our production achievements was the redesign of the product into modules," explains Lakes. "That was critical to our goal of compressing lead time during manufacturing. The modular design permitted each cell to take responsibility for the production, quality, and product improvements of a module. We assigned process engineers to the cells, and several cells have engineers who actually do some of the design work there, depending on how far the cells have evolved to building stand-alone products. ◎ **The result is a transformation of a linear, sequential production concept into one performing things in parallel with drastically compressed lead times.** That helped us to pare order-to-shipment lead time by 50 percent, which played right into the requirements of our Wolfpack strategy."

In 1985, when middle managers Harold Faig and Bruce Kozak first formulated Wolfpack for the Plastics Injection Machinery Business, the goals were more specific than design for marketability and design for manufacturability. "We were aiming for cost reductions of 40 percent, machine functionality improvements of 40 percent, and paring the usual two-year product development time to 270 days," recalls Faig.

"The Wolfpack goals are predicated on setting up a plant floor agile enough to support quick delivery as a marketing strategy," says Lakes. Thanks to strong vendor relations, computer-driven flexible part making, and cellular assembly techniques, the operation can now match or beat the delivery times of offshore competition, even those that are shipping their machines from warehouses. Adds Lakes, "With quick delivery right off the plant floor, we gained two advantages over the competitors that warehoused machines—avoidance of inventory carrying costs plus the ability to customize injection-molding machines to our customers' needs."

Gaining those benefits required confronting two obstacles: a plant that had undergone numerous additions since its construction in 1969 and a reengineering of the plant's traditional culture. "Decades of additions to the facility had severely compromised work flow at the Batavia plant," says Lakes. "With each surge of business, it seems we would add new floor space, but we usually never had the time to optimize the layout. Each new addition brought new burdens of production compromises. For example, a bay that was ideally equipped with cranes and utilities to handle heavy assembly was being used as a weld shop.

◎ **"The solution was to take the time to go at it from a 'clean sheet' viewpoint. No sacred cows—just an analysis based on 'what do we need to do?' "** The result is a cell-based layout that encourages communication between supplier and customer cells, a move that enhances the continuous-improvement process. "Suppliers are next to customers, and that results in an easy way to communicate and coordinate product and process improvements," adds Lakes. By decentralizing and reorganizing the work flow, the Batavia plant doubled production capacity and gained an additional 20 percent of floor space, which is now being used to assemble other product lines.

◎ **Achieving the world-class manufacturing goals of Wolfpack started with a reengineering of the cultural process, and the effort is a continuing one.** First, the traditional approach to product and process development in which functional groups work sequentially and are isolated from one another during the development process was eliminated. Now the organization works in teams to synergistically combine their experience. Hands-on involvement of teams is evident throughout the organization. For example, the Q-5 team of engineering, manufacturing, marketing, materials, and service representatives meets weekly to review and resolve customer problems relating to quality and performance. Product design and updates are the special focus of Wolfpack teams that include both suppliers and customers. Strategic issues are explored and reviewed by the cross-functional China 18 team.

"Wolfpack marked the rebirth of manufacturing for us," summarizes Martin Lakes. "When we started Wolfpack, we realized that winning with that strategy was only possible if we reinvented the way we did things on the factory floor."

—JOHN TERESKO

Cincinnati Milacron, Plastics Machinery Group

Contact: Dave Shardelow, Quality Manager

4165 Half Acre Rd., Batavia, OH 45103

Telephone: (513) 536-2000

Employees: 1,200

Major products: Plastics injection-molding machinery

COHERENT, INCORPORATED

Auburn, California

EMPLOYEE INVOLVEMENT
TECHNOLOGY

Be willing to offer incentives to employees for performing tasks, duties, and responsibilities typically handled by management.

—DIANE BREEDLOVE, DIRECTOR OF OPERATIONS

In 1848 a band of intrepid pioneers stumbled onto gold in a ravine in what is now Auburn, a city some 30 miles east of Sacramento in the foothills of the Sierra Nevadas. The discovery helped to trigger the historic California Gold Rush. Nearly a century and a half later, the Auburn-based Electronic Products Division of Coherent Incorporated found gold of a different kind.

The 1994 Best Plant award-winning Electronic Products Division mines printed circuit boards (PCBs), wiring harnesses, laser head assemblies, laser magnets, and power supplies, most of which are assembled into finished products at other plants operated by Coherent, a Santa Clara, California–based manufacturer of laser equipment for scientific, medical, and commercial markets.

By adapting flexible JIT manufacturing concepts to a high-mix, low-volume environment, and supporting its revamped approach with ingenious homegrown computer systems, the Auburn plant has achieved some remarkable gains. Productivity is up more than 60 percent. Overhead costs have been sliced by 58 percent. And order-to-shipment lead time was slashed by 90 percent.

In Coherent's quest for improvement, however, there were a few uneasy moments. Particularly trying were the first three months after the division took a bold plunge into self-managed work groups. In late 1990, production supervisory posts were eliminated, and employee groups assumed responsibility for hiring, firing, daily planning, performance reviews, training, discipline, and other duties formerly handled by supervisors. "What made it difficult was trying to do so many things at once," operations director Diane Breedlove recalls. "We made a mistake in not allowing adequate time for training. . . . In the first three months, our productivity went down the tubes. And delivery went into the tank. We were asking people to handle functions that they had no training for—that they had to learn as they went along." Fortunately, they proved to be quick learners, and soon the performance curves began heading upward again.

Developed and implemented during a six-month period, the self-management approach grew out of discussions that began when employees complained about the plant's performance review system. In considering alternatives, Breedlove became intrigued with the idea of peer evaluation, where employees would review their coworkers. She began holding a series of weekly meetings with employees, two from each work group, to explore the idea. Three months later, the discussions expanded to include authority over almost all management responsibilities, which meant there would no longer be a need for plant supervisors. (Ultimately, eight supervisors were reassigned.)

For communication and coordination purposes, the work group members take turns serving as "group representative," with temporary oversight responsibility for scheduling, shipping, and other duties. No one is permitted to hold the rotating position for longer than three months, Breedlove explains, "because we wanted each person to experience what it meant to be in that seat—to have to deal with the pressures."

"It keeps you on your toes," agrees Scott Hoppe, an employee in the power-supply group. "When you are the group rep, you have to know what is going on in the whole department."

◎ **Adopting the gainsharing plan in 1991 helped to ease the pain of the early struggle to adapt to the new system, Breedlove recalls, "because it let the employees know we were willing to compensate them for the new responsibilities they were taking on."** Under the gainshare formula, employees get 50 percent of identified savings. The base amount is tied to productivity improvement, with a bonus/penalty adjustment based on quality and delivery performance.

Several computerized systems have helped propel the continuous-improvement process. ◎ **One of the plant's most intriguing innovations is its Auto-Que system, using software developed in-house, which establishes real-time production priorities.** Auto-Que takes over where the MRP system leaves off. "The MRP output is based on our master schedule, which is based on forecasts," notes Patrick Yoho, manager of purchasing and materials. "But forecasts are typically not reality. So we tried to come up with something that was closer to reality, based on consumption."

Working with the MRP projections, Auto-Que calculates a recommended kanban quantity for each component or subassembly. It then assigns production priorities based on current inventory levels, which are constantly updated by back-flushing the computer system at key points in the process. If, for example, the recommended kanban size for an item is 10, but only 5 units are in finished goods or in process, then the kanban is considered 50 percent empty. That item would get a higher build priority than one whose kanban is only 20 or 30 percent empty.

That's the basic approach, but it can be overridden by hot customer orders. "If something in a sales order is not covered by finished goods or work in process and it is due within five days, that will automatically move to the top of the priority list regardless of the percentage off the kanban size," Yoho explains. In effect, the Auto-Que priority ranking, which shows up on computer screens at the start of each production line, becomes the daily production schedule.

The plant also uses electronic method sheets that give step-by-step instructions to assembly workers at their terminals. By eliminating use of paper drawings and providing clear instructions, the system has contributed to a 49 percent reduction in scrap and rework. And a computerized replenishment system automatically creates purchase orders when parts inventories fall below a certain level. Suppliers electronically retrieve the purchase orders by accessing a computer holding file.

The Auburn plant's innovative mix of participative management and technology earn it Best Plant status. It also gives new meaning to the proverbial miner's yell: "Thar's gold in them hills!"

—JOHN H. SHERIDAN

Coherent, Inc., Electronic Products Division

Contact: Diane Breedlove, Director of Operations

12789 Earhardt Ave., Auburn, CA 95602

Telephone: (916) 888-5032

Employees: 110

Major products: Laser assemblies

CONTINENTAL GENERAL TIRE, INCORPORATED

Mount Vernon, Illinois

EMPLOYEE INVOLVEMENT

TECHNOLOGY

My shoes are my office. We practice being on the floor, observing, assisting, and making corrections when necessary. We're very visible to people on the floor.

—JIM DOLWICK, PLANT MANAGER

Truck-tire finishers at the 1994 Best Plant award-winning Continental General Tire facility in Mount Vernon had been vexed for several years by a continuing source of operational frustration. The trim knives used to remove "rubber whiskers" and cosmetically finish commercial truck tires prior to shipment were heavy and poorly designed for repetitive use. They sometimes marred the final product. The knives, inherited from joint-venture partner Japanese tire maker Yokohama Rubber Company Limited, underwent numerous trial-and-error changes. But not until the involvement of shift superintendent Jim Morgan and a team of 25 tire finishers in March of 1994 did anything change.

In August 1994, Morgan presented his case before the monthly meeting of the STAR (Saving Time and Resources) board of directors, the reigning decision-making body at this highly acclaimed facility. He made his no-nonsense pitch to gain approval to fabricate the team-designed knives in-house, provide a pair of knives for each of 25 commercial tire trimmers, and get financial approval for an inventory of spare parts. His five-minute presentation is bolstered by the inclusion of sage comments from Robbie McGovern and Scott Johnson, two tire finishers who took a special interest in the project.

The 16-member STAR board, the top-of-the-pyramid representative body of some 67 employee-involvement teams and nearly 2,000 plant workers, gives its stamp of approval to the project within minutes. The cost will total about $3,000. Clive Norton, a maintenance technician who designed and fabricated the prototype trim knives, will make the final product during unscheduled maintenance time.

◎ The STAR gainsharing program has elicited more than 11,000 such suggestions since being implemented in June 1985, resulting in $30 million in

cost savings, and helped create more than 1,000 new jobs in nine years. No layoffs have occurred at the Mount Vernon plant since 1980, while 51 other tire manufacturing plants have shut their doors in the last 15 years.

Hubertus von Grunberg, chairman of the executive board of Continental AG of Hanover, Germany, parent of U.S.-based General Tire Incorporated in Akron, Ohio, praises the facility as the benchmark for all international operations.

The STAR system keeps all employees focused on business objectives and eligible to share in cost savings on a monthly basis. It is an irreplaceable part of the plant's culture and gives management and workers an opportunity to work together to improve plant operations, solve problems, lower costs, reduce waste, and improve safety.

"STAR," says Jim Dolwick, plant manager, "is probably classified as a gain-sharing program, but it's really the skeleton of the Mount Vernon facility. It gives us the structure to allow for information to flow upward, downward, and sideways. It's a chance for employees to participate in the management of the plant. And it gives employees a chance to share in the cost-saving fruits of their suggestions." Since mid-1985, employees have shared $20 million in plant performance bonuses based on a split of 70 percent for employees and 30 percent for the plant.

While the STAR system is the plant's mantra or hymnbook, it also fosters mutual trust between managers and the people on the floor. "Everybody shares in the monthly bonus, except for the plant manager," says STAR coordinator Sam Lewis. "So we are, in that sense, together. It has forced us to be more communicative with people on the floor and with salaried personnel. It has forced us to be very open and share a lot of information. We are extremely responsive to people on the floor. You can't work with someone you don't trust, and this system promotes trust."

What is the central reason behind the plant's success? "We've been progressively participative," observes Dolwick, who spends five hours of his 13-hour day in the manufacturing environment. "I know it's a buzzword from 10 years ago. But that's what has helped us be successful. It's about getting people involved in their work, where they know their job better than anyone. When people walk out of this plant at shift change, they can say, 'I was a success today.' That's important, because everyone wants to be a success. We help them be a success."

◎ **Employees can question management decisions.** "Management is not infallible, and we encourage our people to challenge us," says Dolwick. "We remove the emotional shackles. We don't have any gag rule. When I make a bad decision, I say, 'I'm sorry, it won't happen again.'

"You can be a soldier's general or a general's general. We choose to be a soldier's general. We don't just listen to staff members tell us what's going on in the plant. We go out there to smell it, touch it, taste it. We want firsthand knowledge. What are the emotions in the plant?" asks Dolwick rhetorically.

◎ **Much of the plant's success comes from its computer-aided hiring practices.** Only one candidate in 100 makes it through the screening process, and successful applicants must show aptitude for teamwork. Once hired, the orientation program does not focus on what their jobs will be so much as on the business challenges the plant faces and how to get things done within the plant's culture.

Once on board, employees cannot be fired without the advice and consent of a peer review process. "What this says to all employees," observes senior vice president of manufacturing Jim Rippy, who managed the Mount Vernon plant for 18 of the last 21 years, "is that no supervisor or member of management can ever take an arbitrary action against them. It creates a system of fairness, objective evaluation of the facts, and helps build plantwide trust." The decisions to terminate or retain run about 50-50.

Looking back on bringing empowerment to the Mount Vernon facility, Rippy, who wrestled with upper management for seven years before it approved the STAR system, says he learned that "you have to crawl before you can walk, and walk before you can run. Sometimes we went too fast in our zeal. To get to world class, you have to do it a bite at a time if you're going to digest it."

—Brian S. Moskal

Continental General Tire, Inc., Commercial Division

Contact: Frank Bruce, Logistics Industrial Engineering Manager

P.O. Box 1029, Mt. Vernon, IL 62864

Telephone: (618) 242-7100

Employees: 2,000

Major products: Radial tires

COPELAND CORPORATION

Sidney, Ohio

Quality
Employee Involvement

We don't analyze [employee] suggestions to death. If an idea looks as if it has merit, we tell them to do it.

—Randy Rose, Operations Director

Copeland Corporation's 1994 Best Plant award-winning scroll compressor plant has turned the market upside down with its high-quality, highly efficient, low-noise products that feature just two moving parts, compared with 7 to 15 in a conventional reciprocating compressor. The scroll compressors are used in commercial and residential air-conditioning units and commercial refrigeration products.

The development of these breakthrough products depended on a team approach. In 1978, almost a decade before the plant began using scroll technology for compressors, management, aware that conventional lathes, grinders, and mills were too slow and too costly for high-volume scroll production, put together employee teams to brainstorm ideas to achieve the manufacturing breakthroughs needed to economically produce and machine the scrolls.

Those teams studied plastics, aluminum, and steel before choosing cast iron as the scroll material. But that was just part one of the breakthrough teams' mission. They also had to simultaneously develop the high-precision machine tools and gauges needed to machine the castings. The reason? The two spiral-shaped scroll elements—one is fixed and the other rotates—must fit together to form crescent-shaped gas pockets. They have an operating clearance of less than one-tenth the thickness of a human hair and must be machined with a precision that is usually needed only for products such as mechanical heart valves.

It was only natural, then, that when manufacturing got under way, the team effort continued. Supplier partnerships provide just-in-time delivery. Work teams on all three shifts attack business and manufacturing costs through kaizen projects. New employees get four days of classroom training. There are team meetings at the beginning of each shift. And there is a mind-set to develop and adopt hundreds of poka-yoke techniques.

◎ **Equally important, Copeland, a wholly owned subsidiary of Emerson Electric Company, has identified and communicated to its 690 employees what operations director Randy Rose calls the "seven wastes of manufacturing."** They are: time wasted waiting to do work, overproduction, unneeded transportation of parts, unnecessary processing, too much material stock or finished inventory, wasted motion, and defects.

As a result, it's almost become a religion for scroll-plant employees to look for ways to reduce costs and improve efficiency. "Sixty percent of the plant looks different in one form or another from two years ago," declares Rose. "The internal change has been so rapid that we have been constantly rearranging the equipment. Some of it has even been on wheels."

Different, in this case, also means better. As measured by worker-hours per unit produced, manufacturing cycle time improved 47 percent between 1989 and 1994 on Copeland's Aspen compressor product line. In turn, that cut the lead time for shipments in half. The market inroads made by Copeland's scroll compressors are equally impressive. Market share jumped from 6 percent in 1990 to almost 25 percent in 1994.

In 1993 and 1994, 42 kaizen quick-step improvement projects reduced factory operating costs by almost $2 million. A shell-fabrication kaizen team, for example, downsized its work area by 30 percent, reduced throughput time from 59 minutes to 13 minutes, and freed up space for suppliers to make just-in-time deliveries.

The workers' willingness to look for ways to cut costs and make manufacturing more efficient can be attributed to several factors. There is a no-layoff policy, so that when a work team proposes changes that cut workforce needs, workers get different jobs, not pink slips. And the contract negotiated with the International Union of Electrical Workers before the plant began operations contains just five work classifications—two in maintenance and three in operations—instead of the typical 60 that exist in other Copeland facilities.

◎ **"We look at the cost of every [component] produced and the breakdown of that cost," adds plant controller Brian Boyd.** If producing a main bearing costs $5, workers will also know the exact cost of the materials, labor, scrap, tooling, maintenance, and fringe benefits that compose that figure. Thus, workers can attack each cost element in the equation. "We look at business costs, material costs, and production costs, not just manufacturing costs," says Boyd. "And we look at all processes, shipping, receiving, manufacturing, and customer services, so we can calculate the impact of a potential change."

◎ **Workers also innovate to make manufacturing as failproof as possible.** "We have over 200 different types of poka-yoke techniques in the plant, and many of them are very low cost," says Rose. For instance, in the shell-fabrication area, where small sheets of steel are rolled and then welded to form the compressor container, there are large, white posters with color pictures of good and bad welds so workers immediately know when a weld is of poor quality and can determine whether it can be repaired.

Another technique was adopted at the suggestion of a worker: If the valve on the inside of the muffler plate—which the operator cannot see—is not in the proper position, a sensor keeps the press from operating. And, in the welding and leak test, a mirror was put on the back side of the line, because the orientation of the product was such that the operator had not been able to see if everything was properly aligned.

One last quality measure: Computers track each step in assembly so that any error can be tracked down to prevent problems in the future. But, as Randy Rose notes, it's not needed often. Thanks to the poka-yoke technique, the compressors have a 99.5 percent first-pass rate. "We believe in quality through ownership. Each operator puts his own mark on the product as it moves down the line."

—Michael A. Verespej

Copeland Corporation, Plant #8
Contact: Gerry Urlich, Plant Manager
1675 West Campbell Rd., Sidney, OH 45365
Telephone: (513) 498-3011
Employees: 1,000
Major products: Air-conditioning compressors

CORNING, INCORPORATED

Corning, New York

You can talk all you want about quality systems, but unless the guy who puts it in the box gives a damn, you're dead.

—ROB HENDERSON, OPERATIONS MANAGER, CORMATECH

CUSTOMER FOCUS
EMPLOYEE INVOLVEMENT

Early in 1989, Corning Incorporated moved its specialty cellular ceramics (SCC) business from its birthplace, a warehouse in one corner of its Erwin, New York, plant, to a brand-new facility of its own on the outskirts of Corning. The move reflected an optimism about the future of that business, which makes ceramic filters to purify molten metals such as iron and steel. But, more important, it represented a fundamental belief in people, a conviction that workers will do an outstanding job if they are treated like intelligent adults.

"It works," asserts Rob Henderson, who as plant manager headed the SCC start-up venture. "You can combine profitability and human dignity."

Clearly, the SCC business is in the vanguard of efforts to explore new sociotechnical systems, that is, organization and work designs that seek to tap the full potential of motivated employees. The concept it adopted was relatively simple: Explain the basic goals and requirements, then give the workers the freedom to decide how the goals can best be met.

Among other things, this 1990 Best Plants award winner abandoned the use of shift supervisors; in effect, workers have no immediate bosses. Instead, it relies heavily on teams and team decision making to keep the plant running smoothly. There are four production teams (the members are called "ceramic associates") and one support team that includes engineering and maintenance specialists. A production team consists of the 16 or 17 ceramic associates who work the same shift.

The teams continually try to improve the production process, notes Norm Garrity, senior vice president for manufacturing and engineering in Corning's Specialty Materials Group. "It is their responsibility not only to control the process, but also to improve it. And they have the freedom to do that. Since there are no supervisors, they don't have to go through any authority to get things done."

In addition, SCC adopted a cross-training policy that made it possible to reduce the number of job classifications to just four, providing a high level of flexibility. During a shift, workers rotate from job to job. ◎ **And, SCC located business support functions such as sales and marketing in the plant as part of a "customer service cluster" to streamline communications and improve response to customers.**

A pay-for-skills program was instituted in cooperation with the union (American Flint Glass Workers), under which the ceramic associates earn wage increases as they progress through four skill levels in various areas. Skill Level 1 consists of the basic knowledge required to do a particular job. By the time they reach Level 4, the associates are considered "experts," qualified to teach others. New workers are required to reach Skill Level 2 in all areas within 24 months.

Further, employees learn not only the specific work skills that they use daily, but, through training, gain a thorough understanding of the business from a competitive standpoint.

Perhaps the most telling evidence of Corning's belief in the value of employees' ideas is this: The ceramic associates and other SCC employees helped to design the new plant, including the layout of the process, the type of cafeteria, and the team structure itself. ◎ **Prior to the move to the new building, a nine-member design team, including six associates and three management people, worked full-time for six months on the project.** Among other things, the team surveyed all employees and asked them to describe, in three words, what kind of plant they would like. That wish list was consolidated into a vision statement for the new plant; and many of the ideas were incorporated into the plant design. The workers also indicated a desire to work fewer days, but on 12-hour shifts. That schedule was adopted.

From a business perspective, the most significant results have been the work teams' contributions to quality improvement, productivity, and customer service. Quality levels, which are measured in terms of customer rejects, have improved dramatically. Between 1987 (when the business was still in the Erwin plant) and 1990, defect rates dropped from 1,800 parts per million (ppm) to just 9 ppm. Customer lead times have been reduced from five weeks to a matter of days. In some situations, the plant can respond within 30 hours.

Meanwhile, process losses were reduced more than 50 percent between 1986 and 1990. And the business is able to meet requested customer delivery dates 98.5 percent of the time. "We used to negotiate on the request date. We don't do that anymore. When the customer says he needs it, that's what goes down on the piece of paper," says Henderson, who has accepted a new position as operations manager for Cormatech Incorporated, a 50-50 joint venture with Mitsubishi that will make catalytic converters to remove nitrogen oxide gases from power plant emissions.

By all measures, the SCC business has been a success, says Henderson. "We've met or exceeded every goal—every process goal, every financial goal." One process goal was to implement just-in-time/continuous-flow principles—which contributed to the huge quality improvement. But the most important requirement for good quality, Henderson asserts, is a committed workforce.

—JOHN H. SHERIDAN

Corning, Inc.
Contact: Christine Nagel, Plant Manager
CE 0101, Corning, NY 14831
Telephone: (607) 974-9000
Employees: N/A
Major products: Ceramic filters

EMPLOYEE INVOLVEMENT

DANA CORPORATION

Minneapolis, Minnesota

The key ingredient for world-class manufacturing is investment at all levels. Give people the tools and the environment they need and you'll get the results.

—BEVERLY KIMBALL, QUALITY MANAGER

The numbers tell the story at Dana Corporation's Minneapolis valve plant, and what a story it is. Between 1987 and 1990, this plant, which makes Gresen hydraulic control valves that are used in heavy construction equipment, has reduced manufacturing throughput time 92 percent and increased productivity 32 percent. The 1990 Best Plants award-winning facility also trimmed customer lead time from six months to six weeks. (In a pinch it can turn an order around in as little as two days.)

And the list goes on. . . . Improved efficiency allowed two plants to be consolidated into one, thus producing comparable output in half the manufacturing space. On-time delivery was boosted to the 95 percent range, quality costs were pared by 47 percent, total inventory was trimmed 50 percent due primarily to huge reductions in work in process (WIP), and subassemblies inventory was slashed by 94 percent. The bottom line: a 470 percent improvement in return on investment and a return on sales of 320 percent.

Oddly enough, that lengthy list of improvements was accomplished with relatively little investment in new technologies. "A lot of what we have done has been low-tech," says Hank Rogers, total quality manager for the plant, a unit of Dana's Mobile Fluid Products Division.

An assortment of tactics contributed to the impressive gains. But, Rogers notes, two strategies were paramount: a streamlined just-in-time (JIT) system built around manufacturing cells and a heavy reliance on a team structure.

The turnaround began with a team-oriented training program launched in late 1986, shortly before the move to cellular manufacturing. Employees were given 40 hours of training over an 18-month period. The plant's EIM (excellence in manufacturing) teams continue to receive ongoing coaching from an EIM facilitator.

The teams are "primarily self-directed," Rogers points out. They elect their own team captains and tackle various continuous-improvement projects, including setup reduction, preventive maintenance, and the adoption of kanban systems.

It took about a year for the team training to have a real bottom-line impact, recalls Rogers, who studied Japanese methods before helping to institute the changes. "Things start to roll after you get about 50 to 60 percent of your people trained."

◎ **But the training did pay at least one early dividend when workers began contributing ideas to improve the original work-cell designs.** The first group that took a look at the engineers' proposed design for its work cell came up with a counterproposal. "They gave us back 40 percent of the space the engineers had allocated for that cell," Rogers recalls. "After that, we let every group do their own cells. We have focused on the teamwork approach in all things—not just the cell design."

◎ **For one thing, the Minneapolis plant has sponsored internal competition among teams in areas such as setup reduction, poka-yoke (foolproofing) methods, and kanban systems.** Members of a team that completes a project earn Dana jackets; winning teams accumulate points that can be redeemed for other prizes.

The combination of quick changeover methods and manufacturing cells produced the 92 percent reduction in manufacturing cycle time and radically shorter customer lead times. "If you put in cells, cut your WIP, and cut setup times, the reduction in throughput time is pretty automatic," Rogers says.

The plant's JIT approach also relies on cross training. Production workers can earn an extra 37¢ an hour by becoming cross-trained. "The result is that 85 percent of our people are now in the same labor grade, which gives us enormous flexibility," Rogers notes.

And, he adds, fewer workers develop back problems. In one six-person cell, two of the workers had histories of back trouble. After petitioning the union (Teamsters) to endorse cross training, they were able to take turns running different machines, using different muscle groups and avoiding long stretches of the same repetitive motions. Since adopting the new regimen, neither worker has experienced back trouble.

A major benefit of the team approach, Rogers points out, is that management listens more closely to what a team has to say than it would to an individual worker. "When you have 10 people saying, 'We have a problem,' you listen to them," he says. "And you do everything you can to resolve the problem."

—JOHN H. SHERIDAN

Dana Corporation

Contact: Beverly Kimball, Quality Manager

600 Hoover St. NE, Minneapolis, MN 55413

Telephone: (612) 623-1960

Employees: 391

Major products: Hydraulic control valves

TECHNOLOGY

DIGITAL EQUIPMENT CORPORATION

Colorado Springs, Colorado

The automation program was undertaken to achieve higher quality and reliability, not to reduce labor.

—BILL STRANG, PLANT MANAGER

The term *enterprise integration* has implications that extend beyond the boundaries of computer-integrated manufacturing. And it's a term that the folks at Digital Equipment Corporation (DEC) take very seriously.

Digital's 1990 Best Plant award-winning Colorado Springs site, a center for the design and manufacture of the RA90 family of disk drives and other advanced storage devices, provides a showcase for the benefits of networking computer and information systems. From the product concept stage through manufacturing start-up to quick response delivery, this plant is fully computer integrated.

◎ DEC's internal networking platform, which supports electronic mail as well as computer-to-computer links, was a vital cog in the concurrent product/process development effort that culminated in the successful launch of the RA90 in mid-1988. And the network, which uses DECnet and Ethernet links, continues to play an important role in coordinating the fabrication and shipment of parts and subassemblies from 40 locations around the globe to two points of final assembly: Colorado Springs and Kaufbeuren, Germany. Process control and quality data are shared across the network to achieve rapid problem solving and continuous improvement.

"We have made extensive use of the network as a tool for all forms of human communication, and for information management and process control," explains Bill Strang, plant manager at Colorado Springs.

But communications and computer technology are only part of the story. One clue to Colorado Springs' determination to stay at the leading edge of world-class manufacturing is the fact that it has established a four-person Office of Continuous Improvement. The unit is charged with ensuring that cross-functional efforts, such as the implementation of just-in-time techniques, go smoothly.

In addition, the Colorado Springs facility has embarked on an ambitious quality program that aims to reduce defects to the six-sigma range. To get there, it is intent on reducing defect levels 70 percent a year. "Our goal," says John Bauer, manager of the Office of Continuous Improvement, "is customer satisfaction through six sigma, which leads to continuous improvement."

At Colorado Springs, the DEC network supports a sophisticated computer system known as CAPS—for computer-aided process support. CAPS plays a multifaceted role, including database management, interapplication messaging, capture of assembly and test data, controlling the movement of in-process material, and interfacing with robot controllers. CAPS also provides traceability of vendor parts to specific finished products.

The heart of the manufacturing operation is a robotic head-and-disk-assembly (HDA) operation, which takes place in a state-of-the-art Class 10 clean room (fewer than 10 particles per cubic foot). Eleven assembly robots and a test station are linked by a clean room–certified conveyor. Completed HDA units are transported to workstations outside the clean room where the final disk drive assembly is done manually.

◎ **One of the more intriguing aspects of CAPS is its material-handling system (MHS), which uses electronic "pull" signals to coordinate material movement.** When a work cell finishes an operation, a MicroVAX controller sends a ready-to-receive signal over the network. The MHS matches ready-to-send messages with ready-to-receive messages, and then issues commands to the workstations to move pallets along the conveyor system. No material moves unless a station has asked for it.

Traditional kanban cards are used to pull material from the warehouse and from local suppliers. When a container of parts from a supplier arrives on the floor, the bar-coded kanban card is wanded, which authorizes payment.

Colorado Springs' flexible production system has also supported efforts to provide quick response to customer needs. For rush orders, a Fast Ship program promises shipment within 24 hours, even if the requested item is not in finished inventory or in the production pipeline.

"We have the ability to turn an order around in 24 hours," says Bill Plante, customer-integration manager. "We can reach into our pull-system pipeline and pull to the finish line whatever the various requirements would be to meet an order."

In 1989, the Society of Manufacturing Engineers honored the Colorado Springs facility with its prestigious LEAD award for "leadership and excellence in the application and development of CIM." Among the achievements cited were a 50 percent improvement in product reliability, 75 percent reduction in installation problems, 50 percent compression in "time to volume" (the time from product conception to full-volume production), 45 percent reduction in staffing, and 50 percent reductions in cost and required manufacturing space.

In October 1994 the Colorado Springs disk drive operation was sold to Quantum Corporation, and in 1995 it was relocated to Malaysia. DEC now operates a new and unrelated computer subsystem operation in this location.

—John H. Sheridan

EDY'S GRAND ICE CREAM

Fort Wayne, Indiana

Practices and systems need to be supported by a common vision. . . . The systems need to complement and support each other.

—KIRK RAYMOND, PLANT MANAGER

TECHNOLOGY
EMPLOYEE INVOLVEMENT

It doesn't take a connoisseur to taste the difference between premium and standard grades of one of summer's most cherished treats. One heaping spoonful of Edy's Grand Ice Cream, arguably one of the richest and creamiest blends made today, will bring a smile to the lips of even the most stone-faced curmudgeon.

But the employees who run this 1993 Best Plant award-winning operation in Fort Wayne believe there is more to the recipe than cream, sugar, nuts, and fruit. To continually reach the level of customer delight that keeps the company on top in this fiercely competitive industry, every Edy's associate manages—and is held accountable for—short-term, bottom-line business results. And that, they contend, can be accomplished only with a solid foundation of values and both eyes trained on the future.

Edy's, which is the jewel in the crown of Oakland, California–based Dreyers Grand Ice Cream Company, was not always this balanced, admits plant manager Kirk Raymond. Like many newcomers to the concepts of empowerment and shared leadership, Edy's stumbled for a few years under the strain of an emphasis that was almost entirely social in nature. Adding insult to injury, the workforce and its production levels were in the throes of exponential growth. (In six years, the plant's full-time employment jumped from 14 to 137.) Missing was an understanding of the business realities that form the basis of sound decision making. It's not that the Edy's workforce was ill-intentioned, says Mr. Raymond. On the contrary: "They had been told stories that they were running a great place, and they were. But business was terrible from a profitability standpoint. You have to understand the business," he asserts, "because that's our future. . . . We're not entitled to sell ice cream—we have to be profitable."

So began a complete organizational overhaul in 1990, developed and implemented internally by teams of associates. (Titles here are largely for the benefit of outsiders. Because the workforce is 100 percent self-directed, managers are referred to not by title, but rather as members of the M-Team.)

◎ **The system's core is comprised of cross-functional teams that are aligned by business unit.** A typical business unit team consists of four or five people who actually make the ice cream (a mix maker, a flavor operator, etc.), a packaging operator, an engineer, a shipper, a palletizer, a maintenance person who cleans the equipment, and so on. Each team is responsible for everything from quality and sanitation checks, meeting individual business goals, internal scheduling, and discipline, to training and career development.

The system works, production manager Scott West believes, because the structure relies heavily on three aptitudes: "Associates need to have a business aptitude that allows them to take process data and make decisions around how they're going to meet their business goals. They also have to have an aptitude that allows them to understand the technical part of our business. And third, they have to have a social aptitude that's really different than in other companies, because they're responsible for career development for the people on their team as well as for discipline. When you work together with somebody for years and you have to sit down and counsel that person on performance, that can be tough."

In addition to business unit teams, there are functional teams and numerous policy-making and activity-coordinating committees on which employees are expected to serve. Being entirely self-managed means more than living without traditional supervisors. ◎ **It means that the associates themselves are responsible for developing and implementing pay rates, hiring and firing, and everything in between.** In these and other matters, Raymond sees his role as one of boundary protector as opposed to internal manager. "Our plant has to be competitive to survive," he says, "and I'm personally accountable for the plant being competitive. But as a plant manager, I have a direct labor [budget], and I don't care how that pie is cut up. In fact, I don't even set the rate of pay at this plant."

◎ **Particularly noteworthy is the pay-for-skills system.** Developed by a team of employees, the system is designed to maximize an individual's control over his or her own career path while creating a breadth and depth of experience and expertise for the teams. The result is a three-tiered system that allows an individual to choose one or a combination of the following paths: first, as an operator, to master the basic skills involved in a module (a unit of tasks within the plant that comprise a "job"); second, to gain in-depth, technical expertise in a module, earning "technician" status; and third, to develop specialized engineering, training, or leadership skills.

If the plant's performance measures are any indication, the systems work. Take the 83 percent reduction in scrap between 1988 and 1993, for example. Material loss is measured and reduced every step of the way. In one case, simple human ingenuity so improved a piece of leased packaging machinery that the supplier modified the rest of its machines throughout the field. "Our convocan department actually makes the cartons that come down the conveyer to the filler," West explains. At one time, though, the side seam was defective. "The filler would force the side seam out, the line would get all jammed up, and you'd get stuff all over the floor. One of our maintenance guys, Kevin Bastion, designed a new side-seam clamp for that machine."

The monthly incentive system, which is paid plantwide, is designed to reward performance in five dimensions (commonly referred to here as the "Cobweb"): people, customer delight, new products and processes, cost, and quality. But, to hear it from the associates, the hard work pays off in more ways than financial. "When you have one of these tough nights," notes Randy Hire, a maintenance specialist, "and you have all this stress, but then you get things worked out, you get the product run, and someone says, 'Good job, thanks for getting the machine going,' you go home and you feel great."

—Tracy Benson-Kirker

Edy's Grand Ice Cream

Contact: Kirk Raymond, Plant Manager

3426 Wells Street, Fort Wayne, IN 46808

Telephone: (219) 483-3102

Employees: 225

Major products: Ice cream, frozen yogurt

ENGELHARD CORPORATION

Huntsville, Alabama

We have a moral obligation to provide an injury-free plant. We feel you can't have good morale without it. So we hammer at safety. And we celebrate safety milestones.

—JOE STEINREICH, GENERAL MANAGER

EMPLOYEE INVOLVEMENT
QUALITY
ENVIRONMENT AND SAFETY

In northern Alabama, the biggest tourist attraction is the elaborate Space Museum at NASA's Marshall Spaceflight Center in Huntsville—it marks the spot where Wernher von Braun and other German expatriates gave birth to the U.S. space program. A stop at the space museum is a must for visitors to the area, insists Joe Steinreich, general manager of Engelhard Corporation's Huntsville site, just a few miles down I-565. On occasion, he will personally escort guests to the NASA facility. But Steinreich much prefers showing off his two plants and detailing the remarkable success story crafted by Engelhard managers and employees.

In the early 1980s, a manned mission to Mars may have seemed an easier undertaking than turning Engelhard-Huntsville into a world-class operation. Employee turnover was in the 150 percent per year range. Morale was poor. Productivity and quality were at unacceptable levels. And the business was losing money.

But a dramatic culture change, which began shortly after Steinreich arrived in 1981, gradually turned the situation around, creating positive attitudes and launching a continuous chain of improvement. By the time this plant earned its Best Plants award in 1991, the business could point to such achievements as a 324 percent productivity gain, superb quality levels, and an incredible safety record—eight years (nearly 4 million labor hours) without a lost-day injury.

The Huntsville site houses two manufacturing plants. The larger one makes catalytic parts for controlling air emissions from automobiles, forklifts, and stationary pollution sources. The second produces silver chemicals. One of the first things that a visitor notices is a triple row of tall, barbed fences surrounding the property. Security is taken seriously because of the high value of the raw materials used in the catalyst plant—platinum, palladium, and rhodium. The precious metals are delivered by armored truck and stored in a bank-style vault. In the production process, they are mixed in glass-lined reactors to create slurries for the plant's catalyst coating line where ceramic-substrate parts are immersed in the solution.

Since a single slurry can contain several million dollars' worth of chemicals, material conservation is a high-priority task. Between 1988 and 1991, employee suggestions helped to reduce slurry-yield losses caused by improper coating levels by 67.6 percent. And the plant also dramatically reduced "accidental" losses—resulting from leaks, spills, and inadvertent slurry cross-contamination.

Accidental losses had been a vexing problem for years. ◎ **So in 1990, in brainstorming meetings, a task team of 20 employees identified numerous ways that loss incidents could be prevented.** "We spent about 30 hours in meetings and got the most incredible payback I've ever seen," says Ken Rogers, manufacturing manager for the catalyst plant. "They came up with 252 ideas." After implementing the most practical ideas, the frequency of loss incidents was reduced from one in every 100 material transfers to one in 1,749.

Other employee-involvement efforts also made a huge difference. One example: A worker on the pack line, where finished catalytic parts are boxed for shipment, came up with an idea—a small reverse-motion belt—that prevents part jam-ups where the conveyor system narrows. The idea, which cost just $100 to implement, eliminated the need to have a worker stand next to the conveyor and poke at the jammed parts with a stick.

Perhaps the most significant change at Engelhard-Huntsville during its decade of change was in management's attitude toward employees. Steinreich, a chemical engineer by training, notes that the facility was once plagued by "some pretty poor theory X management." Upon arriving, one of the first goals he set was to eliminate the high turnover and make Engelhard "a place where people enjoyed coming to work." With that in mind, he instituted a supervisor training program to foster a more people-oriented management culture. Supervisors who balked at changing their ways were replaced. Over time, the new supervisory climate has paid many dividends. The devastating 150 percent turnover rate has been trimmed to a respectable 3.9 percent. In turn, this has enabled Engelhard to be more selective in its hiring practices.

The emphasis on employee involvement contributed to a 324 percent productivity leap in the catalyst plant, which in 1990 turned out more than 40,000 catalysts per employee, compared with 8,134 in 1980. The primary productivity boost, Rogers believes, has come from a strong quality program—which slashed rework in the catalyst plant from 12 percent to less than 0.5 percent. Productivity also soared in the silver-chemicals plant.

◎ **A Total Quality Management effort—including the use of Taguchi methods and FMEA (failure modes and effects analysis) in process development, as well as operator involvement in statistical process control—has earned the plant an impeccable reputation for quality.** A major turnaround in quality occurred in the early 1980s when the plant shifted its emphasis from detection to prevention, says Jim West, quality assurance manager. "In 1980, we had people checking people, trying to detect problems. That doesn't solve the problem and it's very labor-intensive," he says. "Now, quality responsibility lies with the person doing the job."

◎ **Of all their achievements, however, the feat most often mentioned by employees is the outstanding safety record—eight years without a lost-day injury.** Extensive training on safety, reinforced by monthly and weekly safety meetings, has played a big part. At departmental staff meetings, safety is always the first item on the agenda.

—John H. Sheridan

Engelhard Corporation

Contact: Lurline Yarbrough, Assistant to General Manager

9800 Kellner Rd., Huntsville, AL 35824

Telephone: (205) 464-6261

Employees: 300

Major products: Environmental catalysts, silver chemicals

QUALITY
EMPLOYEE INVOLVEMENT

EXXON CHEMICAL COMPANY

Baytown, Texas

Data is a gift to a highly motivated organization. Widespread participation in the gathering, analysis, and use of data is a very powerful tool for understanding and progress.

—RAY FLOYD, SITE MANAGER

With a quick nod toward an array of yellow sheets of paper on a bulletin board, Davy Crockett declares: "It's amazing what you can accomplish with small ideas." This namesake of the coonskin-capped pioneer is a machinist in the rubber-producing Butyl Polymers unit of Exxon Chemical Company's Baytown chemical plant, which has adopted a pioneering spirit of its own. And the object of his remark is no ordinary bulletin board, but what the folks at Baytown universally refer to as a "quality station."

◎ **Visitors to the 3,500-acre complex encounter a quality station—a focal point for improvement activities—at practically every turn of a corner.** Some are fancier than others. But, by design, they all serve the same basic purposes: to convey overarching business goals to each individual work team in a way that is meaningful at that level, to provide a mechanism for stimulating ideas, and to track progress toward the goals for that unit.

The 1993 Best Plants award-winning Baytown operation, near the Houston Ship Channel, has four product divisions and a "core services" unit, including finance and human resources. It produces six billion pound of product annually and gets most of its raw material from an adjacent Exxon refinery. (Together, the Exxon operations in Baytown comprise the largest petrochemical complex in the world.)

Throughout the chemical plant, nearly 2,000 people—over 1,200 full-time Exxon employees and some 700 people who work for outside contractors—are actively involved in the goal-oriented improvement process. It is a major reason why Baytown far outstrips the industry average in such key productivity measures as revenues and profits per employee. "We've got 2,000 people out there making improvements—an average of 12 improvements per person each year," beams site manager Ray Floyd, the chief architect of the quality-station process. "It's a huge number of improvements—and they are additive and compatible."

That is not true of all continuous-improvement programs. "If you simply turn people loose to make improvements," Floyd points out, "it is very common for one person to make an improvement that takes away the advantage created by another person." But through its system of layered goal translation, which cascades key sitewide goals down to the work units, Baytown keeps everyone headed in the same direction.

A full listing of the plant's achievements would consume an encyclopedia-size report. But here is a sampling of the activities that have enabled a world-class facility to get even better.

A series of small improvements enabled the Polymers Division's polypropylene production unit to coax 110,000 metric tons of product annually from a system designed to produce just 70,000 metric tons per year. The technology was originally licensed from a Japanese company, which has been baffled by Exxon's ability to so dramatically increase the output.

Statistical quality practices, adopted in the 1970s and enhanced with the latest real-time, computerized, process control systems, have enabled the Baytown complex to achieve an average process capability of 1.7 Ppk. (Ppk is a measure of process variability.) The Paramins and Basic Chemicals divisions are now operating at 2.0 Ppk or better—a six-sigma performance.

And, in 1993, the plant, which monitors 80,000 valves and other pieces of equipment annually to prevent chemical air emissions, had already exceeded its "Clean Texas 2000" target of reducing toxic chemical releases by 50 percent from the baseline year of 1987.

Undergirding the Baytown plant's improvement efforts are several major initiatives. One is a pacesetting campaign to translate just-in-time and "lean" manufacturing concepts, typically oriented to discrete manufacturing, to the chemical process industry. In the Elastomers Division, which produces synthetic rubber material, cycle time has been reduced by 74 percent, and in some cases production is coordinated so closely that the plant can ship hot material to customers, enabling them to skip their first process step—reheating.

◎ **Another initiative is a "human diversity" program, a pilot project within Exxon Corporation that emphasizes valuing individual differences, equalizing opportunity, and cultivating an atmosphere in which each employee can contribute to his or her fullest ability.** In addition to plantwide awareness training, a team known as the "Diversity Pioneers" implements programs and procedures to systematize the culture change. One subcommittee developed a new sitewide process for recruiting new technicians, in which candidates are screened by a team of technicians who represent the plant's diverse population.

◎ **A heavy emphasis on employee training is bolstered by a central training organization that has 12 full-time trainers, a 6-person library staff, and a fully equipped videotape production studio with the capability to create interactive multimedia training packages.** Employees average about 20 full days of training yearly. But in the Basic Chemicals Division, which is pursuing sociotechnical work redesign, process technicians now receive even more training by rotating off shift one full month out of every five. The stepped-up training grew out of the work redesign effort, which is being overseen by a cross-functional steering committee that includes the presidents of each of the four plant unions.

—JOHN H. SHERIDAN

Exxon Chemical Co., Baytown Chemical Plant
Contact: R. C. Floyd, Site Manager
P.O. Box 4004, Baytown, TX 77522
Telephone: (713) 425-1009
Employees: 2,000
Major products: Chemicals, elastomers, polymers, paramins

FISHER-ROSEMOUNT, INCORPORATED

Chanhassen, Minnesota

AGILE MANUFACTURING
CUSTOMER FOCUS

The best opportunities are in the least glamorous areas. Management visibility can make an area glamorous.

—AMY JOHNSON, OPERATIONS MANAGER

When Emerson Electric's Rosemount Measurement Division opened its 340,000-square-foot plant in Chanhassen in January 1990, it wanted the world's largest pressure-transmitter manufacturing facility to be more than just a leader in world-class technology. The plant, which can produce over 10^{18} different assembly configurations of its modular-designed transmitters, had to have maximum manufacturing flexibility.

In terms of architectural design, that was easy enough. ◎ **There are no inside walls in the Chanhassen factory, and machinery is mobile, not fixed in concrete.** Thus, switching to a new technology—whether it be the inside "brain" of a transmitter or a different product-building technology (welding or joining, for example)—can be done in a "couple of hours, not a couple of weeks," says Edward Monser, vice president for pressure operations for the Measurement Division.

The other half of the equation, though, was a gamble. If Rosemount Chanhassen was to become a world-class plant operating on the principles of continuous improvement, managers had to put the power for changing and improving operations into the hands of the workers. That meant changes in management and manufacturing systems, extensive training, and a different culture for the workers who moved to Chanhassen from other Measurement Division plants.

For starters, says Monser, "we knew that we needed people to know the entire process from beginning to end and that cross training would be necessary to become flexible and to be a world-class competitor." Workers could no longer be dedicated to just one task. So the 1993 Best Plant award-winning facility cross-trained workers and gave them 19 two-hour training sessions in the use of statistical process control. By 1993, roughly 150 cross-functional process improvement teams (PITs) of 6 to 12 employees each were at work. And thanks to the cross training, "you no longer have people protecting their own turf or taking the attitude that that is how it's always been done," says instrument builder Donna Brown-Rojina. "It is now, 'What can we do to improve the process?' because all of us can see how each step in the process affects the next step."

In 1992, for example, on its Model 1151 pressure transmitter alone, Rosemount Chanhassen implemented 216 PIT suggestions that reduced the dollar amount of scrap by 43 percent, improved out-of-box quality 80 percent, cut order lead times from 12 weeks to 6 weeks, and, with the help of a cellular approach in final assembly, cut cycle time 75 percent.

Among the specific changes: An instrument builder—who rotates through work areas where helium in pressurized bottles was used for high-pressure sensor tests, as a cover gas in laser welding, and to check welds—noticed that the helium bottles were being discarded when the gas reached levels that were no longer usable in one area, but were adequate for use in another. At his suggestion, workers began moving the bottles between work areas. This saves $40,000 a month. "It was a change that never would have been made if we didn't have a person rotating, connecting how work was done," says Monser.

The changes are not confined to manufacturing. A PIT team cut order-processing time by 46 percent and after-shipment accounts receivable adjustments by 50 percent. It also found that orders often were sent to the sales office, then reentered a second time at the factory. All orders are now are entered at the factory.

◎ **Since the plant opened, Chanhassen has altered its final assembly area so that workers, instead of being in a straight line, are in a cellular manufacturing configuration in which teams of five to eight employees have all the parts needed for assembly at their workstations and work in a U-shaped flow that enables them to communicate more easily and respond faster to both production and material changes.** Chanhassen also developed supplier alliances that between 1991 and 1993 improved on-time delivery from suppliers from 90 to 95 percent and raised supplier conformance to specifications from 70 to 90 percent.

◎ **In addition, a 24-hour National Response Center handles 100,000 calls a year.** When a customer calls the center, Chanhassen promises a callback from a service technician within 30 minutes—it's usually 15 minutes—and on-site help, if requested, within 24 hours. How has the National Response Center helped? In late August, a Chanhassen customer scheduled a plant shutdown, then realized it did not have all the instruments it needed. The customer called at 10:45 A.M. on a Saturday in need of four transmitters. Workers built them that day, and Chanhassen's production manager flew to California, hand-delivering the parts in just over 10 hours. "It wouldn't have happened [without] a way to get in to get the problem solved," says Monser.

Chanhassen also improved customer service through design-team partnerships similar to PITs—except that customers are involved. "We used to design to meet a broad range of applications, despite the fact that no two chemical plants, refineries, or breweries are alike," says Dave Broden, director of R&D for the pressure group. "Now, while we still use a building-block concept, we design based on customer feedback, not to a sheet of paper."

It is also routine for customers to come to Chanhassen to talk directly to the people in the factory who build the product. Workers will ask the customer how they use the product and what can be done to make it better. "Workers now hear it directly from customers instead of interpreting what was said to engineers," says Broden.

And that applies to internal customers as well. "Before, we used to just build parts and let them stand," says Phil Kelaart, production supervisor with the sensor-manufacturing group. "Now we respond to what our [internal] customer needs. . . . We know our customers are the grinders and we build to their needs, not for inventory. And when there is a problem we all sit down as a group and come up with a solution. Everyone has a chance to speak up."

—MICHAEL A. VERESPEJ

Fisher-Rosemount, Inc.

Contact: Kelly Hoffman, Vice President, Pressure Operations

8200 Market Blvd., Chanhassen, MN 55317

Telephone: (612) 949-7000

Employees: 1,144

Major products: Pressure transmitters

FORD ELECTRONICS & REFRIGERATION CORPORATION

Lansdale, Pennsylvania

SUPPLIER RELATIONS
CUSTOMER FOCUS
ENVIRONMENT AND SAFETY

A detailed vision has been thought out, written down, and communicated to accomplish making perfect products and creating customer delight.

—DUDLEY WASS, PLANT MANAGER

Ford Electronics & Refrigeration Corporation's North Penn Electronics Facility (NPEF) does not want to satisfy its customers anymore. From now on, delighting them is all that will suffice. Delight, explains Dudley Wass, plant manager of the Lansdale unit, is quality and customer satisfaction writ large. In essence, he says, "we want people to tell other people how good we are."

A case in point: When Mercury owner Tom Fauntleroy's car clock failed and local dealers could not repair it, Fauntleroy removed the clock himself and on it discovered NPEF's address. "I was contacted by [NPEF engineer Deborah] Keller after writing to the indicated address," Fauntleroy reported. "She advised me repair was no longer available at that location and took the time to explain that it could be merely a simple problem of bulb failure. After [providing] instructions for bulb removal, she promised to send a couple of bulbs to try, and [asked] to know the results.

"My point," continued Fauntleroy, "is that 'normal service' today would consist of her sending me a form letter with a statement like, 'We no longer service this product at this location. Contact your local Mercury dealer.' " Instead, Fauntleroy got his bulbs from North Penn and installed one successfully.

"At the core, we're trying to be an agile manufacturer," says Wass, who oversees the employees in the 703,000-square-foot facility. "We're not all the way there yet, but we have a pretty good idea what it takes to achieve it." The 1993 Best Plant award-winning facility, the direct descendant of a Philco Corporation operation that Ford acquired in 1961, is a crucible for world-class manufacturing techniques. And it is a crucible with a timer attached. Almost every advancement at NPEF, from automated troubleshooting and repair on the floor to a cutting-edge videoconferencing system tied to other division plants worldwide, shaves more and more time off the production process.

In the world of agile manufacturing, time flies. So can profits if every aspect of the process isn't monitored and continuously improved. "We're trying to instill a mind-set on time," says Diane Darrow, supervisor of the agile manufacturing group. "If equipment is down, that translates into so many dollars of product lost." Because time is production, explains Wass, "NPEF has aggressively implemented continuous-flow manufacturing techniques for each product line in the plant. Since 1991 our efforts have reduced manufacturing cycle time by 78 percent—from 7.8 days to 1.7 days [in 1993]."

To delight its 76 customers around the world, NPEF puts a new and positive spin on the term *clock-watchers.* In June 1993, North Penn inaugurated the use of the Flexibility Enabling Manufacturing System (FEMS), a state-of-the-art computer-integrated manufacturing (CIM) system. Among the advantages of FEMS is its ability to let operators, using real-time data, detect and correct quality and throughput problems.

Another critical time beater is NPEF's new software-based Manufacturing Design Rule Checker (MDRC). Raising the axiom "Measure twice, cut once" to the high-tech level, MDRC allows craftspeople and designers to determine design feasibility before a design goes to production. The result: increased first-pass design success on many of 25 new products, including one that NPEF calls "the most advanced engine controller in the world."

And NPEF doesn't just raise the bar for its own performance. As a plant that consumes six million parts each day, it recognizes that its overall success in getting products to market lies in raising the bar for its 132 supplies as well. ◎ **To do that, NPEF's Total Supplier Improvement Group, initiated in 1992, scrutinizes the operation of the supply base.** Within a year, the group's effort paid off with a 66 percent reduction in supplier in-house defects, and quality rejects of delivered parts were cut by 25 percent—to 163 parts per million.

As it expects world-class service from its suppliers, the plant also delivers world-class service to its customers. ◎ **A North Penn liaison engineer is assigned to each major customer, resolving problems and generally acting as a champion for that customer.** Wass explains: "Liaison engineers provide the customer with fresh eyes to observe their processes and our product applications, as well as make recommendations on material-handling practices, scrap reduction, installation procedures, and throughput improvement."

Those are the champions the customers know best. But NPEF has many champions, from the production worker to the manager at this unionized (United Auto Workers) plant. "We are the best," boasts Delores Herbert, a coordinator on one of the more than 100 self-directed work teams at NPEF. That is no empty claim. Consider that one team in a test-and-repair area reduced its work-in-progress inventory from 4,000 units to 600 units.

◎ **North Penn also is cutting deeply into the area of waste. Its returnable-packaging programs have already eliminated one million pounds of cardboard, pallets, and plastics per year.** And where it once consumed 350,000 pound of Freon as part of its post-wave-soldering process annually, it now uses a no-clean (inert gas) process that eliminates the need for all Freon cleaning agents.

From forming teams to reforming material-handling practices, NPEF aims to keep its flagship status as a producer of Ford electronic power-train and vehicle-control products. To do that, the management team relies on its monthly quality operating system reviews and its advanced quality planning. NPEF also uses an eight-year technology plan to identify emerging technologies and link them to specific products. "North Penn's direction is clear," says Wass.

—JOSEPH F. MCKENNA

Ford Electronics & Refrigeration Corp., NPEF

Contact: Ted Davenport, Manufacturing Design and Development Manager

2750 Morris Rd., Lansdale, PA 19446

Telephone: (610) 584-7112

Employees: 2,000

Major products: Automotive power-train and control electronics

FORD MOTOR COMPANY

Wixom, Michigan

Manufacturing flexibility is a long-range challenge that all auto companies face.

—PAUL R. NOLAN, PLANT MANAGER

EMPLOYEE INVOLVEMENT
SUPPLIER RELATIONS

Ask manufacturing experts where to find the slickest automotive assembly operation in the country and chances are they will cite Ford Motor Company's assembly plant in Wixom, located a few miles west of the company's Dearborn headquarters.

The mammoth 3.95-million-square-foot plant, which produces over 1,000 cars each day, is a complex wonderland of mechanization and automation, including state-of-the-art robotics and a maze of conveyor systems that move through the plant at different levels.

What is particularly impressive is that Wixom builds three entirely different automobiles—the Lincoln Continental, the Lincoln Town Car, and the Mark VII—all on the same line. And, the differences between these cars go far beyond mere sheet-metal wrinkles. The Continental is a unitized, or unibody, front-wheel-drive automobile; the Town Car is a frame-supported rear-wheel-drive car; and the Mark VII is unitized and has rear-wheel drive. Among other things, that means the engine/drivetrain assemblies must be installed differently. For the rear-wheel-drive cars, the engines are lowered into the chassis from an overhead conveyor; on the front-wheel-drive Continental the engine must be installed from the bottom by a worker in a pit.

It all adds up to a major manufacturing challenge—one which has brought out the best in Wixom's 3,200 employees. Despite the complexity and the fact that when the facility won its 1990 best Plants award it had been running flat out with maximum overtime for five straight years, the plant has an impressive record of quality improvement.

The capability of producing three different types of cars gives the Wixom plant unique flexibility in responding to changes in the marketplace.

At Wixom, cars are built according to a preprogrammed model sequence. But the plant is not locked into a particular model mix; in fact, the sequence changed 11 times between January 1988 and August 1990.

It would be an understatement to say that the Wixom plant is quality conscious. A huge sign hanging above the final inspection department at the end of the 15-mile-long assembly line reminds workers: OUR CUSTOMERS ARE OUR NEXT INSPECTORS.

"We are very customer-driven," says Nolan, noting that meetings are held twice a month in the plant's large conference room to go over the results of Ford's competitive new-vehicle-quality (CNVQ) studies, which provide detailed reports on what customers liked or did not like about the Wixom-made cars. The meetings, involving about 100 employees, last anywhere from four to seven hours.

◎ **Wixom was one of the first U.S. auto plants to embrace employee involvement (EI), in the early 1980s, as part of a union-management partnership.** The EI program is guided by a joint union-management steering committee that meets twice a month. More than 30 EI teams meet weekly to tackle quality problems and other matters.

The EI philosophy includes soliciting workers' input on the design of the cars built at Wixom. During the design of the 1990 Lincoln Town Car, for example, 110 prototypes were built right on the assembly line, each mixed in with the assembly of current production models. In the process, workers helped to tweak the design to improve quality and ease of assembly. No fewer than 2,600 suggestions from workers were incorporated into the final design.

As an assembly plant, Wixom relies heavily on its 670 suppliers to ensure that high quality standards are met. And suppliers who are not performing up to snuff quickly get the message.

◎ **Once a month, the plant's five worst suppliers—based on sampling inspections of incoming parts—are summoned to a meeting in the large conference room and asked to document what they are doing to solve their quality problems.** "We meet with the CEOs and presidents of the supplier companies," Nolan says. "It makes the top executives of those companies realize that they have problems that maybe they were not aware of. It can be pretty embarrassing. And if they don't make instant improvement, they may be invited back the following month."

The quality of the cars that roll off the Wixom assembly line is also enhanced by advanced robotic technology and a $500 million state-of-the-art paint shop. The paint facility—a huge 850,000-square-foot building completed in 1990—is connected by a tunnel to the main assembly plant.

One of Wixom's more remarkable workstations features a robot that installs carpeting in cars that already have their doors attached. Since the robot must contend with three different body types, a sophisticated computer-vision system is employed to identify the type of car and to position it properly for the installation.

—JOHN H. SHERIDAN

Ford Motor Company
Contact: Ezra Carter, Human Resources Manager
28801 Wixom Rd., Wixom, MI 48393
Telephone: (810) 344-5431
Employees: 3,200
Major products: Automobiles

THE FOXBORO COMPANY

Foxboro, Massachusetts

AGILE MANUFACTURING
TECHNOLOGY
ENVIRONMENT AND SAFETY

There is minimum overhead employed to operate the plant. This has been accomplished via a total redesign of our business systems. . . . Simplicity is legislated.

—V. S. VENKATARAMAN, PRODUCTION MANAGER

Back in 1986, when Massachusetts was still enjoying its economic miracle, one of the state's old-line manufacturing companies was embarking on a bold new initiative designed to see it into its second century. There was little to cheer about at The Foxboro Company, a worldwide supplier of industrial-automation-control equipment that began trading in 1906. Market conditions in 1986 were poor, and layoffs were impacting the company, which depends greatly on process industries, particularly the energy sector, for its bread and butter. The pain would continue into the 1990s, following the company's takeover by Siebe of the United Kingdom. Total employment dropped from 6,900 to 4,500.

When plant manager Henry Metcalf was called in by the "big boss" in 1986, he was not sure whether he was going to be "praised or fired," he laughs. At the time, he was running a Foxboro manufacturing facility in nearby Bridgewater, Massachusetts. Instead of being fired, he was told about a new product line, the Intelligent Automation (I/A) series of process control systems, and was charged with creating a new operation to produce it. The I/A facility was to be located in a building the company already owned that stands near the center of the town of Foxboro.

The I/A series was a complete departure from the company's existing products, which had been mainly electromechanical in the past. The new line of equipment was taking Foxboro into the world of electronics. The 75,000-square-foot building was required to allow the formation of an assembly plant containing surface-mount equipment (900,000 placements are made per week), processor modules, power suppliers, cables, peripherals, enclosures, communications modules, and other associated equipment. Behind the whole project lurked an even higher ideal, though. The I/A facility was "created with the intention of serving as the vehicle that would change Foxboro's approach to manufacturing and achieve world-class operational status."

"We started with four walls, a floor, and a ceiling," recalls Metcalf. "For three years we worked around the clock to put this together—to the point where wives and husbands hated I/A. ◎ **We traveled all over the country looking at the ways other companies did business.** Hewlett-Packard [a Foxboro rival] went out of its way to be helpful to us, as did companies such as IBM, Texas Instruments, Motorola, DEC, and Allen-Bradley." Armed with this information and a variety of ideas and views from other sources, including its own people, Foxboro spent some $60 million creating the new division.

The first realization that struck many people was that the 1992 Best Plant award-winning facility would be operating on a high-mix, low-volume production schedule. That went against the grain, says Metcalf. "We blew the high-volume production theory out of the water. It even bothered our president at the time, who had been a GE man and who adamantly believed in high-volume production techniques." Control of production at Foxboro's I/A plant now comes down to one item—a customer order. Without that, says Metcalf, nothing gets made. "That can be very difficult for people who were used to making for stock in the past. But there is no warehouse as such. The entire plant is run on JIT." ◎ **There is no paper, either. Everything operates across an electronic networked information system.** As a consequence, there is not a clerk to be found in the whole manufacturing area.

Foxboro concentrates on three main areas in their efforts to achieve world-class status. Their aim is to provide the highest-quality hardware in the marketplace at speed of response that beats all comers—and all this at minimum cost. Minimal overhead was a major goal of the cost structure. It breaks down this way: material, 70 percent; overhead, 27 percent; direct labor, 3 percent. "All programs within our long-range plans fall within one of these three categories. Commitment, along with total employee involvement, will make them successful," says Metcalf. "[The operating platform] is the foundation that doesn't change but serves as the basis for our continual-improvement programs."

Employee involvement and intensive training programs are a way of life at Foxboro, and all employees participate in continuous-improvement programs. But this does not mean that autonomy rules. "The management style is goal-oriented, hands-on, and anticipatory in its approach. Self-directed work teams are a figment of people's imaginations," contrarian Metcalf says. "Everyone should be involved."

Supervisors—and there still are some—meet regularly with the people in their departments and conduct information sessions. This activity can relate to day-to-day work or involve new ideas. Team members work in a variety of departments, coming together to focus on continuous-improvement tasks or other aspects of the plant's life. ◎ **One example of the latter occurred when it was proposed that in this environmentally sensitive world the company should be shipping its products in unbleached, rather than bleached, cartons.** The idea found its way onto the agenda of a team. It drew in engineers and designers as well as shipping and manufacturing employees. Their recommendation: Ship product in unbleached cartons.

To obviate problems with its suppliers, Metcalf meets regularly with them to explain the system that Foxboro operates and what the company expects in the way of quality, cost, and speed. The company uses well-tried methods to help attain its continuous-improvements goals. For example, it employs statistical process control and has spent a considerable amount of money on bar coding that not only helps to track products as they move through the factory, but also enhances the "no-paper" dictum.

"We truly believe, and are on a mission to prove," concludes Metcalf, "that American manufacturing can compete with and surpass anyone in the world."

—Brian M. Cook

The Foxboro Co., I/A Manufacturing Div.

Contact: Ray Webb, Manager of Employee Activities

33 Commercial St., Foxboro, MA 02035

Telephone: (508) 549-3747

Employees: 246

Major products: Electronic process control systems

GE FANUC AUTOMATION NA, INCORPORATED

Charlottesville, Virginia

EMPLOYEE INVOLVEMENT
AGILE MANUFACTURING

[Continuous improvement] is not a campaign to eliminate people, but a program to liberate people . . . from the nonsense of useless paperwork, meetings, and reports that we cause ourselves.

—DONALD BORWHAT, SENIOR VICE PRESIDENT, HUMAN RESOURCES

Can we know the child simply by knowing the parents? Not in the case of GE Fanuc Automation North America Incorporated, Charlottesville, the joint venture of General Electric Company and Japan's Fanuc Limited. Productwise, GE Fanuc is much more focused than either parent, primarily producing programmable logic controllers (PLCs) and computerized numerical controllers (CNCs) for the factory automation marketplace.

The plant is the brainchild of Jack Welch, GE's CEO, and Dr. S. Inaba, CEO of Fanuc, who constructed a new, combined strategy to reach the automation market. A primary issue was the changing nationality of the machine-tool makers who purchased GE's CNCs and PLCs. American machine-tool makers were being replaced by the Japanese competition. Hence the new strategy: Reach these new customers with a joint venture bearing two familiar names. Technologically, the emergence of the new company in 1987 combined their best strengths—Fanuc in CNCs and GE in PLCs. Less autocratic than GE organizations of yore and far less authoritarian than Fanuc's reputation in Japan, GE Fanuc is headed by CEO Robert Collins, an enthusiastic executive who has moved the company from a top-driven, dictatorial style of management to what he calls a "coaching" technique.

Visitors to the plant first see architecture and a building layout that is in keeping with the current Japanese business preoccupation with an edifice complex, but a closer look reveals a more practical set of values. A building that once housed group executives in handsome office space now serves as meeting rooms for the 42 teams that run the plant.

From a performance standpoint, the 1992 Best Plant award-winning operation was also strengthened by a campaign for empowerment that CEO Welch initiated in the late 1980s. The program was designed to maximize inputs from associates and put managers in roles as enablers—counseling teams, eliminating barriers, and encouraging employees to pursue improvement efforts. At GE Fanuc, the effort began early in 1989 when Collins addressed all the plant's associates, 30 to 40 at a time, at an off-site location. Assisted by a consultant from the University of Virginia, the associates were introduced to a management technique called *work-out*. Described as a way to improve methods, work-out includes best practices and process mapping.

In this definition, best practices differs from benchmarking by researching a broader question—the overall excellence of other organizations—as opposed to excellence in such functional categories as production or distribution. Process mapping, the second part of work-out, is a study of asset deployment—an analysis designed to reveal the actual steps in the manufacturing process. The goal is to find places where improvements should be made.

◎ **Work-out is designed to draw the associate into the decision process via team problem solving.** Constant improvement is the focus. Five basic questions form the work-out test: Will this task contribute to our mission? Will this serve our customers? Is this the fastest way? Is this the simplest way? Do we have the self-confidence to do this? When a continuous-improvement target is found, employees apply the following process: Identify and contact all stakeholders; assess importance of target to each; search for a better way; get approval for a change; and finally, communicate results to stakeholders.

Installing the cultural change brought some surprises. "For example," says Robert F. Wayand, senior vice president of manufacturing, "some of our best performers under the old system had problems adjusting, while some former 'problem' performers have done outstanding jobs under the new system." The shift in values may be responsible. "Star performers under the old system tended to be the employees who would follow directions; they would find out what production quantity was expected and then try to surpass it by some amount," Wayand observes. "Under the new system, they had to become self-directed. In contrast, some of the 'problem' employees were undoubtedly placed in that category because they were always self-directed—even though under the old system they were told not to be."

By 1992, GE Fanuc's continuous-improvement journey had yielded impressive results. A 52 percent reduction in organizational layers led to a collapse of the traditional management pyramid. Empowerment has resulted in a 25 percent reduction in the bureaucracy of seeking approvals. Other benefits include a $150,000 reduction in scrap and rework and a 74 percent reduction in the time needed to develop new products—down from two years to as little as nine months in some cases.

Although GE Fanuc has turned against automation for automation's sake, plenty of automation is evident in its revised factory of the future—even CIM. ◎ **The effort to rationalize the production technology started in 1986 with a complete re-layout of the factory into a U-shaped continuous-flow configuration.** That change reduced cycle time 25 percent and pared work-in-process inventory. Another significant initiative is the design-for-manufacturability program that ensures that designs fit the production scheme. As a result, 98 percent of more than 300 different types of circuit boards fit into a standard 8.7- by 11-inch form factor.

The recovery of the Charlottesville plant was not simply a matter of revising the emphasis from technology to cultural factors, but rather one of shifting the focus to the process. With that understanding, Collins believes the facility is finally on its way to becoming the true factory of the future.

—John Teresko

GE Fanuc Automation NA, Inc.

Contact: Dave Friesema, Senior Vice President, Manufacturing

P.O. Box 8106, Charlottesville, VA 22906

Telephone: (804) 978-5494

Employees: 1,149

Major products: Programmable logic and computerized numerical controllers

GENERAL ELECTRIC

Auburn, Maine

You have to provide [employees] information so they can make effective business decisions.

—KEITH SCHAFER, PLANT MANAGER

TECHNOLOGY
ENVIRONMENT AND SAFETY

If you walk the factory floor at 1995 Best Plant award winner General Electric Company's Electrical Distribution & Control Division (ED&C) plant looking for Mona Dunn, you might have trouble finding her. A cell operator, she normally works at the station that produces lug-kit assemblies (a component of circuit breakers), one of an astounding 5,000-plus variations of plated stampings, machined parts, welded/brazed assemblies, and electrical components the facility supplies to 50 sister plants in the division.

But you could just as easily find Dunn, a 25-year veteran of the plant, at a different cell—one that produces, say, circuit-breaker stab assemblies. Or maybe she'll be in the electroplating area, or, for that matter, anywhere else on the Auburn plant's 124,000-square-foot floor, which features more than 100 separate manufacturing processes. "I move to a different job at least once a day, and sometimes three or more times," says Dunn. "I'll go wherever I'm needed—to any area of the plant that needs help."

And she's no different than most of the plant's 283 employees. ◎ **Cross-trained to perform any job in the facility, they use computer terminals located throughout the plant floor—part of a homegrown computer system called ON-TRAC—to learn the priority and status of each order in real time, as well as workload, material availability, and quality requirements.** Armed with this information, employees make their own decisions about where to work at any given time. There are no supervisors on the floor.

"The main thing is to get the work out," says Dunn, reflecting an employee mind-set that has enabled GE Auburn to offer customers a remarkable degree of customized service. The facility, a 100 percent make-to-order business, is known as "the Burger King of ED&C." Its customers can "have it their way"—

in a dazzling variety of lot sizes, tolerances, or materials. Special one-shot orders aren't unusual. Neither is overnight delivery.

Between 1990 and 1995, this plant has achieved a sparkling 116 percent improvement in productivity. (Shipments have grown from $36 million to $48 million, while the number of employees has dropped from 472.) Scrap and rework have been reduced 68 percent during the span; customer rejects, 35 percent; cycle time, 90 percent; order lead times, 40 to 50 percent; and cost per unit, 27.5 percent. Inventory turns have improved 182 percent, from 10.77 to 28.95. And despite the 5,000 variations of its customized stampings and machined parts, first-pass yield is a near-perfect 99.995 percent.

But empowered employees alone don't explain the plant's success. Another vital factor, indicates plant manager Keith Schafer, is management's emphasis on team measurements. "Our measures help drive behavior," he says. "It's not enough simply to tell people they're 'empowered' and to go out and do what they think best. You have to have them grounded and focused."

GE Auburn's employee empowerment stems from a 1988 reorganization, as do other aspects of the plant's turnaround. Parallel decisions were made then, for example, to implement the ON-TRAC computer system, to adopt kanban systems (which now control 75 percent of the plant's output), and to use an "infinite load" scheduling approach that maximizes throughput of highest-priority needs. The latter replaced a "finite" approach, that source/service manager Tom Nezwek explains, "presumed predictable performance levels, fixed process, available material, and adequate personnel." Now, he says, the plant "continuously reschedules itself" and has the ability "to pull out all the stops" to meet rush orders or changes in ship dates.

Nothing gives GE Auburn more pride, though, than its accomplishments in the area of environmental health and safety (EH&S). During the last three years, the plant has cut its water consumption by 30 percent and its air emissions by 85 percent, as well as dramatically slashing or eliminating its use of chemicals. Its safety performance is equally glittering. Workers' compensation costs have plummeted from $1 million to $25,000 and its OSHA-recorded accident rate has dropped 50 percent from 1990 to 1995. By September 1, 1995, the facility had logged 1,108 consecutive days, over three years, without a lost-time accident.

"You don't have a good health and safety record by accident," stresses Anne Paradis, environmental-health-and-safety engineer. ◎ **Much of the credit goes to a hardworking safety team of five hourly and three management employees and a rotating team of 11 safety observers.** So sterling have this plant's achievements been that it has won two prized awards: In June 1995, it collected one of the new Governor's pollution-prevention awards from the state of Maine, and in 1994 it earned ED&C's Golden Chair Award honoring the division facility with the best EH&S record.

—WILLIAM H. MILLER

General Electric, Electrical Distribution & Control Div.
Contact: Tom Nezwek, Service/Sourcing Manager
135 Rodman Rd., Auburn, ME 04210
Telephone: (207) 786-5100
Employees: 300
Major Products: Made-to-order electrical components

GENERAL MOTORS

Grand Rapids, Michigan

The biggest lesson I've learned is that there's nothing that can replace having personal contact on the floor with the teams. I used to dream about doing it. I used to talk about doing it. Now I do it.

—RON KORTE, PLANT MANAGER

CUSTOMER FOCUS
EMPLOYEE INVOLVEMENT
QUALITY

Ricardo Garcia has been to General Motors Corporation's Ramos Arizpe engine plant in Mexico seven times in the last two years. The production-line assembly operator acts as a customer service ambassador for GM's Delphi Energy & Engine Management Systems plant in Grand Rapids. An hourly employee who was born in Los Nogales, Mexico, Garcia literally speaks his customer's language. He has helped explain the nuances of the Spanish-language product manual, how his plant's hydraulic valve lifters are made, how to identify product defects, and how to store and handle the valve lifters without causing damage.

During one trip Garcia solved a vexing problem. When Delphi's returnable plastic shipping containers came back empty to Grand Rapids from Mexico, they had tiny sand particles on the inside. (Sand and engine parts don't mix.) The bilingual Garcia discovered through casual conversations with Arizpe employees that the shipping containers were stored outside the plant near a sand pit where gusts of wind would blow sand into the containers. Now the containers are wrapped in plastic. Problem solved.

"GM Mexico had a hard time getting answers to some technical problems because of the language difference," points out Ron Korte, manager of this plant, which is a division of GM's $22 billion Delphi Automotive Systems Components Group. "Before Ricardo began going to Mexico, conversations and communications were conducted only in English, and that created an instant language barrier," he adds.

At Delphi Energy, the customer service commitment goes well past linguistics. "We take a 'quality with a vengeance' approach," says Rich Erickson, plant quality manager. "We'll even manually sort to protect the customer from manufacturing problems. If it's a concern to our customer, we don't argue. ◎ **Each**

year we list the top five customer complaints and eliminate them the following year." Labor and management work hard to create the shortest line of communication between those who make the product and the customer who uses the product.

When the GM plant in Saint Catharines, Ontario, called Delphi Energy to report a quality problem, eight employees jumped into a van. By the time they reached the Canadian plant seven hours later, the customer plant had traced the bad part to another supplier. But before heading home to Michigan the Delphi customer service team stayed around to sort seven pallets of problem stock.

When Suzuki Motor Corporation complained about noise coming from Delphi's hydraulic valve lifters, a cross-functional team of Delphi employees boarded a plane and spent three days in Japan tracing the problem to the oil pump—made by another supplier.

Delphi is a maverick facility that has thrown out the book on some tried-and-true manufacturing principles when they interfere with customer satisfaction. The plant scrapped materials resource planning and replaced it with a customer-oriented pull system of production scheduling. A complex and costly computerized automatic-storage-and-retrieval inventory-control system was scuttled in favor of a placard-based kanban system, and as a result inventory float has been reduced from 5 to 10 days to virtually zero.

Of 30 to 40 basic manufacturing action strategies emphasized by GM corporate to improve quality, Delphi Energy concentrates on just five. "We picked five key strategies—workplace organization and visual controls, lead-time reduction, a pull manufacturing system, identification and elimination of waste, and planned maintenance—because they made the most sense for our plant and our customers," says Korte. ◎ **To assess how well the strategies are working, each Friday Korte devotes his day to meeting the plant's teams. He spends about 30 minutes with each team, and he meets with every team in the plant once every five weeks.** "It helps me understand their roadblocks and obstacles and how I can help them overcome them," says the plant manager. "And unless you make a commitment, have appointments and schedules to get it done, it doesn't get done."

Also, the independent-thinking plant leadership doesn't buy into computer-integrated manufacturing, because it produces only a handful of product lines in very high volume and at high speed. The plant's 1,179 employees make about 500,000 precision-machined hydraulic valve lifters a day in the 1.4-million-square-foot facility. ◎ **To test the plant's systems, rogue (defective) master parts are purposely run through its manufacturing systems every day.** "We want to make sure our errorproofing system is functioning properly," says Erickson. "Our assembly-line operators begin the day's production with known defective parts," he says.

One result: In 1992, 1993, and 1994, the plant received the Toyota Excellent Award for Quality Performance. During an 18-month period in 1994 and 1995, it shipped 1.5 million valve lifters to Toyota Japan with a zero defect rate.

—Brian S. Moskal

GM, Delphi Energy & Engine Management Systems
Contact: Peter Gallavin, Personnel Manager
2100 Burlingame SW, Grand Rapids, MI 49509
Telephone: (616) 246-2294
Employees: 1,241
Major products: Hydraulic valve lifters

GENERAL MOTORS CORPORATION, CADILLAC

Detroit, Michigan

EMPLOYEE INVOLVEMENT
NEW PRODUCT DEVELOPMENT

If a person has to make a certain connection in the assembly process 456 times each day, five days a week, it soon becomes apparent that the way the connection is designed and laid out has a direct impact on the person and his quality.

—ROY ROBERTS, MANUFACTURING MANAGER

Given the genesis of GM/Cadillac's Hamtramck Assembly Center, its success, including a 1990 Malcolm Baldrige National Quality Award, is nothing short of amazing. This 1991 Best Plants award-winning facility was designed for Detroit's Polish neighborhood of Hamtramck in the late 1970s. GM's original vision was for a competitive victory that was to come from a new plant using tomorrow's technology to build the Cadillacs of today.

But when the plant opened in 1985 the reality was somewhat different. In her book *Rude Awakening*, automotive analyst Maryann Keller told what she saw: Robots designed to spray-paint cars were painting each other; a robot meant to install windshields was found systematically smashing them; factory lines were halted for hours while technicians scrambled to debug the software; robots were smashing into cars, demolishing both the vehicles and themselves; and computer systems were sending erroneous instructions, leading to body parts being installed on the wrong cars. So much unproved and misapplied technology had been installed that just removing it improved production and efficiency, remembers Larry Tibbitts, Hamtramck's plant manager.

Tibbitts describes his problems with an automated robotic system for applying seam sealants: "Despite the transponders, sensors, and electronics that guided the 12 robots, it seemed that we always needed people to follow the automatic applicators and make sure the seams were properly sealed. Then last year we tried applying the sealants manually. To our surprise, we found that that required fewer people." Being a working laboratory for advanced technology was not a cost-effective role for the plant. Instead, Tibbitts points to employee involvement as the key ingredient of the progress that Hamtramck has made.

For example, the plant's employees achieved a 73 percent reduction in discrepancies per vehicle between 1986 and 1991. They also recorded a 65 percent improvement in productivity for the same period in all measures of quality, customer satisfaction, and other measures of competitiveness. These include a 50 percent decrease in warranty repairs and a 50 percent reduction in problems per hundred vehicles as measured by a blind GM survey of new-car purchasers after 90 days of ownership.

Employee involvement is Cadillac's obvious success in this 77-acre plant. One work team member, a rather shy Shirley Green, strengthened that interpretation: "Before [the worker-involvement program], engineers and management didn't talk to us, and they certainly didn't want to hear any ideas we had about our work. Now we have an opportunity to contribute our ideas and produce better cars."

◎ **One example of that change is the assembly-line effectiveness center (ALEC).** Roy Roberts, Cadillac's manufacturing manager, explains how ALEC represents a profound shift in Cadillac's thinking and in the way it does business: "We wanted to get the voice of the assembler into the product design process—way before the new product was introduced. So [in 1989] we carved out a 20,000-square-foot section of the plant, surrounded it with a privacy fence, and gave regular assembly workers an early chance to evaluate and make suggestions. When the assemblers first looked at the new Seville and Eldorado . . . almost 300 modifications were made. These were focused on quality, productivity, and cost advantages for the product and the people."

◎ **Cadillac also made major improvements in new designs with its design-for-manufacturability (DFM) program.** The bumpers of the Seville and Eldorado are a good example. Nuts and bolts were replaced with snap-in fasteners. Where the 1990 Seville required 217 fasteners in the bumper system, the 1992 model required only 34. Engineers calculate that some 19 minutes were eliminated in assembly time, not only by reducing parts—and they've been reduced by nearly 50 percent—but also "by examining every movement that is used in bringing material to the line and in the movements the operator makes in the assembly process."

Other significant process improvements include statistical analysis of fits—door, hood, and deck lid—using data from a vision system and the use of computer simulation to predict the effect of model changes on the assembly line.

In 1987, a 12-member simultaneous-engineering team (engineers, assembly technicians, and supervisors) began a continuous-improvement process to reduce electrical-system problems. With input from 170 electrical-system assemblers the number of defects was reduced 90 percent.

The plant's winning performance comes from its team approach, and Tibbitts expects even better results as the plant moves up the learning curve. But the broader goal for this plant is to lead a *paradigm shift* from mass production to lean manufacturing, a new, more flexible way to run a manufacturing plant. The term, coined by the Massachusetts Institute of Technology's International Motor Vehicle Program, combines the best-of-craft-production with the best-of-mass-production techniques in a low-cost, highly flexible way. The implications, says the International Motor Vehicle Program: "Half the business effort in the factory, half the manufacturing space, half the investment in tools, and half the engineering hours to develop a new product in half the time."

—JOHN TERESKO

Cadillac Motor Car Div. of GMC
Contact: Ben Ippotito, Assistant Director
2500 E. General Motors Blvd., Detroit, MI 48211
Telephone: (810) 575-1928
Employees: 3,800
Major products: Automobiles

GILBARCO, INCORPORATED

Greensboro, North Carolina

Don't limit improvement activities or TQM to just the plant environment; establish broad policies and practices to encourage excellence in all business activities.

—Jerry Smith, Manager, Manufacturing Strategy

Quality
Agile Manufacturing
Employee Involvement

The 1994 Best Plant award-winning Gilbarco, Incorporated, creates and manufactures gasoline dispensers for the Information Age, as well as the control systems and software programs that station owners use to manage them. The company, which employs 1,400 people in its Greensboro plant, has a long history stretching back to 1865 and the era of gas lighting. More than 100 years later, in the early 1980s, Gilbarco introduced the revolutionary multiproduct dispenser, doubling the company's market share from less than 25 percent to 50 percent and driving it to marketplace leadership.

As so often happens, success brought with it the seed of destruction—overconfidence. "We were sitting here fat, dumb, and happy," ruefully laughs president Mark Snowberger. The competition quickly caught up with Gilbarco, and by 1985, market share had slipped back to 32 percent. Compounding matters, long-time owner Exxon Corporation cut the company out of its corporate navy, leaving it to brave rough seas alone until it found a new berth with the United Kingdom's General Electric Company PLC in 1987.

The turnaround started in early 1986 when, after an extensive analysis of the company's competitive position, Gilbarco rolled out a new manufacturing strategy called CRISP (Continuous Rapid Improvement in the System of Production). CRISP centered on the plant floor and incorporated total quality control, focused factories, just in time, and total preventive maintenance.

◎ **Today, CRISP has evolved and spread into every corner of Gilbarco's 500,000-square-foot headquarters—from new-product design to accounts receivable—as EDGE (Everyone Dedicated to Gilbarco Excellence), which calls for the eventual achievement of six-sigma quality in every function.** Thus far, the results of the program are impressive. The use of cross-functional teams, concurrent design, design for manufacturing and assembly, and end-of-line configuration have reduced the product development cycle by 70 percent and increased reliability by 65 percent, as measured on the company's newest product line, The Advantage.

At the same time, a reorganization into focused factories of self-directed work teams and the successful application of a virtual litany of world-class techniques raised productivity per employee by 69.2 percent. The company's achievement of 0.07 percent scrap as a percent of sales led Best Plants judges to wonder if a decimal point had been misplaced on the application form. The company's market share has rebounded to 47 percent.

◎ **Despite the continuing growth of the company, manufacturing space requirements are significantly lower today than in 1985.** More than 150,000 square feet of space has been freed up during the company's improvement journey, allowing the sale of a 64,000-square-foot office building that once housed engineering and the consolidation of a 60,000-square-foot warehouse, 25,000 square foot of leased offices, and a small training facility. The benefits have been twofold: reduced costs and much-enhanced communication between manufacturing and engineering. Also, purchasing, planning, manufacturing engineering, and quality functions have all been moved to the factory floor.

Management at Gilbarco is quick to identify people as the company's real competitive advantage. All of the plant's 750 production employees and 75 office employees (mainly in the drafting and financial areas) are working in permanent self-directed work teams. The ratio of employees to supervisors has improved from 18:1 in 1989 to 47:1 today, and half of the company's organizational layers have been eliminated. At the same time, union grievances (manufacturing employees are represented by the Teamsters) have dropped 88 percent since 1986.

The list of accomplishments of manufacturing teams is long—more than 300 process-improvement initiatives have been launched since 1990. Teams have reduced setup times on the 100-ton press from 45 minutes to 5 minutes, reduced costs in the graphics shop by 60 percent, and cut cycle time by 55 percent, saving $1.8 million annually. Emergency orders can now be built and delivered for shipment in two hours versus 7 to 10 days previously. ◎ **Their work is supported by a corporate commitment to education that includes, among other programs, 57 hours of initial training, a computerized learning center offering 25 self-directed educational courses for employees and their families, and a goal of 2 to 3 percent of employee time devoted to training.**

Lest manufacturing garner all the acclaim, Gilbarco's administrative departments are writing some success stories themselves. ◎ **Applying TQM improve-**

ment techniques and teamwork to traditional white-collar jobs has fueled many of the company's most recent gains. Process improvements generated by self-directed teams in drafting, for example, created an express lane for minor engineering changes that cut overall drafting cycle time by 67 percent while doubling volume. Similar stories are told in the controller's department, where work-flow redesign enabled a 10 percent staff reduction (attained through attrition) and an almost nonexistent billing error rate. In customer service, a telesupport center operates 24 hours per day, year round, fielding 23,000 phone calls each month.

To what end is all this work being applied? Nothing less than marketplace leadership. "We used to think of ourselves as a pump company, but we are really a marketer in the forecourt of a gas station," declares Mark Snowberger. "We are going to be the technological leader in this industry. We lost it once, but we got it back, and we are going to keep it."

—THEODORE KINNI

Gilbarco, Inc.
Contact: Jerry Smith, Manager, Manufacturing Strategy
P.O. Box 22087, Greensboro, NC 27420
Telephone: (910) 547-5407
Employees: 1,500
Major products: Fuel dispensing equipment

HEWLETT-PACKARD COMPANY

Roseville, California

QUALITY
NEW PRODUCT DEVELOPMENT

One thing we've realized is that if you want to get to be very responsive in getting products to market—and also get your cost structure low—your quality needs to start approaching perfection.

—WADE CLOWES,
NCMO OPERATIONS MANAGER,

Hewlett-Packard Company (HP) has long been regarded as a high-quality manufacturer of computers and computer peripheral equipment. But in the early 1980s it learned a valuable lesson—that there was still room for improvement.

HP's "teacher" was its Japanese subsidiary, Yokogawas-Hewlett-Packard Limited, which applied total quality control methods to stage a dramatic turnaround that culminated in the subsidiary's winning the Deming Prize for quality in 1982.

The following year, CEO John Young called for a companywide total quality control program and set a "stretch" objective of reducing the failure rate of HP products in the field by a factor of 10 in the decade of the 1980s.

By 1990, the 10× goal had been "largely met," says Joseph Mixsell, a computer manufacturing engineering manager at the firm's 500-acre Roseville manufacturing complex, where the dust was still settling following a major consolidation that involved the relocation of manufacturing operations from Cupertino, California.

Among the units based at the 1990 Best Plants award-winning Roseville facility are: the Networked Computer Manufacturing Operation (NCMO)—which builds HP's line of multiuser computer systems, including the HP 3000 and HP 9000 series, as well as networking products such as I/O devices; and the Roseville Personal Computer Operations (RPCO), which makes HP's high-volume line of PCs and display terminals.

By 1990, both the 3000 and 9000 lines met the 10× goal, Mixsell notes. One result is that the reliability of HP's bigger computer systems has improved dramatically. For example, the mean time between failures on HP 3000s had become 25,000 hours—about eight times longer than on older versions.

Impressive gains had also been made on the personal computer side. In 1990, for example, terminals produced by the RPCO unit were 30 times more reliable than they were at the beginning of the 1980s, notes Neal Wagner, quality manager. And during that period the selling price dropped from $7,000 to $500.

"Our customers want PCs that don't fail," says Steve Helland, RPCO operations manager. Helping to enhance quality in the RPCO facility is a state-of-the-art surface-mount technology line in the printed circuit assembly area. A computer-integrated manufacturing system guides operators who assemble and test system processing units, the heart of the personal computers.

◎ **In addition to using problem-solving teams to achieve continuous improvements in quality, cost, and manufacturing cycle time, the Roseville site has adopted a hoshin strategy.** (*Hoshin* is the Japanese term for a breakthrough advance.)

"Basically, the philosophy is that you can do only so many breakthrough kinds of things at one time," explains Helland. So a structured planning technique "focuses the whole organization" on a limited number of breakthrough targets.

In the NCMO unit, one hoshin effort dealt with cost-of-complexity issues. Among the objectives: (1) reduce manufacturing costs by considering product design alternatives, and (2) ensure that the selling price of products reflects the true costs (including quality-related costs).

◎ **The cost-of-complexity program builds on a pioneering effort in activity-based costing (ABC) launched in 1985 within NCMO.** The initial focus was to identify the "true cost drivers" that impact various cost centers.

Among other things, the analysis found that the choice of components for a printed circuit assembly can influence production costs. For example, the cost to insert a part on an automatic insertion machine might be 15¢, while a "backloaded" part—added manually after the wave-soldering operation—might cost $1.50. "If you can change a design to make a part autoinsertable and avoid the backload process, the process becomes less costly," notes Chris Barmeir, NCMO controller.

"ABC accounting, directly allocating costs to a product, has a lot of implications in terms of better pricing decisions and make-or-buy decisions," Barmeir adds. "But our focus has been on giving the right kind of messages back to our lab designers."

Within NCMO, quality has been enhanced by a homegrown brand of employee empowerment and the use of continuous-improvement (CI) teams. Whenever a problem occurs, operators are authorized to stop the process and put a particular part or assembly on hold by entering a brief description of the problem into a terminal, which is connected via a broadband local area network to monitors throughout the plant. The screens show which processes have been halted—and why. This alerts engineering or procurement staffers who can help to solve the problem.

CI teams can be formed on a temporary basis by any employee in order to tackle a problem or explore an opportunity for improvement. One CI team, including a process engineer, operators, and a supervisor, came to grips with a wave-soldering machine that was creating an excessive number of solder bridges, which cause short circuits on a PC board. "They collected data for two or three months, and tortured the data until it confessed," recalls Mike Mickey, NCMO manufacturing manager. "When it confessed, it turned out that the flux vendor had changed recipes. So we worked with the vendor to go back to his original recipe and the problem disappeared."

—JOHN H. SHERIDAN

Hewlett-Packard Company

Contact: Public Affairs Department

8000 Foothills Blvd., Roseville, CA 95678

Telephone: (916) 785-8000

Employees: 3,300

Major products: Computers, display terminals

HONEYWELL, INCORPORATED

Phoenix, Arizona

AGILE MANUFACTURING
EMPLOYEE INVOLVEMENT

. . . management support was a key factor. It is important that all levels of management demonstrate an ongoing commitment to become world class.

—GALE KRISTOF, MANAGER, WORLDWIDE MANUFACTURING PROGRAMS

Honeywell's Industrial Automation & Control (IAC) business unit designs, manufactures, and configures the sophisticated systems that enable its customers, including refineries, chemical plants, and paper mills around the globe, to achieve world-class process control capability. Since the early 1980s, IAC's manufacturing facility in Phoenix has been learning to harness the potential of teams in a focused factory environment. In the mid-1980s, it began to develop strong "Supplier Alliance" partnerships and to cultivate an outside-in perspective on customer satisfaction.

Such an organization might be expected to demonstrate a solid grasp and a full deployment of world-class manufacturing principles. Yet, despite its prior technical and business success, in late 1989 the management team began to harbor some nagging doubts about its performance. "We posed the question: Has there been a paradigm shift around quality?" recalls Gayle Pincus, who was then vice president for manufacturing. "That question caused us to open our eyes and see what was happening in the world."

For the next five months, a dozen of this 1993 Best Plants award-winning facility's top manufacturing staffers turned their attention outward, visiting Japan, exploring what other companies were doing, talking to consultants, and even doing book reports. ◎ **Each of the 12 managers read the latest tomes on subjects such as just-in-time and Japanese manufacturing systems, and they shared what they had learned.** The exercise gave birth to a three-year world-class manufacturing (WCM) program that established ambitious goals in three primary areas: defect reduction, short-cycle production, and materials management. The goals included slashing defects tenfold and cycle times by a factor of five by year-end 1992.

To support the effort, a WCM Program Office was created to provide resources and to take a systemwide view. But ultimate success, Pincus and her colleagues agreed, would hinge on recruiting as many managers as possible to become change agents. And, clearly, the WCM campaign also needed the energetic support of the 700-plus plant workforce. ◎ **So, on April 4, 1990, Pincus shut down the plant and took everyone to an off-site meeting at another Honeywell location.** "We spent the six hours trying to articulate the need for change and then explaining what the change would likely be, in broad terms," she recalls.

Three years later, defect rates had been sliced by 70 percent, the cycle time on parts was down 72 percent, inventory investment had been reduced 46 percent, and customer lead times for the Local Control Network (LCN) and Process Manager (PM) products had been slashed by 72 and 78 percent, respectively.

These achievements didn't occur without struggles. One had to do with "managing the white spaces," the gaps between different links in the internal supply chain. In 1991, the push to reduce cycle time was bogging down. So Randy Harris, director of manufacturing for the IAC systems business, accepted a special assignment to research the problem. He concluded that the various teams in the value chain for each product line "had a tendency to suboptimize the total supply chain because they were primarily focusing on their own areas."

Producing a system component such as the LCN involves three key stages: building printed wiring assemblies; assembling them into modules; and "racking and stacking" modules into cabinets. Each stage is handled by a different team. To get the teams to think in unison, Harris took the three team managers aside and told them that they had to function as a team. "I said, 'The three of you are responsible for the total product line. And at the end of the year, you are going to be measured on how you perform as a product line.' "

For a decade, Honeywell IAC has been learning the secrets of teaming, often through trial and error. ◎ **One thing it learned is that teams need to have control over all of the things that impact their performance.** "The first team we created in 1981 failed because we didn't give them all the tools," says Gale Kristof, manager of worldwide manufacturing programs. "They had 80 percent of the ingredients, but we left out two critical ones, scheduling and inventory. You have to schedule to be able to control your destiny."

The team culture is also bolstered, he notes, by the 1987 decision to convert to an all-salaried workforce. In addition, the creation of a Productivity Placement Panel has removed psychological barriers to continuous improvement.

The panel serves as a matchmaker to find positions for people whose jobs are eliminated by advances in productivity.

IAC's strong Supplier Alliance program has been developing "custom supply solutions" to help synchronize material flow from suppliers with manufacturing demand. In 1992, a team known as the "Paradigm Busters" created a matrix to classify purchased items according to the predictability of need as well as the dollar amounts involved. Using that matrix, the team then developed a series of the supply alternatives, such as shared scheduling, use of "resident" suppliers, kanban, EDI (electronic data interchange), and supplier stocking suitable for each commodity type. The Phoenix plant now has two on-site resident suppliers, one of whom monitors the usage and inventory of items stocked by his company and writes his own purchase orders for replenishment.

Ed Szkaradnik, who succeeded Pincus as vice president of manufacturing, says, "Materials represent about 70 percent of our cost structure, and I think we're at a point where we need to understand materials much more intimately than we do today to take the next steps in cycle time, the next leapfrog."

—JOHN H. SHERIDAN

Honeywell, Inc., Industrial Automation & Control

Contact: Gale Kristoff, Manager, Worldwide Manufacturing Programs

16404 N. Black Canyon Hwy., Phoenix, AZ 85023

Telephone: (602) 313-3200

Employees: 1,350

Major products: Automation and control systems

IBM CORPORATION

Rochester, Minnesota

The ultimate defect is to have produced something the market doesn't want or need.

—LARRY OSTERWISE, GENERAL MANAGER

CUSTOMER FOCUS
TECHNOLOGY
EMPLOYEE INVOLVEMENT

It is not difficult to understand why IBM Corporation's sprawling development and manufacturing complex in Rochester was chosen as the computer firm's most promising entry in the 1989 and 1990 Malcolm Baldrige National Quality Award competition.

At IBM Rochester, a multipronged quality strategy is evident, ranging from the use of sophisticated design tools to statistical process control (SPC) methods enhanced with artificial intelligence. And, the quality push includes an extensive effort to capture the "voice of the customer," to ensure that products are designed, packaged, and serviced in a manner that will keep customers happy.

"Our goal is to have 100 percent customer satisfaction," says Larry Osterwise, general manager at Rochester and also a director of the firm's Application Business Systems (ABS) unit which makes small and intermediate general-purpose computer systems.

Indications are that the 1990 Best Plants award winner has hit the mark. Says one manufacturing consultant: "They've done a good job of linking people together to serve customers." The largest IBM facility in the world under one roof, the Rochester complex occupies 3.6 million square feet of space on a 586-acre site. It is the principal development and manufacturing site for both IBM's Storage Systems Products Division and its ABS line of business. Chief products include the hot-selling AS/400 computers and hard disk storage devices known as DASDs.

One measure of Rochester's success in quality efforts is that from 1984 through 1990, write-offs for scrap and excess inventory were reduced 55 percent. In addition, when the AS/400 line was introduced in 1988, IBM extended its warranty period from three months (on the predecessor Systems 36 and 38) to a full year. Mean time between failures has more than doubled since 1983, and engineering changes to designs have been slashed 45 percent.

The reduction in engineering changes "is a pretty good indication that you are driving your defects back early into the design process and removing them there," says Roy Bauer, manager of engineering and operations. Quality, Bauer points out, starts with design quality. ◎ **To get customer input early on, IBM has created Customer Advisory Councils—groups of about 20 customers who meet several times a year to let the folks in Rochester know what they think about the product line.** "We'll disclose what our plans are for two or three years in advance," Osterwise notes, "and ask them: 'What do you think of that? What do you want that is different from that?'

"We've gone from early manufacturing involvement to product development," he adds, "to early and continuous customer involvement. We want to ensure that we not only produce a defect-free product, but also a product that they want."

Among the methods IBM Rochester uses to ensure top quality are sophisticated design tools that verify conceptual product designs and then automatically translate them into the detailed descriptions needed for the production of computer chips. This eliminates errors that may occur in manual translation. In addition, a design tool known as EVE (for early verification engine) enables a high-level simulation of an entire system. To ensure an error-free manufacturing process, design information is electronically translated into instructions for such equipment as the automatic-insertion machines that build printed circuit boards.

Because the AS/400 computers were designed in a variety of configurations, ranging in price from $18,250 (including software) to about $1 million, the plant's CIM system helps to ensure that a customized system is built properly. "A customer order from a branch office is transmitted to manufacturing and is used to electronically control the build sequence," says Bauer. Bar codes on the computer case and on inserted parts are wanded at each operation. Since the bar codes contain the process routing, the CIM system will notify operators if the proper build sequence is not followed.

◎ **On-line SPC technology is used in critical operations.** Monitors indicate at a glance the status of process parameters and also provide a snapshot of the process conditions over the last 12 hours. Enhanced with artificial intelligence, the SPC system analyzes the data being tracked and notifies an operator if a process is beginning to drift out of control.

IBM Rochester has several programs to help ensure that its employees are in tune with customer concerns and requirements. ◎ **For example, manufactur-**

ing personnel are given temporary assignments of three months or so in which they make customer partnership calls—telephoning customers who have made recent purchases to inquire how they like their new computer systems.

And, in an Advocates program, employees from various functions, including manufacturing, spend six weeks in marketing branch offices around the world. "They are given about two weeks of training on interacting with customers," Osterwise notes. "The idea is that they will be advocates for our products, but they also learn about the true uses of our products in the customer's establishment."

IBM Rochester has also implemented its version of JIT production, called *continuous-flow manufacturing*. Among the results: a reduction in the assembly and test cycle from three weeks for the System 38 to about two days for an AS/400.

Lead-time reductions are aided by good SPC methods, Osterwise points out. "Simply stated," he says, "the fewer defects you have, the faster the process flows. And the faster the process flow, the fewer defects you have. The two are tightly coupled and they are both based on process analysis."

Moreover, he adds, the better things get, the easier it is to get even better. "It may be a little counterintuitive, but the more you get into defect reduction and cycle-time reduction, the easier and faster it seems to become—because you get people not accepting the status quo and [who focus] on improvement. As you focus on increasing participation and empowerment, you bring the full power of your workforce to bear."

—JOHN H. SHERIDAN

IBM Corporation
Contact: Steve Hoisington, Quality Manager
3605 Highway 52 North, Rochester, MN 55901
Telephone: (507) 253-9000
Employees: 4,800
Major products: Computers

JOHN CRANE BELFAB

Daytona Beach, Florida

To meet or exceed customer expectations is just the minimum price of admission. Act like your customer . . . and respond instantly.

—DOUGLAS FOCKLER, PRESIDENT

AGILE MANUFACTURING
QUALITY
EMPLOYEE INVOLVEMENT

At 1993 Best Plant winner John Crane Belfab, employees like to reminisce about the January day management "rolled the holy hand grenade" into the infrastructure; President Doug Fockler recalls "nuking" functional departments; and "getting blasted out of the box" is a positive event every employee is encouraged to experience. Lest this leave the impression that the company resembles one of the smoking wrecks occasionally dragged from the track of nearby Daytona International Speedway, consider this: Revenues (before acquisitions) more than doubled between 1987 and 1993, domestic market share rose from 12 percent in 1985 to 22 percent in 1992, cycle times are down 75 percent, and annual inventory turns have increased from 1.7 up to 30 in some product lines.

The company, which produces edge-welded bellows and bellows seals, generates $20 million in revenues from four semiautonomous, product-based business units housed within its 36,000-square-foot headquarters. It is a cellular structure designed to turn on a dime in response to customer needs. ◎ **Each product group encompasses the entire business process—from product development to shipping—and is staffed by a self-directed workforce.** A simple, unstaffed shipping station that every colleague is trained to use is located in the center of the building and serves as the common end point of each product group.

Not even the walls are fixed. Instead, they were specifically constructed to "float" for easy and inexpensive renovation as product units expand and decline in response to the marketplace. "I don't know if I'd go that far again," reflects Fockler. "So far, our market hasn't changed that fast." But, for the most part,

Fockler delights in the efficiency of the layout. "It's 65 feet from one end of the business to the other," he says, pointing out the Seals product group.

As might be expected, the phrase "close to the customer" takes on a higher meaning at Belfab. ◎ **An internal "no-use-as-is" policy prohibits the product units from asking customers to accept even minor deviations in quality.** Internal benchmarks, including zero defects, 100 percent on-time delivery, and a commitment to "always be twice as good as the competition" set lofty standards. And an emphasis on fulfilling customer expectations in addition to their needs "raises the price of admission," according to Fockler.

Two of the company's product groups are dedicated specifically to single customers. Customer logos and colors are used in the interior design of the workplace and product teams are encouraged to visit their customer/partners. "One new employee, Jessie James, thought we *were* our customer," chuckles product unit manager John Kidder. "He was answering the phone using their [company] name and causing all kinds of confusion. They couldn't find a Jesse James on their payroll." Blurring the lines between buyer and seller has paid off in a big way. Since 1990, sales volume from two of Belfab's largest customers has increased 188 and 1,280 percent, respectively. In 1991, the company was named top supplier of the year by Applied Materials and, in 1992, earned elite supplier certifications from Pratt & Whitney, Rockwell, and Allied Signal Engine Division.

It was no near-fatal business catastrophe that propelled Belfab down the track to fundamental change. Instead, the story begins with the simple desire to reduce inventory. "About one-quarter (8,000 square feet) of this building was devoted to inventory," says Doug Fockler, "and we knew that represented a huge cost." The quest to reap the savings of inventory reduction led the company to JIT, and JIT has, in turn, led to a virtual litany of best-plant practices.

Today, business units that generate almost half the company's revenues occupy former stockrooms. Inventory levels are down 51 percent from 1987 and, in most cases, are stored on a few shelves within production areas. Only the Seals product group still maintains a inventory of the 6,000 different replacement parts that are almost always needed on an emergency basis by its customers. Fockler, who is passionate about the elimination of inventory, as well as trashcans and file cabinets ("they're places to hide waste"), vows to do away with that also.

In addition to restructuring the business's processes, the traditional boundaries between management and worker had to be completely redrawn. For Doug Fockler, who thinks of the Belfab workforce as "who really pays my salary," that meant eliminating privilege and class distinction in all forms. Gradually, five staff levels, including 11 hourly pay grades and 35 job descriptions, were reduced. ◎ **Only two levels of management remain, and today's workforce is all salaried.** "We've always had good relations with our union [Sheet Metal Workers International, Local #15] but there was still that salary/hourly wall," says Fockler. "Now, everyone gets paid when they are sick and everyone gets the same benefits package."

—Theodore B. Kinni

John Crane Belfab, a member of the TI Group

Contact: Douglas Fockler, President

P.O. Box 9370, Daytona Beach, FL 32120

Telephone: (904) 253-0628

Employees: 237

Major products: Welded bellows and bellows seals

JOHNSON & JOHNSON MEDICAL, INCORPORATED

El Paso, Texas

You've got to trust, reward, and recognize your employees. You can't restrain them; you can't put handcuffs on them.

—MICHAEL LEWIS,
DIRECTOR, MEXICO OPERATIONS

EMPLOYEE INVOLVEMENT
CUSTOMER FOCUS
QUALITY
COMMUNITY INVOLVEMENT

In April 1993, when Mexican farmers protesting low crop prices blocked the Zaragosa Bridge between Juarez and El Paso, the traffic and customs departments of Johnson & Johnson Medical Incorporated's (JJMI) border operations shifted into quick response gear. The JJMI Artcraft plant in El Paso, which transports thousands of cases of finished goods twice a day from Mexico into the United States, was awaiting finished surgical products produced by three sister *maquiladora* plants in Juarez. The bridge closure threatened to disrupt not only the just-in-time manufacturing operation, but also its 99 percent–plus on-time customer delivery rate.

To prevent that, the benchmark-worthy traffic and customs team (U.S. Customs has made JJMI's traffic system the envy of the twin-plant industry and rewarded Artcraft with a special security agreement, reducing inspection by 90 percent) of 24 associates rerouted all U.S.-bound trucks to the still-under-construction Santa Teresa border crossing in nearby New Mexico. And when the Santa Teresa crossing became clogged, traffic and customs came up with another innovative, self-directed solution. The Bridge of Americas remained open as a border crossing, but because of extensive renovations, it could not handle the tonnage of fully loaded trailers. So the JJMI team decided to break up full trailers into smaller loads that would not exceed the bridge's weight limits. Once again, JJMI freight was the first to cross the border, ahead of hundreds of other *maquiladora* shipments into the United States, with minimal disruption in the flow of priority finished goods.

"These guys [customs and traffic associates] were flexible and fast, which is one of our four critical success factors," recalls Jack Morrison, Artcraft plant manager. "We beat the competition getting across the border. Our customers did not suffer, because we kept product in the pipeline. We respond to our customers, both internal and external, in a way that nobody else can."

The 1993 Best Plant award-winning Artcraft facility has achieved some remarkable milestones since it seriously began applying innovative manufacturing and teamwork techniques in 1989. Before 1989, border management was more punitive and authoritarian. Unionized production workers were not invited to the annual Christmas party and were "counseled" for being 30 seconds late to work. Now, U.S. and Mexican operations, 3,000 associates strong, are run on the basis of respect between salaried and hourly workers and a common bond built around a quartet of critical success factors: customer-driven quality, fast and flexible processes, lowest cost, and total associate involvement.

As a result of such cooperation, the Artcraft operation was the first domestic Johnson & Johnson facility, as well as the first in the medical products industry, to receive ISO 9000 certification. Its border operations have been benchmarked for its fabric-cutting techniques, customs department, and the uptime and efficiency of its cobalt 60 sterilizer, the largest in the United States. (In 1989, the sterilizer was ready for action about 60 percent of the time; by 1993 uptime was 97 percent. Each percentage point of increase represents a savings of $100,000.)

The Artcraft plant and its three Mexican assembly plants truly have flexible manufacturing systems; nothing is bolted down—except plastic protection guards on production equipment. Some workstations are formed with pie-shaped sewing tables; others are laid out in a straight line. Thousands of square feet of workstations can be reconfigured at the drop of a sombrero. In one modular assembly area, a simple telephone call from sales or marketing can activate the design and manufacture of a brand-new product within hours.

◎ **Flowcharts hang on walls everywhere—waste, costs, defect rates by shift and product line, and other chartable information are visible to every employee.** "Every associate on the border understands that our goal is to be the lowest-cost producer in our product line," stresses Artcraft's Morrison. "Quality—consumers the world over demand it. You won't get a second chance if you lose a customer. We flowchart and measure everything. We measure quality in parts per million and perhaps by [1993] year's end in parts per billion."

The Artcraft facility has translated customer expectations into product standards and specifications. ◎ **A cross-functional, value-analysis team, headed by the technical-services manager, brings in customers, such as nurses and doctors, to determine customer needs.** "In some cases, we have gone to hospitals to watch surgeries in progress to determine how our products are used, how they need to be folded, how they must be in the right sequence so that everything can remain sterile," says Tim Thomas, director of quality assurance for the border operations.

◎ **Artcraft completed some 35 benchmarking projects between 1990 and 1993.** Honda in Marysville, Ohio, was benchmarked for its just-in-time deliveries by suppliers, Federal Express for on-time delivery and customer satisfaction, and Milliken & Company for quality of design and manufacturing. "We always come away with two or three things that other people are doing better," says Thomas. "You are almost foolish not to go out and see how talented people in different environments and cultures are doing things."

◎ **On both sides of the border, the J&J facilities are socially responsible.** For example, the Artcraft facility sponsors at-risk students at Canutillo Middle School through a $6,000, two-year grant program. "We pay [potential dropouts] to tutor first- and second-grade students in the three Rs," says Morrison. "It gives them a sense of self-worth and builds up their self-esteem." For the last six years, associates in Mexico have rehabilitated a school each year in neighborhoods heavily populated by Juarez employees. (The Mexican government builds schools, but doesn't maintain them.) "We get $20,000 to $30,000 together and 100 to 150 associates, and they fix the plumbing, the lights, the electrical, paint, and even replace desks," says Morrison.

—BRIAN S. MOSKAL

Johnson & Johnson Medical, Inc.*

Contact: John Morrison, Director of Operations

7850 Paseo Del Norte Dr., El Paso, TX 79912

Telephone: (915) 833-2711

Employees: 242

Major products: Sterile medical apparel and devices

* The El Paso Artcraft facility is scheduled to close in December 1996. The operation is being merged with its sister plant in Mexico.

JOHNSON & JOHNSON MEDICAL, INCORPORATED

Sherman, Texas

EMPLOYEE INVOLVEMENT
NEW PRODUCT DEVELOPMENT
AGILE MANUFACTURING

The success of this business depends on every one of us. It's the most important thing we have because that's the way we're going to feed our families, that's the way we're going to educate our children, and that's the way we're going to be secure in our old age.

—TIM MCGRAW, PLANT MANAGER

The line of suspicion and distrust between union leadership and plant management at Johnson & Johnson Medical Incorporated (JJMI) was once as deeply etched as in any U.S. manufacturing plant, maybe even deeper. But 12 long years and a lot of hard work on both sides have so obliterated that line that today there is but one team at this 1991 Best Plant award-winning maker of products for medical and personal use. A shared vision and seldom-before-seen cooperation also keep this facility profitable for its management, union members, and Sherman itself, the small town 60 miles north of Dallas which is, in large part, supported by the plant.

◎ **The linkage between JJMI managers and union officials is so tight, in fact, that Billy Mayo, president of Local 513 of the United Textile Union, keeps a full-time office on the plant premises.** And he sits in on weekly meetings with Tim McGraw, the plant's manager, and his staff. Not only has this unique partnership engendered deep and lasting trust, it has also introduced into the plant a pride of ownership that is immediately palpable. Throughout the sprawling facility (the plant spreads out across more than 13 acres of gently rolling hills) are reminders of both internal success stories and external realities.

Take the Wall of Honor, for example. Here a visitor catches a glimpse of employees who have identified prices of nonconformance over $25,000. Elsewhere, the story is told on a bulletin board of one of 275 Quality Improvement Teams (representing a 70 percent involvement rate), complete with photos, Pareto diagrams, and other problem-solving tools it employs.

These demonstrations of internal accomplishment abound. But the attention is not entirely focused inward. ◎ **One hallway features what has come to be**

known as Competitors' Corner. Here, JJMI employees can examine and compare their products with those of direct competitors.

These signs of awareness of what it takes to be a world-class competitor are matched by both widespread commitment and technical expertise. Pulling two rolls of gauze bandages out of sequence, machine operator Ruth Ann Mihalakis shows a visitor how she conducts statistical process control quality checks four times throughout each shift. That is not unusual, notes Leroy J. Wiltz, production manager. In fact, 99 percent of all plant operators are IQS (Integrated Quality System) certified.

Technical know-how is not incidental to JJMI's mission of total employee involvement. The concept of "enabling" is critical, says Jim Burke, information-services manager. "The whole goal-setting process and the training process are critical to making empowerment work." Those sentiments are backed by an in-house Quality Education System (QES), which delivers 16 hours of quality improvement and problem-solving training to every JJMI employee. And that is only one part of a larger plantwide training program. Wiltz estimates that, including the QES courses, employees received an average of 40 hours of training in 1990. Included are the principles culled from quality masters such as W. Edwards Deming, Joseph Juran, and Philip Crosby.

◎ **JJMI also initiated a partnership with Grayson County College. The idea was to design a one-week program, which began in 1990, to certify qualified JJMI employees as industrial instructors.** Upon certification, these instructors are authorized to teach and then perform IQS certification throughout the plant for new operators. Within a year, some 35 instructors had passed the course.

Continuous improvement efforts by QIP (Quality Improvement Process) teams throughout the Sherman plant during 1990 generated $2.6 million worth of savings. A high-speed machine generating nonsterile sponges, for example, is now operated by two people. The cell produces roughly 3,100 pads per minute. In the past, six operators ran the machine.

Hundreds of small success stories so convinced plant management of the competitive value of teams that the plant's first completely self-directed work team was finally sanctioned in 1990. Comprised of 61 employees across three shifts, the team links mechanics, operators, and supervisors. Part of the team's magic involves the modular overhaul of a critical one-step production machine. In an effort to reduce downtime, the team's mechanics broke the large and complex machines into some 50 or 60 modules. With a complete spare of every module, mechanics can replace an entire nonworking unit with a working one.

The malfunctioning part can then be repaired without major production interference. The goal is to limit downtime for major repairs to less than four hours.

◎ **JJMI's cardless kanban system has eliminated the transactions required to support material planning and control systems.** Moreover, the pull system used by the fiber-finishing division, for example, has reduced work-in-process turns dramatically by producing only what is demanded by the plant's finishing floor. That was accomplished, in part, by eliminating incentives for operators to keep producing regardless of demand, notes Joan Simonetti, department manager. Advanced manufacturing has literally changed the face of this 29-year-old facility. Just-in-time and flexible manufacturing, inventory reduction, work-cell design, continuous flow, and self-inspection have all combined to reduce the need for warehousing space by 70,000 square feet.

"The success of this business is very important to these people [in Sherman] because this is their home," concludes plant manager Tim McGraw. Yet, even though the JJMI team has made tremendous strides thus far on the road to becoming a world-class manufacturer, McGraw feels compelled to add, "I personally don't feel we've accomplished but 20 percent of what we're capable of." So, stay tuned.

—Tracy Benson-Kirker

Johnson & Johnson Medical, Inc.

Contact: Joe Cernero, Quality Improvement Process Administrator

P.O. Box 9100, Sherman, TX 75091

Telephone: (903) 868-9000

Employees: 642

Major products: Wound-care sponges, tampons

KENNAMETAL, INCORPORATED

Solon, Ohio

Once [employees] were put in control of the whole process, they came up with their own ideas about how to make it better and reduce costs.

—PAUL CAHAN, DIRECTOR OF OPERATIONS

AGILE MANUFACTURING
QUALITY
EMPLOYEE INVOLVEMENT

Deftly pulling a freshly finished tool-handling part from a giant creep-feed grinding machine, operator Marianna Frusteri passes it to a visitor. "Aren't they beautiful?" she asks enthusiastically. Then, grinning at the surprised response of her visitor, she says without a trace of embarrassment, "This is my job, and I'm proud of these." It is this fierce pride that distinguishes the employees at Kennametal Incorporated's 1991 Best Plant award-winning Solon operation.

In modern plants, automation and other advanced manufacturing technologies have eliminated much of the backbreaking manual labor. But they have also stripped away some of the pride of craftsmanship—the greatest reward of blacksmiths and lathe operators from eras gone by. The challenge that Kennametal has successfully faced is finding ways to put the pride of craftsmanship back into the manufacturing process.

"We've had a big culture change here," explains unit manager Richard Monday. "[Employees] really want to make good products. They care about quality. It's just a matter of us working together to get closer to that goal." Workers openly agree. "Things have changed a lot around here—and all for the good," 17-year-veteran-machinist Bruce Swindell Jr. shouts over the whine of his five-axis milling machine.

Flexibility and good communication help Kennametal management ferret out ways to put pride back into the workplace. Plant employees are not afraid to try out new approaches or to go back to the old methods when new approaches do not work out. And they do not pretend to know all the answers.

Paul Cahan, director of operations, admits that in a few cases Kennametal actually obtained greater benefits than it had anticipated when it put some new practices into place. ◎ **"One of the key benefits of our moving toward the**

work-cell method of production that we didn't initially appreciate is that this was the first time these people saw the entire product made from start to finish in one area," he says. "Before, they would do one process and then pass it on. No one saw it all the way through. No one took responsibility for the finished product. But in the cells, employees take ownership. It's their product, and they can influence how it turns out."

Kennametal also uses self-management approaches to boost the sense of ownership in the production process. Employees are given a great deal of autonomy to develop the best way to organize and set up the machines they run. Instead of instructing employees on the one and only way they should do a certain job, management stresses the goals they should work toward and then lets the employees find the best way to bring about improvements in those areas.

◎ **Hanging above each manufacturing cell is a set of colorful graphs that display these goals as key performance indicators and specify how each work team has performed in the top-priority cost and quality measures.** For one workstation, these include on-time performance, tooling cost per hour, tooling disbursements, percentage of rejects (to take a positive view, Kennametal labels this the "right first time" indicator), and setup ratios.

"The key performance indicator charts show how each group is meeting or exceeding its targets," says Monday. "We post these because they communicate what goals are important. Then we work with employees to draw the connection to how they actually can impact performance. Keeping these objectives visible, and explaining what's behind the figures and how employees can influence the variables, is very important."

Cahan stresses that Kennametal's team approach encourages employees to work smarter and identify ways to improve in key performance areas. "Working hard is not enough alone," he says. "Even before we had this major culture change, we always worked hard—very hard. But our problems were a result of the way we did things, not because we didn't try hard." By creating a system that is designed to track and expedite rush orders automatically, the chaos has been eliminated and employees' efforts are directed at improving the system rather than fighting fires.

Open communication between management and nonmanagement employees helps Kennametal build continuously upon past improvements. Says Cahan, "You won't see a big division between management and workers here. You can talk about anything to anyone at any time." And a stroll around the plant proves the claim is true. Employees greet the plant manager casually and openly, providing follow-up answers to questions that Cahan and the employees had obviously discussed earlier.

Kennametal's success is the product of expert use of both its capital resources and its human resources, with the stronger emphasis on the human resources. "The fact that we have a new facility isn't what makes us a good plant," says Cahan. "You can have a beautiful plant and all the latest technologies and still have employee morale problems, quality problems, and on-time-delivery problems.

◎ **"The whole world-class thing comes down to how you treat and manage your employees," he says.** "You can't be successful unless you work as a team. If you systematically keep trying to resolve problems and treat your employees well—treat them fairly, train them, work as a team, and provide a clean and positive environment—then they'll do the best work they can."

This approach has led to significant performance gains. The company has improved its on-time delivery performance on standard products from 82 percent in the second quarter of 1989 to 99 percent two years later, and from 53 to 95 percent on specialty design products during the same period. Lead times have dropped from 42 days in the first half of 1989 to 27 days in 1991. The plant has achieved more than $1 million in cost reductions in 1990 and 1991.

—DAVID ALTANY

Kennametal, Inc.
Contact: Steve Anderson, Plant Manager
6865 Cochran Rd., Solon, OH 44139
Telephone: (216) 346-5510
Employees: 580
Major products: Tooling products

LOCKHEED MARTIN CORPORATION

Moorestown, New Jersey

Communications is a key enabler toward achieving a real cultural change. Communicate . . . communicate . . . communicate!

—BILL ADAMS, MANAGER QUALITY PROGRAMS/SYSTEMS

EMPLOYEE INVOLVEMENT
AGILE MANUFACTURING

More than 90 percent dependent upon sales of shipboard combat systems and radars to the Navy (mainly the vaunted Aegis weapons system), Lockheed Martin's huge Government Electronic Systems (GES) plant in Moorestown faced a bleak future in 1991. Not only were orders sharply dwindling as a result of cuts in the defense budget, but many of the systems components that the plant had been producing in-house were scheduled for outsourcing. More layoffs, which already had cut employment at the facility to 3,000 from as many as 5,000 in the halcyon days of the 1980s, appeared inevitable.

Then, the 1994 Best Plants award winner took its future into its own hands. Aided by union-management cooperation unique in its size and scope, GES achieved a remarkable culture change that has enabled it to protect jobs and become a tough, efficient, committed competitor. To be appreciated, the transformation needs to be seen and sensed. Yet a striking array of statistics attest to its reality. For starters, the plant, which already was a good performer, boosted productivity by 64 percent. It cut inventory by 80 percent, manufacturing costs by 25 percent, and achieved 100 percent on-time delivery.

Tour the plant floor and you hear employees enthusiastically describe their accomplishments. For example, in the plant work center that builds water-cooling equipment for the Aegis system, Tom Howarth, an engineer, points out how an employee team trimmed manufacturing cycle time from 253 to 126 days. The effort enabled the plant to bid on, and win, contracts to build outsourced components in-house. Instead of a 40 percent production cutback that had been scheduled, the work center has increased its production by 80 percent.

Meanwhile, in the work center that builds antennas for radars in the Aegis system, electronic tester Mike Monahan tells how a team redesigned the work area, cutting floor space needed for the same amount of production by 50 percent. Another team halved cycle time from 58 to 29 days. And in the work center that builds electronic modules—a component that had been targeted for outsourcing—teams reduced "touch labor" by 25 percent, cycle time by 20 percent, scrap by 57 percent, and defects by 54 percent. "Now, the work will remain in Moorestown," says Steve Ibbetson, a microcircuit specialist and International Union of Electronic, Electrical, and Salaried Machine and Furniture Workers (IUE) shop steward.

Obviously, change this profound did not just happen. The story goes back to 1989, when the plant's then-parent, General Electric Aerospace launched its outsourcing program. During the next two years, half of GES' production was gone, along with some 1,100 jobs, to other GE units and outside suppliers. When plans were announced in November 1991 for 400 more layoffs, concerned leaders of IUE Local 106 proposed that management and the union cooperate to save jobs.

Thus was born the GES/IUE Competitive Initiative. Although similar cooperative efforts are common elsewhere, the effort at Moorestown is unique because of the size of the plant, the size of the local (it is the IUE's largest bargaining unit), and the bitterness of previous labor relations. ◎ **Within weeks a joint delegation of union and management officials from the plant visited GE facilities at Lynn, Massachusetts, and Schenectady, New York, that had been through similar trauma.** They also attended a Best Plants conference in Chicago, at which winners of the 1991 competition shared their secrets of success.

To achieve the change, the plant undertook a variety of initiatives. Among them, it flattened the organization from nine layers to three; boosted flexibility by reducing union job classifications from 26 to 4; created 16 work centers backed up by intense cross training and statistical process control; implemented MRP II; launched a performance measurement system featuring eye-catching displays of current data on the shop floor; strengthened supplier relationships, cutting the number of suppliers from 8,000 to 714; and built a "Competitive Initiative Room" that serves as a nerve center of the effort.

But if there was any one thing that helped turn the corner, management and union officials agree, it was the emphasis on communication. The plant uses all the traditional communication tools. ◎ **More important, however, is the focus on face-to-face meetings, featuring a "48-hour flow-down" process in which key messages about the business are transmitted from management to each employee.** All managers are trained in listening skills.

The only factor that can slow the momentum, both management and the union fear, is one beyond their control—further cuts in the Navy's budget. Vice president of manufacturing (now company president) Joe Threston, who started his career at the plant 35 years ago as an entry-level engineer, foresees a "continued downward trend for three or four years" in the defense business. As a result, GES is positioning itself to cope with further U.S. defense cuts through a diversification strategy. As part of its recently developed vision for the year 2001, the plant is emphasizing international sales; it already has won military contracts with Japan and Germany. It's also working to expand its U.S. non-government, nondefense sales.

Still, the plant's prime customer is, and will remain, the Navy. And it is a pleased customer. "The plant's performance has been consistently outstanding and keeps getting better," praises Rear Admiral George Huchting, the Navy's Aegis program manager. He singles out GES for "solving problems up front and worrying about contracting details later."

—William H. Miller

Lockheed Martin Government Electronic Systems
Contact: Carmen Valentino, Manufacturing
P.O. Box 1027, Moorestown, NJ 08057
Telephone: (609) 722-6147
Employees: 3,000
Major products: Combat and radar systems

QUALITY
AGILE MANUFACTURING

LORD CORPORATION

Dayton, Ohio

[The employees] said we were too heavy-handed, too authoritative. Their response was almost: "Get the hell out of our way so we can do our jobs."

—DAVID LICHTINGER, PLANT MANAGER

At the end of lunch in a Chinese restaurant in Dayton, Mike Rogers cracks open his fortune cookie. "You tend to believe the incredible while doubting the improbable," it reads. Those are appropriate words for Rogers. As employee-relations manager at Lord Corporation's Dayton plant, he has firsthand experience with the incredible.

For example, Rogers has seen the Dayton plant return from the brink of demise in 1985 to become the company's most productive plant in 1991. He has seen manufacturing cycle time for one product (helicopter tension-control spools) reduced from 75 days to 7; a 50 percent return of components shipped to in-company customers reduced to zero; overall scrap and rework costs fall 85 percent; productivity increase 30 percent; and typical setup time fall to one hour from four to eight hours in 1985.

Most important, Rogers witnessed a workforce that felt hindered in the past become an all-salaried, fully self-directed, and highly motivated workforce. Workers in the "new" Dayton plant schedule their own work. They perform all jobs within their manufacturing cells. They hire their own team members. They select and order their own equipment. And they have been promised there will be no layoffs caused by fluctuations in workload or increases in productivity. Indeed, the improvements are astounding at this nonunion plant that makes vibration-isolation mountings for airplanes.

The situation could hardly have been worse at the plant in 1985, says Rogers. That was the pivotal year. "The only way it could have gotten worse was if we were sold or closed." It was frustrating, agrees Dave Lichtinger, plant manager. "I couldn't understand why a small facility shouldn't be the best place to work. Why wasn't this a dream job for me and everybody else here?" he asked himself.

With nothing to lose but their old way of managing, Rogers and Lichtinger decided to go to the experts—the workers—for help. ◎ **"We asked them, 'What would you change to make this place perfect, to be able to tell your friends that this is a good place to work?' " Lichtinger remembers.** What the two men heard from their employees surprised them. In no uncertain terms, the workers told them that although they were preaching empowerment to the workers, they were not behaving like the enlightened managers they thought they were. Rogers and Lichtinger took that response seriously and began putting together involvement teams. At first, the teams worked on problems at the plant, such as reducing scrap or setup time.

Then, in the summer of 1987, the first work-cell team was formed in order to make a pitch-control shaft product for helicopters. "We got a lot of conversation going . . . and got the team to really think things out," says Lichtinger. "When they came up with ideas or decided they wanted to do something, we let them do it. If they said they wanted to move equipment, we got riggers in and we moved equipment."

◎ **It was important that the employee team put the cell together themselves, he insists.** "They laid it out. They took a piece of paper and put the equipment where they wanted it. They measured the floor. They helped the engineer program the jobs. Everything." Almost immediately after the team began operating, it produced miracles, he says. With no supervisor looking over their shoulders, team members reduced scrap from $350,000 a year to virtually zero in 1991. They reduced work-in-process inventory from $3.5 million to $150,000. And, they slashed lead time from 160 days to 32 days.

Seven more teams followed, which were organized according to the work: tension-control spools, short-run miscellaneous products, heavy-precision repetitive machining, bonded products, assembly, quality support, and maintenance support.

Even though the plant and people were changing quickly, as dollars were salvaged spirits began lifting, and trust developed. "That was the big thing," says Rogers, "not only our trust in them, but a trust in us that took a lot of work." That trust was clear to Phil Hilderbrand, a team member in the short-run cell, when the two managers told the workers how much money was being wasted in scrap. "They'd say, 'We could use this money for better equipment and benefits," he says. "There used to be hoppers filled with scrap parts. And now there's only a dozen or so [parts] in a 55-gallon barrel."

When the improvements occurred, he notes, the plant did bring in the new technologies it needed—five Toyoda machining centers, now busily churning out products, for example. Researched and purchased by a team, these were important to the plant, says Lichtinger, because of their accuracy and automatic features that free the operators to do other tasks in the cell.

◎ **Of course, behind this transformation were "thousands and thousands of hours of training," he remembers.** He and Rogers read W. Edwards Deming's books and attended seminars. And every operator has received the technical training needed to be able to do all the work within a cell, instead of operating just one machine. For example, each operator is a certified quality inspector, which requires 200 hours of training. The old inspectors are now auditors and ensure that the plant conforms to customer, government, and internal specifications.

Lord's Dayton plant would not have made it to the 1991 Best Plants winner's circle without also using some of the best technologies as well as the best management practices. Real-time computerized statistical process control, computer-integrated manufacturing, manufacturing-resource planning, and electronic data interchange are all important elements of the plant's operations. But these are icing on the cake. The technologies are in the plant, reaping rewards, because every person at the plant has changed in order to be successful in the new world of industry. "It's motivated people who make [technology] work," insists Lichtinger. "Otherwise, you're wasting your money."

—THERESE R. WELTER

Lord Corporation, Aerospace Products Division

Contact: Mike Rogers, Employee-Relations Manager

4644 Wadsworth Rd., Dayton, OH 45414

Telephone: (513) 278-9431

Employees: 120

Major products: Aircraft vibration-isolation mountings

MARLOW INDUSTRIES, INCORPORATED

Dallas, Texas

AGILE MANUFACTURING
EMPLOYEE INVOLVEMENT
QUALITY

We no longer have a lot of non-value-added effort going into meetings and bureaucratic proceedings. We have freed up the phantom organization that you don't realize exists.

—RAYMOND MARLOW, CEO

Employees at 1993 Best Plant award-winning Marlow Industries Incorporated think back to how the Dallas firm used to make its thermoelectric coolers and components before it shifted decision making to employees in 1988, and they shake their heads in disbelief. "We had no standards, no specifications," says Peggy Holmes, supervisor in the thermoelectric material growing area (where ingots are formed). "We used to check the specifications and the rods' properties after they were made. The good stuff we would cut up and ship to customers. And we would just toss the bad. So the yields were poor, and we had to work a lot of overtime to fill customer orders."

Today, her department is a shining example of the progress made throughout the entire company as a result of empowerment, work teams, and the creation of minifactories. "There are far fewer variables now," says Holmes. "When we improved the process, we increased capacity and yields." (Yields in 1993 approached 95 percent, compared with yields barely above 60 percent in 1987.) "We now produce three times as much with the same number of people and the same floor space."

Overtime is a thing of the past, she adds, and "we have learned to recycle." In the old days, out-of-specification ingot rods were scrapped. Now they are brought to specification and fed back into the process. "We have eliminated landfill costs and lowered our costs at the same time." And the minifactory's mission has changed: "Now, we sell what we make," says Holmes, "and make what we sell."

The stories are similar throughout the Marlow plant. Employee teams, given responsibility over their operations and the mission of customer satisfaction, have dramatically changed the way the company makes its coolers—solid-state electronic devices that cool, heat, or stabilize temperatures of electronic components. The plant has won six major quality awards from customers, as well as the distinction of becoming the smallest company ever to win a Malcolm Baldrige National Quality Award (in 1991).

Give credit to the organizational changes introduced in 1988 that broke manufacturing down into three materials minifactories and five assembly minifactories, each with its own process engineer. Employees were given responsibility for quality and the minifactories were focused on customer and project needs, not functions, through market-segment teams. ◎ **"We call it kit-to-ship," says plant manager Chris Witzke. "Each minifactory has its own—not a corporate—overhead. Each team manages its own factory, improves its own process, performs its own maintenance, and recycles or disposes of its own materials."**

Employees, not management, control the process flow, the way components are made, and how products are assembled. They create their own statistical process control charts to track quality levels and have designed new equipment to increase efficiency. At any given time, more than half of the 170 employees are on teams solving problems that they, not management, have found.

As a result, by 1993, manufacturing costs were 37 percent lower than in 1990. Manufacturing cycle time was sliced from 15 days in 1988 to 10 days in 1992, and order-to-shipment lead times are down 37 percent—from 16 weeks to 10 weeks. Since 1987, productivity improvements have averaged 10.6 percent annually. Waste disposal costs are down 57 percent from 1988.

Overall, however, the transition was not entirely smooth. The early bottom-up attempts—including CEO and company founder Raymond Marlow's walk-around chats with workers and "Marlow Mottoes," such as "Quality before quantity, but with quantity not far behind"—didn't do the trick. And supervisors had not bought into the program, so change did not really take hold until Marlow began to educate managers and implement quality expert Philip Crosby's 14 quality steps.

Still, employee teams did not realize how willing management was to change. "You know there is a communications gap when the workers think something is a barrier and you don't," says Witzke. ◎ **"We had to tell every team that they could spend up to $300 without anyone else's signature.** That freed up the thinking. We hadn't realized that employees were spending two to three weeks trying to justify [to management] a $250 expenditure."

After that, he says, "you knew things were working well because you would hear employees ask: 'When are we going to do something about it?' instead of 'When are they going to fix it?' " Mable Campbell, a line leader in the assembly area, agrees. When Marlow began to use worker teams, "it was the first time that I felt that I could take the time I needed to perform the job," she says. "Before, there was always a push to ship product, and it felt like a sweatshop. We have found that by not rushing we get a better product out."

Employees also revamped customer service. Instead of having representatives take calls randomly, responsibility is broken down by market segments. Each rep has the authority to resolve problems up to a cost limit of $20,000. "We have cut the telephone calls in half," says Witzke, and the customer service staff can now "do more proactive things," such as calling customers to let them know when an order has been shipped.

Ironically, while quality was the original driver behind the changes at Marlow, quality as a separate program was eliminated in 1991. "Eight of our ten strategic objectives were all quality-related," says Marlow, "so we merged the two. **◎ We have no separate quality plan. Instead, we have one strategy, one mission. Our business goals embody quality goals. We want total business excellence."**

—MICHAEL A. VERESPEJ

Marlow Industries, Inc.

Contact: Pam Jennett, Human Resources Manager

10451 Vista Park Rd., Dallas, TX 75238

Telephone: (214) 340-4900

Employees: 300

Major products: Thermoelectric coolers and systems

MEMC ELECTRONIC MATERIALS

Saint Peters, Missouri

EMPLOYEE INVOLVEMENT
AGILE MANUFACTURING

By having [manufacturing] feedback quickly and getting this information back, we can identify where our weaknesses are, and how we can correct them.

—JONATHAN JANSKY, PLANT MANAGER

Silicon wafers look simple enough, but their production is a capital-, labor-, and technology-intensive process. A handful of wafer makers control the world market, with just one major supplier in the United States, 1995 Best Plants award winner MEMC Electronic Materials Incorporated. The company headquarter's plant in Saint Peters occupies 500,000 square foot and is expanding to 550,000. It caters to customers in the booming semiconductor industry that are forever pushing the technology envelope. "One of our customers has a sign in its cafeteria that says, 'Obsolete in 18 months,' " says plant manager Jonathan Jansky. "Their product cycle is that sharp, and that comes down to us immediately."

Graham Fisher, director of manufacturing technology, says, "If we're to respond to what our customer needs, we've got to understand very clearly what the customer wants within a two- or three-year time frame in terms of manufacturing and then a three-, five-, seven-year time frame for technology." Building that understanding is the plant's strategic-customer quality team which develops, through strategic customer interaction and annual surveys, a three-year outlook defining service and product parameters. A research-and-development planning council looks at the three- to seven-year time frame for technology development plans.

Underlying these plans is the plant's vision to be the best supplier of silicon wafers in the world through continuous improvement. ◎ **Employee teams have identified tools to realize that vision, the four Cs (customer satisfaction, corporate citizenship, cost control, and culture), and six plant systems (technology, people, structures, reward, renewal, and information) that must be continuously improved to achieve the vision.**

Quality improvement, training programs, and empowered work groups drove the Saint Peters plant and MEMC to profitability, according to Jansky. From 1989 to 1994 MEMC net sales have more than doubled and net income after taxes rose to $38 million. The Saint Peters facility feeds 45 percent of MEMC's total net revenue.

The plant's quality initiatives began in the early 1980s. "It wasn't a fad," Jansky says. "We put in programs that are still here today. They were part of the process, not the program of the month." They began with a statistical process control program, and in 1985 a quality improvement system was established to provide training. That same year the plant switched to just-in-time (JIT) manufacturing. "Immediately, our yields went up, our inventories went down, our cycle times just crashed from 50 days down to 3 days, and we had some huge successes in the process," says Jansky.

◎ **Because the plant runs 4-, 5- and 6-inch silicon wafers over the same equipment, quick changeover techniques, many automated, have also been established.** "It literally used to take us a shift to change things out," Jansky says. "Now we can do it in just 10 minutes." The variety of wafers is also enormous. MEMC customers require various and changing parameters for their wafers—thickness, flatness, chemical makeup, and data quality. In 1994 the plant produced 2,000 different wafer types, 90 percent of which did not exist three years ago.

The plant moved toward empowered work groups in 1989 and seeks employees who combine high technical skills with interactive skills: ◎ **"We have 56 hours of team-skills training that focuses on issues such as trust, consensus building, win-win, confrontation, effectively conducting a meeting," says Brad Eldredge, director of human resources.** The 100 percent team-based structure includes a supplier-improvement team, cross-functional work teams, corrective-action teams, a policy-advisory committee, a tactical-operations committee, and total-machine-performance teams. One cross-functional TPM team redesigned product flow to integrate in-line inspection within the process, saving $550,000 per year in material and 12 hours per cycle.

If empowerment and quality improvements brought the plant to the end of this century, information technology may well lead it into the next. As wafers move through manufacturing, each customer order is traced. Operators throughout the process collect detailed parametric information and update computer documents at terminals. The system enables operators to alert engineers downstream to deviations that may be occurring. As important, it provides to the customer, via a PC and modem, an information encyclopedia of wafers as they move through production, collecting information from 20 databases.

"We're getting validation of the fact that we've done some very good things over the years," reflects Jansky of the Best Plants award. In 1994, the plant won the Missouri Pollution Prevention Award, Missouri Governor's Energy Conservation Award, and the Missouri Quality Award.

—GEORGE TANINECZ

MEMC Electronic Materials, Inc.

Contact: Tim Kolp, Custom Quality Assurance Specialist

501 Pearl Dr., St. Peters, MO 63376

Telephone: (314) 279-5000

Employees: 2,000

Major products: Polished and epitaxial silicon wafers

METTLER-TOLEDO, INCORPORATED

Worthington, Ohio

[Work cells require] much training and coaching, but the payoff is enormous.

—ROBERT CLEGG, PLANT MANAGER

AGILE MANUFACTURING
QUALITY

Columbus is perhaps best known as the town where the Ohio State Buckeyes play a pretty classy brand of football. Just a few miles down the road, in Worthington, the folks at Mettler-Toledo can brag about some pretty classy teamwork themselves.

In the Worthington plant, which produces some 1,200 weight indicators and 40,000 printed circuit boards (PCBs) each month, the emphasis is on self-managed work teams and a pay-for-skills program that encourages cross training.

These concepts go hand in hand with the plant's advanced just-in-time and cellular manufacturing methods. "To get the required effect and benefit from work cells throughout a factory—not just the initial volunteers—flexibility through cross training and a willingness to learn new skills are mandatory," says Robert Clegg, who served as plant manager at Worthington from the inception of its manufacturing improvement program in 1985 until leaving early in 1990.

At Worthington the payoff has included dramatic reductions in the plant's throughput time. In its focused factory PCB area, the manufacturing cycle was reduced from two weeks to three days; for the weight-indicator devices, the cycle time was collapsed from two days to as little as 30 minutes. The plant also reports 67 percent reduction in work-in-process inventory, an 85 percent drop in defect rates in the PCB line, and on-time deliveries in the 99 percent range. There was a 24 percent increase in productivity in the first two years of its improvement program, based on value of shipments per employee.

The productivity increases have continued, reports John Lucas, who succeeded Clegg as plant manager in time to receive the facility's 1990 Best Plants award, "but we quit tracking it. We're holding manpower fixed, and our output is still climbing."

◎ **The plant, which schedules its production based on customer demand, uses a pull system and kanban signals to maintain its continuous-flow system.** On the weight-indicator side, the process begins with the entry of an order into the computer system, which signals a printer in a work cell to create a delivery label (including bar-coded information on the model type). "The work team will then build the unit, attach the delivery label, send it to the shipping dock, and it's gone," explains Lucas. Because the plant does not build anything without a customer, its finished-goods inventory has been reduced dramatically.

Lucas, who spearheaded a similar improvement program at the company's Systems Division plant in nearby Westerville, Ohio, before taking the Worthington post, believes the most impressive thing about the plant is its pay-for-skills program and self-managed work cells. As employees complete training and certification in a new skill, they earn pay increases. For the company, the benefit is a workforce with the kind of flexibility needed for a successful cell-based operation.

In the self-managed work environment, each cell has 8 to 12 members and functions as "a minibusiness" within the company. The members of each cell have control over their own work hours, skill training, quality control, and certain expenses.

◎ **In addition to learning new job skills, workers can also earn extra money by becoming cell "coordinators."** Each cell has five coordinators who tend to the business side of their operation. There is a coordinator for quality, training and scheduling, customer-supplier relations, cost performance, and "workplace organization," which includes safety, housekeeping, and maintenance.

By 1990, 80 percent of the Worthington workforce had completed training for at least one coordinator task. Since the number of qualified coordinators exceeds the demand, the positions are rotated, so, in a sense, everybody gets a shot at being a manager.

The self-directed team concept has been well accepted by the workers, says Lucas. "In the past, some people had limited horizons. But now they are able to grow, and they feel they have some control over their own destinies."

—John H. Sheridan

Mettler-Toledo, Inc.

Contact: Dave Platt, Plant Manager

1150 Dearborn Dr., Worthington, OH 43085

Telephone: (614) 438-4390

Employees: 179

Major products: Scales

MILWAUKEE ELECTRIC TOOL

Jackson, Mississippi

. . . the basic position we took with the workforce was very honest and upfront: "We want you to be business owners with us."

—Harry Peterson, Plant Manager

Agile Manufacturing
Employee Involvement
Technology

In October 1990, plant manager Harry Peterson stood by as Jerry D. McCormick, vice president of manufacturing, announced that Milwaukee Electric Tool's Jackson plant would implement just-in-time (JIT) production using a cellular organization and a workforce focused on continuous improvement, or kaizen. "It was a very courageous move, pretty gutsy, to say, 'We're successful, we're going to change radically to cellular manufacturing with all the attendant changes that go with it,' " recalls Peterson. "It was a courageous choice by the local folks, my boss [McCormick], and the president of the company [Richard Grove]."

By May 1991, this 1995 Best Plant winner was fully cellular and demand-flow implemented throughout. The productivity improvements were immediate, dramatic, and sustained, enough to allow the company to delay scheduled new plant expansion until January 1995. Jackson slashed order-to-shipment lead times from five months in 1990 to two days, and inventories fell 33 percent. Customer fill rates improved to 98 percent, up from 70 percent in the 1980s, and in-warranty activity fell 48 percent from 1990 to 1995.

◎ **The plant also changed to a process-hour accounting system from a standard-hour cost-accounting system.** "We threw out traditional measures and began to focus on broader measures," says Peterson. "In the old system, we'd encourage you to run parts, keep the machines running, keep the efficiency and utilization up. It's important, but today we focus on having the line run what the market needs."

That move pushed the business departments to quickly adapt, and an equal if not greater burden was placed on the workforce, who were asked to function as self-directed teams. The entire production workforce now participates in the teams, and the employee-to-supervisor ratio has risen from 18:1 to 50:1. This flattened the organization and drastically reduced the communications chain; no more than two levels exist between cell members and plant manager. Although a few employees initially left the plant, annual job turnover is now less than 1 percent.

"We go after 10 percent productivity improvement per year. That's part of our mission, vision, goals, and objectives," says Peterson. Actual improvement has been 65 percent in the last five years, while costs have dropped 15 percent. "You can get a good bit of that with a business change from batch to cellular, but there is a point in time where those returns are diminishing. You're still improving, but the opportunity for 1 percent per month gets tougher. So there have to be other areas, and those areas are problem solving that makes the problem go away forever; world-class technologies; and the right use of the technologies that are out in the world. These require a workforce that's really energized, because we literally expect our hourly team members to drive a lot of that, especially problem solving."

The plant develops its workforce by emphasizing education. It is committed to 40 hours of annual training per employee, including kaizen workshops. A new Milwaukee Electric Tool World-Class College will provide an additional 40 classroom hours over a 10-week period.

◎ **The plant's pay-for-knowledge (PFK) system encourages employees to achieve new job certifications.** Peterson cites two reasons for PFK: "We want highly motivated, highly flexible employees. And you have to pay for that. That gets your agility. A second issue, we wanted our employees to continue to grow their base of knowledge so that as we bring in world-class technology we don't [need to] bring in people to run it."

Under the PFK system employees are involved with determining their training schedule, pay rate, and daily work assignments. PFK job responsibilities include setup, operation, and troubleshooting of equipment; preventive maintenance; job rotation and recertification; and adherence to cycle, or *takt* time (in repetitive operations, the cycle time between completion of units), quality expectations, and safety requirements. PFK jobs must be performed once in an 8- or 10-day period, says Peterson, and some employees carry up to nine certifications. Employees mark their job activities on team boards in the cells and are encouraged to decertify themselves if they're not meeting requirements. Employees can also be decertified if they show a pattern of failing to meet PFK obligations, which other cell members can note during peer evaluations.

◎ **Employees log onto a PC on the plant floor and confidentially evaluate cell members.** The ratings are then tabulated by management, which also prints hard copies for a review process. The cell teams can also recommend corrective actions, but their suggestions come short of discipline. Peterson acknowledges that peer evaluations are a trying process for cell members, as were the many other changes they've encountered and those that are likely to come.

—GEORGE TANINECZ

Milwaukee Electric Tool Corp.
Contact: Harry Peterson, Plant Manager
4355 Milwaukee St., Jackson, MS 39209
Telephone: (601) 969-3033
Employees: 415
Major products: Handheld electric power tools

MOTOROLA, INCORPORATED

Boynton Beach, Florida

Manufacturing advances toward world class . . . invoke new levels of competency in suppliers while providing customers with unparalleled value in product and services.

—MARK PESTA, DIRECTOR OF QUALITY

AGILE MANUFACTURING
SUPPLIER RELATIONS
NEW PRODUCT DEVELOPMENT

Even before Motorola Incorporated won the Malcolm Baldrige National Quality Award in 1988, the firm's "Bandit" facility in Boynton Beach had attracted considerable attention. The totally automated production line, set up in an 8,000-square-foot corner of Motorola's pager manufacturing plant, represented a breakthrough application of CIM technology. Among its more notable achievements were dramatic quality improvements due in large part to the precision and repeatability of robotic assembly and its quick response capability.

The 1990 Best Plants award-winning Bandit facility, which produces Motorola's alphanumeric radio pagers in lot sizes of one, has the remarkable ability to begin producing customized pagers within 20 minutes after a salesperson enters a rush order (relayed by a computer at Motorola's headquarters in Schaumburg, Illinois, to the main computer in the Florida plant). The actual production of the BRAVO pagers, customized to the proper radio frequency, takes less than two hours, including manufacture of printed circuit boards, final assembly, testing, and packaging for shipment.

◎ **By integrating a surface-mount PC board line with back-end assembly on a U-shaped continuous-flow production line, the manufacturing cycle was reduced by 85 percent compared with previous methods.** In Bandit's "paperless" manufacturing environment, 30 Seiko robots, controlled by a bank of Hewlett-Packard A400 real-time computers, handle all assembly tasks. And most of the inspection, including laser-based machine vision to check the positioning of components, is automated.

From start to finish, the only touch labor is performed by a human inspector at the end of the line who checks the pagers for cosmetic defects. The 40 employees assigned to the Bandit line include cell managers who monitor the process, support staff, and workers who make sure the robotic stations are loaded with all the components they need.

The Bandit line (which took its name from the development team's willingness to "steal" good ideas wherever they found them) yielded big quality gains, including a 250 percent improvement in out-of-box quality and a 350 percent improvement in field reliability.

Among the factors contributing to the improvements were a simultaneous engineering effort in which the traditional BRAVO pager was redesigned for robotic assembly. ◎ **Also contributing is a rigorous supplier-selection process which reduced 300 candidates to just 22 best-in-class sole-source suppliers, chosen for their willingness to commit to extremely high quality levels.**

Automated assembly is enabled by flexible robots that can change their own "end effectors" (grippers) within five seconds of receiving a computerized instruction. The vision-aided placement of components "gives you repeatability that a human will never be able to match," notes Len DeBarros, director of manufacturing for Motorola's Paging Division.

The line boasts computer-controlled statistical process control (SPC) technology that provides real-time feedback on quality data. Real-Time Quality Management software from Automated Technology Associates was adapted to run on the Hewlett-Packard computers that collect data from the robots and other devices. The line controller that processes the SPC data can automatically shut down a unit if it detects a problem. Real-time SPC goes "beyond counting mistakes to preemptively trying to avoid them," explains Scott Shamlin, DeBarro's predecessor who spearheaded the start-up.

◎ **One Motorola objective in developing the Bandit technology was to accelerate its learning curve and then adapt the knowledge to other manufacturing operations.** That has happened. The migration of technology from Bandit to the older BRAVO manufacturing operations in the same building has been completed, reports DeBarros. Many of the back-end processes on the older production line have been automated (the plant now has a total of 150 assembly and test robots).

For example, on the Bandit line, bar codes on the pallets that carry the pagers through the production process give the robots the specifications for each pager. On the older line, workers involved in assembly steps wand bar codes on the pager housing; in turn, the host computer downloads customer-specific information to their terminals. "We have integrated all the pockets of excellence from Bandit—those with a real cost impact—into the BRAVO facility," DeBarros notes. "Bandit is the benchmark from which we're launching all of our other factories of the future." That includes such elements as automated testing, lot sizes of one, build-to-order manufacturing, and the paperless manufacturing concept.

The older BRAVO line can now match Bandit's ability to produce pagers in less than two hours. And both operations met Motorola's interim goal of a 10× quality improvement—that is, a 90 percent reduction in defects. In 1990, the Boynton Beach facility had already reached the five-sigma level (fewer than 200 defects per million). And some of its lines have achieved a zero-defect performance for several months in a row.

—JOHN H. SHERIDAN

Motorola, Inc.
Contact: Mark Pesta, Director of Quality
1500 Gateway Blvd., Boynton Beach, FL 33426
Telephone: (407) 739-2360
Employees: 3,200
Major products: Alphanumeric radio pagers

NEW UNITED MOTOR MANUFACTURING, INCORPORATED

Fremont, California

EMPLOYEE INVOLVEMENT
AGILE MANUFACTURING
QUALITY

Nobody knows the situation, the problems, better than the team members. Management's role is to help and support them.

—OSAMA KIMURA, PRESIDENT

Near the end of the 1.3-mile-long final assembly line in the New United Motor Manufacturing Incorporated (NUMMI) plant, at the point where overhead conveyors deliver seats to the line on a just-in-time (JIT) basis, you will meet "Christine." Christine installs seats in the freshly minted Toyota Corollas and GEO Prisms as they inch their way along the conveyor system. A tough job for a lady, you say? Well, Christine is no lady. She's a robot, a rather unusual automaton, who toils right alongside humans who do the same work.

But the line workers do not see Christine as a threat. In fact, the employees at NUMMI—they're called team members—assisted the plant's assembly-maintenance crew that built Christine from scrap parts.

One reason the workers do not resent the robot is that it promises to eliminate one of the most backbreaking jobs in this 1990 Best Plant award-winning facility. Another is that they are confident that no jobs will be lost, even if a battery of similar robots should appear. A provision in NUMMI's agreement with the United Auto Workers makes that clear: "The company agrees that it will not lay off employees unless compelled to do so by severe economic conditions that threaten the long-term financial viability of the company." ◎ **The contract goes so far as to pledge that alternative measures—including "the reduction of salaries of its officers and management"—will be taken before resorting to layoffs.**

That pledge is only one of the things that makes NUMMI special. The now-famous joint venture, created by Toyota and General Motors and incorporated as an independent California company, was launched as an experiment to find out if Toyota's famed manufacturing system could be implemented successfully with unionized American workers and U.S. suppliers.

The answer seems clear. NUMMI—which went into full production in November 1985 in a facility and with a workforce that had been abandoned by GM—has not only been a commercial success, it has also demonstrated the potential of American workers in a nonadversarial environment that emphasizes teamwork, mutual trust, and respect. For one thing, the productivity and quality achievements of the NUMMI team members prompted a decision to invest $300 million in a new line that began building 100,000 Toyota trucks annually in August 1991.

The team structure is central to NUMMI's success. The workforce is divided into teams of four to six people; each team has a team leader, an hourly worker who earns a small premium over their regular hourly wage.

Teams are expected to support the plant's kaizen (continuous-improvement) efforts and to participate in the Japanese-style consensus decision-making process. Team members also have a major voice in matters that affect their work areas, including the content of individual jobs.

◎ **To keep the just-in-time (JIT) production line flowing smoothly, a "standardized work" system is used.** All tasks must fit into a 60-second takt time; since the plant is geared to produce one car every minute—about 220,000 a year—workers must be able to complete their production tasks efficiently in 60 seconds or less without undue stress.

The key to cultivating a team spirit is "to involve all team members in everything—that means quality, cost, safety . . . everything," says NUMMI president Osamu Kimura, a former executive with Toyota in Japan. "It is important to ask them to think and to make a plan by themselves."

One result of the NUMMI approach is that, in 1989 and 1990, the plant implemented an average of 10,000 employee suggestions annually—about four per year per employee.

The NUMMI production system is a modified version of the Toyota system. ◎ **Within the plant a kanban pull system averts the waste of overproduction.** When a team member in the body shop needs a new supply of side panels from NUMMI's contiguous, state-of-the-art stamping plant, he or she pushes a button that activates a light on the kanban board in a storage area near the stamping operation. As forklift drivers zip by, they pull kanban cards from the slots next to the turned-on lights and promptly deliver a rack of the requested stampings.

Another cornerstone of the Toyota/NUMMI production system is *jidoka,* the principle of quality at the source: Defective parts are not to move from one step in the process to the next. In some areas, machines shut themselves down automatically when they sense abnormal conditions.

◎ **And there is the *andon* system—a combination of lights and music that alert supervisors that a problem has occurred.** A team member who spots a quality problem pulls a cord, which initiates the slightly irritating musical sounds. (Each work area has a distinctive melody, making it easy to identify where the problem is.) If the problem cannot be corrected within the 60-second takt time, a second pull on the cord brings the line to a halt. The result is that corrective action is promptly taken.

Productivity is one of NUMMI's strong suits. The reasons include dedication to continuous improvement, elimination of waste, low absenteeism, and old-fashioned hard work. "Nobody here has an easy job," says Gary Convis, a former Ford executive who is NUMMI's vice president for manufacturing and engineering. "Even the guy who drives the forklift truck has a job based on standardized work. . . . The team members will tell you that they work harder now than they did in the GM days."

—JOHN H. SHERIDAN

New United Motor Manufacturing, Inc.

Contact: Community Relations Department

45500 Fremont Blvd., Fremont, CA 94538

Telephone: (510) 498-5769

Employees: 4,800

Major products: Automobiles

NIPPONDENSO MANUFACTURING, USA, INCORPORATED

Battle Creek, Michigan

QUALITY
AGILE MANUFACTURING

If a radiator fails, the customer will not be impressed with our statistics on quality achievements. One problem for him represents 100 percent failure.

—ART LEARMONTH,
VICE PRESIDENT OF MANUFACTURING

How many plant tours begin with a vice president insisting that the visitors see the locker rooms first? In an ordinary plant even the employees try to avoid it. But to Mineo "Mike" Hanai, the condition of the locker room is an important indicator of how the rest of the plant is being managed. And judging by the tour of the factory, as well as the condition of the locker room, this is no ordinary plant. For starters, the performance of this auto-parts supplier is beginning to be competitive with its Japanese parents.

The workplace environment is a hybrid—not American by American standards, but not Japanese by Japanese standards either, says Stan Tooley, vice president of human resources and administration. But Nippondenso Manufacturing USA is unquestionably world class. Quality is extremely high. In 1991, when the company earned its Best Plants award, customer defects ran just 1.2 parts per million. The line inspection defect rate was 0.036 percent and the scrap rate was a mere 0.16 percent.

When these people talk quality, they really mean it. Art Learmonth, vice president of manufacturing, tells how they handled that rarest of rarities—a defect that got to the customer: "An auto dealer from New Jersey called us and told us of an error of omission, a radiator hadn't been tapped. At an executive management meeting, we discussed the options. One was the creation of a warranty department with a budget that would reimburse the dealer for correcting the factory error." President Mineo "Sam" Kawai suggested another approach: "Have the man who made the mistake fly to New Jersey and tap the hole so that he can understand the shame of what happened."

Learmonth remembers that the American management contingent was somewhat skeptical, maybe the trip would be considered more reward than punishment, but it was done anyway. "It was a good decision. Both the dealer and the customer were impressed that someone would fly from the factory to handle such a minor problem. It was also perceived by people in the plant that the company is serious about quality." Indeed, the warranty department and its budget are still nonexistent.

While quality attainments of this magnitude might tempt some firms to relax, Nippondenso views even one failure as critically as the customer would. To support that concern, the company has evolved a defect-corrective-action system. "All defects found by a customer are evaluated by a team from manufacturing, engineering, and quality assurance. Once countermeasures are determined and assigned, the quality assurance department audits the process," says Learmonth.

That is just the beginning. ◎ **Next comes *yoko-nirami,* a process for broadening the preventive measures to catch similar potential for defects.** "It comes into play once individual root causes of problems have been identified. In this process, management steps back from the problem, opens up its vision, and tries to see where else the solution could be applied," adds Learmonth. He describes a steplike process that goes from the point where the defect originated to other lines producing the same part, lines producing similar parts, even lines involving completely dissimilar parts, and finally to the entire plant itself.

Then, twice a month, a meeting is convened with senior management to critique the corrective actions and to assure that nothing is overlooked. "The whole system is designed to prevent problems from recurring," says Learmonth. "Our customers know that every time a defect has been found, our president and others on the executive staff have approved the corrective action."

Quality also describes how the company regards its associates. Learmonth explains: "We consider our people as part of the company, not as resources like machinery or materials. Not only is Nippondenso their company, they, not management, run it. All we [the executive staff] do is remove the barriers and act as facilitators." Influenced by traditional Japanese labor practices, Nippondenso does not subscribe to the view that labor is just another input to production to be hired, fired, or laid off at will.

Certain factors keep Nippondenso from guaranteeing lifetime employment in the United States, however. "One is that Americans do not have the social custom of going from school to a lifetime career with a company," explains Tooley. "Another is avoiding the risk of litigation." ◎ **Nevertheless, Tooley says he cannot conceive of ever laying off anybody.** "We don't even have a policy addressing the issue. It's simply not part of our system. Many alternatives would be explored first." Some options, such as executive pay cuts, are typically Japanese. President Kawai says he would take personal responsibility by resigning if the company came under extreme economic duress.

◎ **Customer response is further sharpened by extensive training and education.** In the second quarter of 1991, for example, the 1,100 employees received 8,000 hours of training in management, quality, safety, computers, and other technical topics. All of the courses and films are designed and scripted internally. To ensure effectiveness, team leaders, at three- and six-month intervals check to see if the new skills are actually used.

In addition to training, the associates are also provided with advanced manufacturing technologies, much of it state of the art. The list includes computer-aided design (CAD), flexible manufacturing systems, robots, electronic data interchange, bar coding, machine vision, and air-float die-changing systems. The CAD system and most of the robots are internally developed by the Japanese parent. The technology is backed up with such practices as kaizen, total productive maintenance, and just-in-time. The result is 100 percent on-time delivery, total inventory turns of 15.69, annual work-in-process turns of 332.79, and a 102 percent productivity gain since 1986.

And the locker room is sparkling.

—John Teresko

Nippondenso Manufacturing USA, Inc.
Contact: Jim Burkheimer, Superintendent
1 Denso Rd., Battle Creek, MI 49015
Telephone: (616) 965-3322
Employees: 1,500
Major products: Automotive components

NORTHROP GRUMMAN CORPORATION

Cleveland, Ohio

AGILE MANUFACTURING
QUALITY
NEW PRODUCT DEVELOPMENT

The cellular operation, with self-directed teams, instills a sense of individual and team responsibility. . . . There is true ownership.

—JIM SCHUSTER, OPERATIONS MANAGER

Going the extra mile took on new meaning at Northrop Grumman's Naval Systems (NS) in Cleveland. That is how much travel distance was eliminated from the assembly of a torpedo guidance component when the manufacturing facility went cellular in 1991. In addition, cycle time on the component was reduced by 75 percent, cost by 60 percent, and productivity was enhanced three to one.

For some 20 years, 1993 Best Plant award winner NS rolled along with little competition as a supplier of torpedoes and related undersea weapons systems to the U.S. Navy. But as competition heated up in the mid-1980s, NS found itself in a war of its own, "reducing prices for torpedoes as much as 25 percent per year—and some 75 percent over five years—just to stay competitive," recounts general manager Wayne Snodgrass. And as the cold war cooled, the bottom all but dropped out of the market. "We were in a price-based, winner-take-all competition against some of the most sophisticated companies in the world, in a shrinking market," says operations manager Jim Schuster. "The only way we could stay in business was with a complete restructuring based on a total quality program involving 100 percent of our people," adds Snodgrass.

To start, NS adopted the Westinghouse (its former parent) corporate total-quality model based on 12 "Conditions of Excellence," such as customer orientation, accountability, motivation, participation, culture, and communications. **◎ In addition, NS began benchmarking other world-class operations, learning about kit carts for better material handling from Texas Instruments, training from Ford, a new approach to subcontracting from Boeing, empowered work teams from Rockwell, and involving union personnel in process improvement from General Motors.** In making changes at NS, "we took the best of what Westinghouse corporate had to offer, the best of what was already in place, and the best of our benchmarking and combined these with our own innovations to get where we are today," notes Snodgrass.

In 1990–1991 NS changed from a traditional functional organization to a product-oriented organization with cellular manufacturing driven largely by self-directed work teams. "In our previous factory structure, the operation consisted of a series of complicated handoffs with little accountability," explains Schuster. "We had a lot of priority problems. We could store enormous amounts of work in process [WIP] that had a tendency to get lost." But, supercharged by some 2,400 process-improvement suggestions from employees, manufacturing operations were simplified via process mapping and eliminating non-value-added activities wherever possible.

The cumulative effect of these efforts is startling, including WIP reduced 80 percent, process steps down 60 percent, cycle time cut by 75 percent, and distance traveled sliced by 88 percent. Total stock was reduced to two-thirds of its previous level in a time when sales volume doubled. In recognition of this performance, NS was named Westinghouse's "most improved division" in 1990 and received its highly prized Signature Award for manufacturing excellence in 1991 and 1992.

◎ **In the process of restructuring, inspection was built into the assembly process.** "We made the decision to integrate quality into the hard-line manufacturing team, dedicated to the cell, so we could get quality at the source," explains Roger Bakkila, quality assurance manager. On more than 400 measurement points on the advanced-capability torpedo, "first-pass yield improved by more than 35 percent," he says.

But the Navy dropped a bombshell of its own. "In the first quarter of 1992 we learned that torpedo purchases would soon end. That was a real demotivating factor," recalls Snodgrass. Faced with the prospect of losing 40 percent of its sales base within two years, NS had to react quickly. While other defense contractors were buying assets to capture new business or selling assets to get out of the defense business, NS mined the imagination and creative force of its employees.

◎ **Through brainstorming, the employees rallied to come up with some 400 new-product ideas, which were boiled down to a top 40 by the summer of 1992.** The result: nine new product and service segments that are expected to represent 60 percent of NS sales within three years. "Nine of my twelve direct reports were also made product managers, each with a manufacturing cell to support his efforts," says Snodgrass. "They were given independence, the freedom to become entrepreneurial, utilizing resources and technologies not available outside of the company. This once-in-a-lifetime opportunity has been a motivator, a driver to excite employees to create their own destiny."

The new businesses exploit selected technologies developed in the torpedo business—"dual-use" technologies in NS jargon. Says Snodgrass: "Now that we have a competitive factory, we can compete with anyone." To prove the point, NS is already producing electronic assemblies for a commercial sprinkler system as part of its new contract-manufacturing business. Drawing upon its expertise in torpedo stealth and structure, it solved a noise problem at a local airport that helped launch Q-jet, a mobile passive-noise suppressor for jet engine maintenance. And Third-World countries could be cleaner and quieter if the plant's E-wheel hits the road en masse. This battery-powered motorcycle is designed for daily commuting in places like India, Thailand, and Malaysia. NS torpedo-propulsion technology, engineering, and manufacturing disciplines combined to complete the first prototype in a mere five months.

But the real power is in the people and the structure at NS. "We've come a long way, baby, organizationally," says Jim Schuster. "We're not only singing off the same sheet of music, we're writing the music together."

—TIM STEVENS

Northrup Grumman, Naval Systems

Contact: David Shih, Operations Manager

18901 Euclid Ave., Cleveland OH 44117

Telephone: (216) 692-5105

Employees: 500

Major products: Undersea weapons systems, commercial electronics

PELLA CORPORATION

Carroll, Iowa

Really, the only way to achieve top quality is to build it right the first time—and that is what we do.

—HILLARD KEENEY,
VICE PRESIDENT OF MANUFACTURING

AGILE MANUFACTURING
QUALITY
EMPLOYEE INVOLVEMENT

How do you improve the productivity of a manufacturing plant by nearly 100 percent in an eight-year period? "You set tough goals," says Jim Booy, manager of Pella Corporation's Carroll plant, which has accomplished just that. "We have a team here that knows how to rise to a challenge."

Pella, a family-owned firm, is well known as the maker of high-quality windows and doors. It is the second largest wood window manufacturer in the United States. The Carroll plant, which specializes in high-volume window lines, is characterized by one consultant as "the top quality and productivity plant in a company where quality and productivity are universally high."

The Carroll facility is much smaller than Pella's main plant in Pella, Iowa. It is also much newer and has more up-to-date equipment, including proprietary machinery designed and built by employees. "The Carroll plant is kind of our test lab facility for trying out new things on the line, new machinery and new techniques," says Hillard Keeney, Pella's vice president of manufacturing.

Booy, who has run the plant since its start-up in 1982, says productivity has climbed 8 to 10 percent annually ever since, for a cumulative gain of 96 percent by the time the facility earned its 1990 Best Plants award. "It now takes half the number of people that it used to take to run some of the areas."

Among the keys to that improvement, he says, have been automation, improved work methods, reduced material travel time, and inventory reduction. Having excess inventory, Booy points out, can sap productivity. "You spend a lot of time maintaining and controlling inventory," he says. "And it doesn't gain you anything."

Here is another startling statistic about the Carroll plant: Between 1985 and 1990, it increased work-in-process (WIP) inventory turns from 12 to 150 a year by implementing continuous-flow production methods. "That was accomplished over a period of time by reducing setup times and by end-lining the operations," explains Keeney.

◎ **By "end-lining," he means tying various operations end-to-end from parts making to final assembly "so that the whole plant simulates one big assembly line."** In the highly vertically integrated plant the process begins with metalworking (including aluminum extrusion) and woodworking operations in cells that feed parts directly to the assembly line at the precise point where they are needed. And employing a just-in-time (JIT) strategy, parts are made just as they are needed.

Setup reduction has played a critical role, since the plant makes windows in 300 different sizes. On one machine, which makes eight different parts, the use of computerized controls reduced changeover time to a mere 10 seconds. The equipment can now be automatically adjusted to make parts for different-size windows. On three other machines, setup time was sliced from four hours to about 10 minutes. "We had to buy some equipment to make that happen," says Booy.

◎ **Much of the setup reduction was a result of equipment research and brainstorming, he adds.** "If there is a problem such as a bottleneck area, the employees in the area will hold a meeting and just brainstorm on what improvements might be made. And when the new equipment comes and it represents a lot of their ideas, they make sure it works."

With the plant's streamlined production flow, lead times to the firm's distributor network have been reduced from three weeks or longer to just five days. And, in 1990, on-time shipments averaged 99.4 percent. In its approach to JIT, the Carroll plant relies not on kanban pull signals but on the ability of workers to think for themselves. ◎ **Work orders for components were eliminated; instead, each cost center and work cell now gets a copy of the final assembly schedule for a given day—the types and quantities of windows to be produced.** "From that, they have to determine what parts will be required to build those units," Booy explains. "They have to pull the drawing and see what parts their particular cell will have to make. Of course, after they get used to it, they don't even have to look at the drawings anymore."

Among the benefits of a low-WIP, fast-throughput system is higher quality, says Keeney. "Probably more important than the inventory reduction is quality improvement. Because the parts are used immediately after they are made, any quality problems show up immediately, and they are solved immediately.

The Carroll plant has no in-process inspectors. "Everybody is an inspector in our plant," says Booy. "We do a quality audit and we have a quality technology manager who works on quality problems. But no person is to operate or run a part that is not to print." Quality problems are tracked with a demerit system that gives points for various types of defects. "We've averaged better than a 10 percent a year reduction in our demerit level," Booy notes. "We're now about at the point where you can't reduce it any more."

A strong quality-circle program has also had an impact. The quality circles typically consist of people who work in the same area. They meet weekly and deal with one problem or improvement project at a time. But their focus is not limited to quality; they also wrestle with a variety of issues, including cost reduction, production flow, maintenance, and quality of work life.

Operations have also been enhanced by dividing the Carroll plant into two focused subplants, each with its own product line. Each subplant is almost a self-contained business unit, with its own support staff, including production schedulers, maintenance, material-handling people, and an R&D engineer. The subplants deal directly with suppliers on the release of materials (though purchase contracts are negotiated by a central purchasing function).

Giving the subplant manager control over the support functions eliminated the excuse making that often occurred in the past. No longer can the manager complain, for example, that a problem was not resolved quickly because the engineering people were slow in responding. "Now they all have the same goals and objectives," says Booy. "And the objective is to produce the number-one-quality window—safely."

—JOHN H. SHERIDAN

Pella Corporation

Contact: Jim Booy, Operations Manager

P.O. Box 247, Carroll, IA 51401

Telephone: (515) 628-1000

Employees: 325

Major products: Windows, doors

ROCKWELL INTERNATIONAL

El Paso, Texas

We live or die via our team concept.

—Jim Lake, Vice President of Operations

Employee Involvement
New Product Development

In 1983 when Rockwell International Corporation's Autonetics Electronic Systems Division plant opened, it was a feeder plant to the North American Aircraft Division, supporting the Air Force's B-1B program. The plant was optimized for a single program, the efficient high-production output of B-1B instrument panels, wire harnesses, and electromechanical power–control assemblies.

"The head count was approximately 1,500 in a three-shift operation," remembers Dwayne Weir, plant manager. And performance was impeccable. "The final program results included 100 percent on-time deliveries, high quality, and an under-budget total cost performance over the four-year production run," adds Don E. Teague, manager of program administration.

However, operational excellence was no protection from environmental change. Since 1986 and the end of the Cold War, Pentagon spending declined by nearly 50 percent, effectively removing more than 220,000 jobs from the defense industry. The 113,000-square-foot El Paso–based plant's survival has hinged on its ability to continuously reinvent itself.

Today, with only 240 employees, this 1995 Best Plant award-winning facility simultaneously supports some 20 Department of Defense (DOD) and NASA programs, all short-run production efforts. Annual shipments are valued at approximately $150 million. "Reinvention" especially applies to its transition to an adaptive, agile, world-class manufacturing facility. El Paso has adopted quick setup methods, versatile tooling, and created the capability to handle very small lot sizes—as low as one unit, says Rene J. Vargas, manager of manufacturing.

The scope of production varies from electromechanical assemblies to circuit cards, cables, harnesses, box assemblies, and consoles. Those products dictate the facility's primary production capabilities—including fully automated surface-mount and plated through-hole printed-wiring assembly lines, multiple mass-soldering technologies (wave, vapor phase, and infrared), automated component and wire preparation cells, semiautomated fine-pitch assembly, in-line conformal coating, clean rooms, and wire-harness and mechanical-assembly stations.

Despite all the modern equipment, a tour of the facility discloses that the winning performance derives from people systems, the workplace culture, and employee morale. ◎ **All employees are organized into self-managed integrated product/process teams (IPTs) and team members select their team leaders.** "That started about a year and a half ago when the 30-member common-process IPT selected . . . their team leader, and it has worked so well that the process is now accepted practice," says Teague.

The IPTs are cross-functional, multiskilled, and vertically integrated with manufacturing, quality, engineering, and production support personnel on the same team. The IPTs are trained in data-collection, data-analysis, and continuous-improvement strategies. Team members are empowered to make design, manufacturing, and business-process decisions. Highly visible performance charts provide feedback on team performance.

◎ **Suppliers become members of IPTs during the design phase of product development and work as partners for the mutual success of the project.** As team members, suppliers are allowed to develop their products in conjunction with Rockwell's, says Teague. To facilitate better supplier relationships, Rockwell is emphasizing long-term contracts and is cutting back on the number of vendors. In the last three years, the plant has reduced the number of suppliers from 1,500 to 500. Rockwell's purchasing function, located at divisional headquarters in Anaheim, California, helps suppliers develop and implement statistical process control (SPC). Once their processes are under control, Rockwell can eliminate incoming inspections.

The enhanced manufacturing prowess of the El Paso facility, especially in quality, flexibility, and market responsiveness, gives its division headquarters important options to deal with a changing marketplace. For example, Rockwell corporate reports a significant increase in U.S. commercial and international markets. Those two areas combined accounted for less than a third of total sales in 1984. Today each contributes approximately one-third of the corporate total.

Rockwell has also repositioned El Paso as a corporatewide manufacturing resource (as opposed to being a manufacturing site for its division). The plant is now considered as a manufacturing site for any of Rockwell's division that wants to bring it work. At the moment, approximately 2 percent of the work being performed at El Paso is for Rockwell's commercial divisions. "We have submitted proposals for programs that could make that an 80:20 [commercial-aerospace] mix, but we don't have a formal goal," says Vargas.

—JOHN TERESKO

Rockwell International, Autonetics Electrical Systems Div.

Contact: Rene J. Vargas, Manager of Manufacturing

P.O. Box 4858, El Paso, TX 79914

Telephone: (915) 755-5000

Employees: 240

Major products: Electromechanical components

SIEMENS AUTOMOTIVE

Newport News, Virginia

We've put the inventory where the customer wants it: in shippable product.

—DOUGLAS GODFREY, DIRECTOR OF LOGISTICS

EMPLOYEE INVOLVEMENT
QUALITY

Individual effort has figured prominently in the collective pull to transform 1995 Best Plant award winner Siemens Automotive Induction, Fuel and Emission Components Division, in Newport News. In the late 1980s, this former maker of automotive electromechanical items—from car radios to brake shoes to air pumps—was losing an average $31 million annually. Now, it is a focused manufacturer of gasoline fuel injectors with yearly sales growth of 35 percent.

Margaret Wilkens, for instance, voluntarily redesigned the inventory management system for the three clean rooms in which Siemens' Deka family of fuel injectors is produced. With a sparkle of pride in her eye, she reports the results: a reduction in work-in-process (WIP) inventory from $450,000 worth to less than $200,000.

Walter Tillman Sr., a manufacturing manager of the components group with more than 25 years of manufacturing experience, would file such initiative and command of work processes under: "We've come a long way since . . ." Remembering when the words "I don't pay you to think, leave your mind outside the door" reverberated on the shop floor, he cites a new spirit of employee involvement for the 40 percent gain in market share that has propelled the manufacturer from a distant fourth in the world market for fuel injectors in 1988 to within spitting distance of market leaders Bosch and Nippondenso.

Shifting from the organization's autocratic management approach required radical measures. The deconstruction of the functional organization into customer-focused groupings of self-sufficient product teams—injectors, fuel systems, components manufacturing, and an operation in Pisa, Italy, each with its own centers of development, production, and finance—"was a huge shock to the system," says William Fodness, director of the injectors group. "We were a very traditional organization of very traditional boxes with defined responsibility and little budgets."

The process of developing the division's mission statement of continuous improvement toward world-class manufacturing status eventually became a vehicle for consensus. ◎ **The real "test of the integrity of the mission statement," says Fodness, has been management's faithfulness to its 1988 promise to avoid layoffs on the basis of productivity gains accomplished through employee-involvement initiatives, as well as to do everything possible to avoid falling into the automotive industry's typical pattern of layoff and recall.**

The company sent a message that "people are not expendable" at a time when most companies were wholeheartedly dedicated to cuts, says Glen Rotella, district director of the facility's union, International Association of Machinists and Aerospace Workers (IAM). What's more, "They were spending money on training while losing money in production."

Human resources director Loretta Conen says, "We developed a 45-hour course equal to three college credits, and 180 people volunteered to take it. That's 180 people off the job when we were losing money. But we did it." Developed in partnership with local Thomas Nelson Community College, the World-Class Manufacturing class grew into a pay-for-knowledge (PFK) program that provides employees who achieve skills beyond their "home" job with a premium wage and the flexibility to work outside the narrow scope of job classes without contractual restrictions.

In a unique agreement with IAM, the PFK system is layered on top of union rules and procedures so as to "continue to respect the values of the union in terms of seniority, job classification, and rights," says Conen. Sixty percent of the hourly workforce has taken the course, and 30 additional offerings have been added since the program's ratification. This sort of training initiative has contributed to the dramatic transformation in workforce skill levels. The 1988 level of 59 percent "nonskilled" workers has sunk to less than 35 percent, while the ranks of the "highly skilled" have swelled from 25 to 37 percent.

The division's "Company Wide Quality" (CWQ) program focuses on six major elements: training and education, communications, improvement planning, employee involvement, productivity, and rewards and recognition. ◎ **One reward program, called SHARES, deposits $50 for every implemented suggestion into a central fund distributed equally to all employees at year-end.** When an idea extends beyond the ken of a single individual, a problem-solving framework called a *small group improvement activity* (SGIA) is available to form a team—often a cross section of hourly and salaried employees, customers, and suppliers—to brainstorm solutions and develop an implementation plan.

◎ **A variation on the SGIA called POWER (process optimization with early results) dedicates teams full-time for a period of three to five days to analyze and resolve specific problems.** In a series of joint POWER teams with Greystone, a plating and heat-treating supplier that relocated the bulk of its operations from Providence, Rhode Island, to Virginia in order to work more closely with Siemens, the division has "completely shifted inventory from WIP into finished injectors," says Douglas Godfrey, director of logistics.

—POLLY LABARRE

Siemens Automotive, I,F&E Components Division

Contact: Sam Byrd, Manager of Marketing

P.O. Box 2302, Newport News, VA 23602

Telephone: (804) 875-7000

Employees: 1,070

Major products: Automotive fuel-injection components

SONY ELECTRONICS, INCORPORATED

San Diego, California

EMPLOYEE INVOLVEMENT

ENVIRONMENT AND SAFETY

During the introduction of the employee involvement programs . . . it was clearly communicated that no employee would be laid off as a result of the programs.

—STEPHEN BURKE,
VICE PRESIDENT OF BUSINESS PLANNING

Blend equal parts apple pie and sushi and what do you get? A combination requiring an acquired taste, no doubt. But, in a nutshell, that's the menu served up at Sony Electronics Incorporation's San Diego Manufacturing Center (SDMC), a 1994 Best Plant award winner and home of Trinitron picture-tube and color television set assembly. Sony Japan provided the initial product and manufacturing automation technology, and the American management approach to employee empowerment has set the table for a multicourse feast.

"All technical, engineering, and equipment changes have limitations in terms of productivity improvements," says Ramesh Amin, senior vice president of manufacturing for Sony Display Tube Company (SDTC). "But, when all the people are involved with their hearts and minds, the improvements are unlimited."

In August 1989, the executive committee of the SDTC, one of four operating divisions at the San Diego location, met to change its basic top-down approach to management and problem solving, an approach that was simply not yielding results. "SDTC was regarded as the worst operating manufacturing facility throughout the Sony system," says Ross Beattie, director of human effectiveness. Symptoms were typical of low morale, including lack of ownership of results, poor communication, low productivity, high costs, and lagging quality. Absenteeism was almost 7 percent, turnover was 4 to 6 percent per month, and the division was unprofitable.

Setting the foundation for change, the SDTC executive committee introduced a program of redesign to a "high-performance work system" based on participatory management. ◎ **The effort began with high-profile trials, attacking two of the most troublesome problems at the time: 100 percent**

per year employee turnover in the chemical mixing area, and jealousies and family problems related to a new shift-assignment schedule. First trained in teamwork skills, employee teams met both challenges.

The success of these initial efforts sparked expansion of the concept, resulting in the formation of a support team of employee volunteers to guide the change. "What the support team really supports is the change itself," says Cassius Zedaker, support-team member. "How do we get from the traditional style of management to a participatory style? That's what our function is. To be the eyes and ears of what's going on in the organization and to help internalize the changes."

To learn how other teams operate, SDTC support-team members visited Clark Equipment Company, Sherwin Williams, McNeil Consumer Products Company, TRW, and the San Diego Zoo. After these experiences, "the team performed a skit for the workforce, comparing an autocratic versus participatory environment for decision making," says support-team member Terry Carrington. The employees of this nonunion shop embraced the participatory style wholeheartedly. Individual process-redesign teams, consisting of technicians, operators, and maintenance employees, were created throughout the division, and the rest, as they say, is history.

The San Diego site has now become "our most profitable and cost-competitive maker of Trinitron tubes," says Stephen Burke, vice president of business planning. "Turnover has virtually vanished, while comprehensive skill-building and team-building efforts at all employee levels have resulted in higher productivity and safety." For instance, in 1990 cathode-ray tube (CRT) production was 1,395 per employee. By 1993, it had jumped to 2,474 per employee.

"The improvement resulted from increased line speeds, reducing dwell time, and a small investment in automation, but primarily it was due to the cooperation of the employees, helping us improve our productivity with the same equipment complement," says Joe McHugh, senior vice president of Sony Display Component Products, assemblers of Trinitron TVs and computer display monitors.

"We don't think of it as a program anymore; it is our way of work now," says support-team leader Judy Dowhan. Teams are responsible for all day-to-day operations, setting work schedules, ordering maintenance and equipment, even establishing career paths and suggesting promotions.

In addition to building morale and trust among the employees, participatory management has delivered results right to the bottom line. For instance, two redesign teams implemented shift-synchronization schedules, increasing profitability by $5 million annually. The mixing department, after solving its employee turnover problems, attacked others. Chemical waste was cut, saving $20,000; process changes cut $142,000 more; a new purchasing strategy netted $107,000 per year; and phosphor reduction yielded a whopping $600,000 return, reports team leader Sam Farinas.

◎ **Growth at SDMC has been complemented by team-based solutions to waste and recycling challenges, significantly reducing impact on the environment while contributing to profitability.** Among the achievements: SDMC was first in the industry to recycle lead-containing waste glass from CRTs, to the tune of 120 tons per month (4 percent of total usage). City water usage has been cut 60 percent since 1990, the result of 36 percent improved efficiency and drilling three wells on site. Through waste reduction and recycling, 4,500 tons per year of nonhazardous waste have been diverted from landfill.

SDMC also participates in a volunteer EPA initiative, begun under the Bush Administration, called Green Lights, saving $314,000 a year on lighting costs by converting to high-efficiency systems. "The program paid for itself in seven months," says Doug Smith, manager of corporate environmental affairs. In fact, Sony Electronics is saving some $2 million a year at U.S. facilities alone with Green Lights systems.

—TIM STEVENS

Sony Electronics, San Diego Manufacturing Center
Contact: Irene Ball-Dumas, Manager, Employee Communications
16450 W. Bernardo Dr., San Diego, CA 92127
Telephone: (619) 487-8500
Employees: 2,500
Major products: Color television sets and components

SPX CORPORATION

Owatonna, Minnesota

We have a saying around here: Change is a threat if it's thrust upon you, but it's an opportunity if you're allowed to participate and to influence change.

—LARRY D. STORDAHL, GENERAL MANAGER

AGILE MANUFACTURING
QUALITY

Glide south by car on I-35, well past the Twin Cities and deep into farm fields that stretch impressively in every direction. But don't get too lost in reveries about the pastoral frame of southern Minnesota. Otherwise, you'll likely miss Owatonna, a small town (population 20,000) that's home to the 1991 Best Plant award-winning Power Team Division of SPX Corporation.

If you miss seeing Power Team, a maker of special high-pressure hydraulic tools, you forfeit the chance to understand how empowered employees and a continuous-improvement workplace have melded into one of America's best plants. Thoroughly mix that empowerment into a just-in-time/total quality control (JIT/TQC) environment and the formula yields world-class results that draw the attention and praise of customers and vendors alike.

After all, who would not be impressed with a 65 percent increase in throughput in the five years since the traditionally organized firm began moving from a batch-mode to a JIT environment? And who could not take notice of internal scrap/rework/warranty costs that have dropped from 4 to 0.3 percent of the sales dollar? Moreover, it is impossible not to recognize such advances as cycle time on critical parts being measured in minutes today when it was once measured in weeks and months. Or that statistical process control is effectively integrated throughout the operation of the 206,000-square-foot plant.

This is another success story that begins without a "significant emotional event." In other words, Power Team was not awash in red ink or facing imminent closing before it began its JIT/TQC odyssey. In 1985, SPX bought Owatonna Tool and soon made a commitment to a new manufacturing strategy. That ushered in divisionalization and the 1987 creation of Power Team to serve various markets with special high-pressure hydraulic tools and work-holding components. "Exceptional growth created the need for a new plant in 1989," explains operations manager James Schultz. ◎ **"We laid out the plant in four small, focused factories, with support staff located to the appropriate fabrication/assembly areas."** But the change was far more systemic than that. "We've been successful at getting the vast majority [of employees] to understand the need for continuous improvement, what just in time is, why low setups are important," notes Schultz. From the start of its journey toward world-class manufacturing status, Power Team developed a strategy that includes the use of Waste Elimination Teams as well as Customer-Satisfaction Teams.

◎ **In 1985 a formal Operator Certification program was started.** To eliminate the need for cumbersome, time-consuming inspections, Power Team certifies its machinists to inspect their own work. Each operator must pass two consecutive error-free audits on product specifications and employee responsibilities. Certification is not a one-time event. Repeat audits are performed to make sure that the "process in control" benchmark of three sigma is maintained. Work falling outside of that prescribed tolerance limit means loss of certification—and, subsequently, the need to regain certification. Michael O'Brien, quality manager, notes that operators themselves established the rules for certification. "They value doing their job right," he says, "and they feel everyone should."

Power teamwork is a way of life here. Team leaders in the plant might even call it the fuel that powers the machine of customer satisfaction. "Recently, we've had a lot of requests for short lead times on products that aren't normally in our line—special service tools, special clamps, things like that," says product manager Dann Rypka. "We've had a lot of opportunities where the customers says, 'OK, here's a small order. We'll give you a big order if you can perform in a certain amount of time.' Last week we had a situation like that. We got together with upper management, people from the floor, and other members of the team. When that meeting broke up, it wasn't everybody just going back to his desk. That group of people broke up into four or five little groups around the room, talking about the way they do this or how they could accelerate that in order to deliver on time."

The payoff of such teamwork is obvious: In an industry in which typical customer lead time is two weeks, Power Team has an order-to-shipment time of two to five days for its PE Series Electric Power Pumps. And, now that Power Team has been able to document its success through teamwork, the company is beginning to share its secrets with vendors. "We waited intentionally until now," Schultz explains, "because we wanted to be a model for them."

Additionally, Power Team is taking on the challenge of globalization through its units in Europe, Australia, and the Pacific Rim. "We've got a challenge to bring those non-U.S. warehouses up to speed at this time," Stordahl says.

But partnering and globalization should prove to be easily met challenges for the empowered team of Owatonna. Even as respected a name as Matsuura recognizes that the power is in the team. A member of the Matsuura family once visited Power Team, which happens to be one of the best U.S. customers of the Japanese machine-tool manufacturer. "We gave him a very thorough tour," recalls Stordahl. "And after detailed discussion with a lot of our people, including machinists and engineers, he made a remark that we're very proud of. He said we were the most Japanese-like company that he had seen in North America."

—JOSEPH F. MCKENNA

SPX Corporation, Power Team Division

Contact: Gary Glienke, Director of Human Resources

2121 West Bridge St., Owatonna, MN 55060

Telephone: (507) 455-7753

Employees: 400

Major products: High-pressure hydraulic tools

STEELCASE, INCORPORATED

Kentwood, Michigan

We are now product-focused. We have looked at what drives customer needs and worked it backward into the organization.

—JOHN GRUIZENGA, GENERAL MANAGER

EMPLOYEE INVOLVEMENT
TECHNOLOGY

Take a state-of-the-art plant on a site that is the equivalent of 16 football fields. Add in the foresight by company management, before the plant was built, to let employees run the day-to-day operations. Then, instead of hiring new employees, transfer to the new plant employees from three existing plants who have yearned for years for the chance to use their brains, not just their hands. Presto, you have the ingredients for a world-class plant that manufactures office furniture systems. Specifically, you have Steelcase and its 1991 Best Plants award-winning Context Division in Kentwood.

A case in point: For 18 years, Jerry Hammond had been a spot welder making components for Steelcase office furniture without even knowing the people in nearby departments who worked on the same products. Now, Hammond, employed for years in a system that forced him to work as an isolated individual, works as part of a team, running as many as six different pieces of equipment with fellow employees.

◎ And his team—not top management—decides how it will run its part of the Casegoods factory (one of four units) in the Context plant, which was built in 1990 at a cost of $90 million, half of which went for state-of-the-art equipment and technology. "We decide what is the best way to do things, what we need, on a day-to-day basis," says Hammond. "Before, you left your brains at the door, ran a die machine, welded a part, or put on a trim." Just as important, he adds, it is now one team whose members work together without barriers between departments. "I worked with these people [in another Steelcase plant] for 18 years, but never knew them. Now I talk to them all the time because assembly and trim, paint and weld, are all on the same team. Their problem is mine, and vice versa."

These changed roles for workers and managers, and new relationship between them, was by design. Steelcase management wanted workers and managers at Context to be one team and insisted there be no barriers between white- and blue-collar workers. That is why the main meeting room at Context is positioned between the offices and the plant, with an entire wall that is a window into the factory.

The Context plant itself is part of Steelcase's grand plan for five plants in Kentwood that connect in a hub-and-spoke configuration to its 1.2-million-square-foot physical distribution center. The well-thought-out internal design of the Context plant, which is organized around products, not processes, must be credited to the employees. "We did a lot of interacting with people," says John Gruitzenga, the plant's first manager. ◎ **Whenever the plant made an equipment purchase, "we brought in the operator, the setup man, the maintenance man, and used teams to decide what type of equipment to buy."** Adds current Context plant manager Dick Bierschbach: "We pulled in operating teams off the floor [of our other plants] and asked them what wasn't working, what would work better, what equipment we should purchase for the new plant, and what the layout should be."

In other Steelcase plants, "we had organized by process," says Gruizenga. "There was a machine shop, a weld division, a paint shop, etc. This plant is laid out by products, and there's one supervisor responsible for the product from raw materials to finished product."

There are 41 self-directed production work teams and four support work teams that tackle day-to-day problems and 14 teams that deal with specific product or assembly areas, as well as more complex plantwide issues. Among them: safety, scrap and waste, paint quality, and shipping. And workers are cross-trained, as time permits, during regular working hours.

The difference in how the Context plant operates compared with traditional Steelcase plants is not just the use of work teams and bottom-up decision making; it's also the technology employed and the fact that fewer people are needed. There is 1 supervisor for every 33 workers compared with a 1-to-12 ratio elsewhere. And whereas a traditional Steelcase plant has 20 to 26 employees per 1,000 manufacturing unit counts, Context has 11. Recently when Steelcase decided to increase production by 4,000 manufacturing unit counts at both Context and its sister plant, Context requested 16 new workers, while the conventional plant requested 80.

◎ **Among the technological differences Context boasts are an automated laminate line that has reduced the time needed to turn a piece of particleboard into a desktop from 20 hours to 30 minutes, a 400-ton die press used**

in the blanking and forming of desk drawers that can automatically change dies in 6 to 8 minutes (instead of the usual 60 to 75 minutes), and, thanks to a state-of-the-art abatement system, emissions levels at Context of one-half the other Steelcase plants. These efficiencies enable Steelcase to turn raw materials into finished goods, for the most part, in three days compared with three weeks. Costs have also been cut. Raw material inventories are 55 percent lower; work-in-process inventories are down 26 percent.

"This plant is doing everything from the bottom up," concludes Gruizenga. "We wanted to build ownership among the workers and give them the tools they need, the accountability, and the responsibility. What we're trying to do is build product platforms so workers have full autonomy on that product and a complete set of tools to meet customer needs. Customer orientation has come to the forefront."

—MICHAEL A. VERESPEJ

Steelcase, Inc., Context Plant
Contact: Dennis Johnson, Quality Manager
5353 Broadmoor SE, Kentwood, MI 49508
Telephone: (616) 698-5500
Employees: 496
Major products: Office furniture

STONE CONSTRUCTION EQUIPMENT, INCORPORATED

Honeoye, New York

CUSTOMER FOCUS
EMPLOYEE INVOLVEMENT

We're not into world-class manufacturing, but world-class customer service.

—BOB FIEN, CEO

This is the story of a small manufacturing company whose products admittedly used to be poor in quality. It found itself consigned to the cellar of its industry—an industry, like many in the United States, that faces heavy competition from the Japanese as well as the Germans. Its home is in Honeoye (Native American for "broken finger"), a small town situated in the Finger Lakes region of New York. It's just the kind of name you would expect for a Cinderella story. Since 1986, the plant, owned by Stone Construction Equipment Incorporated, has been making dramatic improvements in quality and productivity. By 1991, it was recognized by the construction industry as one of the best in its field and earned its own Best Plants award.

Stone makes light construction equipment: cement and mortar mixers; concrete finishing equipment; contractor pumps; and handheld, ride-on, and walk-behind compactors. The plant is 100 percent employee owned. Although president and CEO Bob Fien does not believe that this kind of ownership is the only kind that achieves the level of improvements that Stone boasts, he has no doubt it helped.

That is because the employees at Stone are the forces driving hard to gain world-class manufacturing status. That does not mean, though, that managers are sitting back and resting on employees' laurels. They, too, are working hard to reach new heights. The combined efforts of management and employees have resulted in a significant list of improvements since Stone launched itself on a journey toward world-class status in 1986. For example, Stone used to take some 10 days to make one of its mixers. By 1991, the mixers were completed in one day. Order-to-shipment used to be the industry average of two weeks; by 1991, Stone had its time down to two days.

The introduction of such techniques as JIT, computer-aided design and computer-aided manufacturing, concurrent engineering, cellular manufacturing, and computer-integrated manufacturing have all played their parts in Stone's renaissance. But there is more to it than that. And, while the company is pleased with its reputation as a quality manufacturing enterprise, executives contend that is not what drives the operation. Instead, Fien describes the company as market-oriented, dedicated to customer satisfaction. ◎ **"Everyone at Stone understands the importance of customers and gets involved with them through customer visits, by fixing equipment at job sites, and having employees sit in on customer complaint committees," says Fien. "This makes customer needs real to employees."**

Stone puts a lot of trust in its employees. Trust with a capital *R*, says Fien—that is, managers have been trained to manage by *Respect* rather than power. Quality inspection has been replaced by process control. Time clocks have been eliminated.

Then comes training through development. Several programs are in place to teach people how to do a job and to show them why the job is done and its impact. The process is also focused on fit. The idea is to make sure all processes fit the company's direction.

A key point, says Fien, is understanding what participation is and what it is not. "It is not management by consensus, nor is it a process where everyone participates in everything. We try to create an environment where the people who are knowledgeable in a particular activity can give constructive input for upgrades and solutions, and where the input is seriously considered," he explains.

Worker input is also treated seriously by the workers. The company initiated an employee-driven manufacturing change request (MCR) system. In a nine-month period, 700 MCRs were received. Of these, 80 percent were implemented. Further, no promise of monetary rewards was required to generate the response. Continuous-improvement projects are initiated by employees, and employees work with design engineers and vendors to improve component quality and working relationships.

A tour of the plant reveals cell manufacturing sites that have been set up by individuals and groups to their best advantage. This has resulted in reduced waste in areas such as material handling, scrap, inventories, and paperwork, as well as setup reduction.

Teamwork is widely evident, as are modern work philosophies such as JIT, poka-yoke, and cross-trained-worker operations. All this team energy has "dramatically increased communication between workers," explains Dick Nisbet, plant manager. In addition, it has helped develop a strong pride of ownership among the workforce.

Using process control rather than final inspection has put the responsibility for quality in the hands of each employee. "Make it right the first time" is the ideology at Stone. A practical demonstration of this is that workers can halt a machine if they see that quality is not going to meet their high standards. ◎ **Customers are not forgotten either. They know exactly who to call at Stone, because each employee signs off completed products.**

All this is fine and dandy, but what about competitiveness? Fien and his managerial colleagues say that Stone's flexibility has helped reduce costs and increase productivity. That places the company on an equal footing with its Japanese and German competitors.

Perhaps the moral to the Stone story lies in the fact that while everyone in industry talks about the need to keep the customer satisfied, the people at Stone live it every day. And how does a company get there? By working on improvement every day and, in the words of Chuck Lybeer, vice president of manufacturing, by "letting change happen."

—BRIAN M. COOK

Stone Construction Equipment, Inc.

Contact: Chuck Lybeer, Vice President, Operations

32 East Main Street, Honeoye, NY 14471

Telephone: (716) 229-5141

Employees: 200

Major products: Light construction machinery

SUPER SACK MANUFACTURING

Fannin County, Texas

We need to tie the amount people will earn to the effort as quickly as possible.

—David Kellenberger,
Vice President of Manufacturing

Environment and Safety
Employee Involvement

The summer before he joined 1995 Best Plant award winner Super Sack Manufacturing Corporation at its plant in Savoy, Texas, Brad Eisenbarth had worked in an old-line manufacturing plant where he witnessed and experienced the all-too-common communication problems between those in power and those on the floor "trying to get things done." He never wants to go back. "It is mind-boggling how much wasted effort there is when you can't work on the same level," says Eisenbarth. "Here it is all one group. It is a phenomenal difference having self-directed work teams."

That sentiment is shared by Brian Suchsland, an industrial engineer at Super Sack's five-year-old sister plant located 12 miles east in Bonham, also in Fannin County. Suchsland had worked in a traditional manufacturing plant before joining Super Sack this summer. "There I just clocked my time," says Suchsland. "Here you have the enthusiasm and attitude you need for a successful business. This place is manufacturing heaven."

There are several reasons why the 325 employees, who manufacture flexible intermediate bulk containers (bags) for the agricultural, food, and pharmaceutical industries, hold their employer in such high regard. First, they have the freedom to perform their jobs the way they see fit. "Each team meets as a unit at the beginning of each shift to talk about what happened yesterday, to review the day's work, to discuss quality criteria, and to go over work orders," says Janet Gott, the plant manager at Savoy.

Second, workers learn to perform every aspect of sack production—and are responsible for safety, quality, and housekeeping as well—and their duties change depending on the needs of the day. Third, there's no threat of team leaders becoming too powerful. Team leaders rotate every three months, with each team member filling the leader's role before anyone gets a second term.

Fourth, team members hire, fire, and discipline the people on their team. Workers can transfer to another team only if that team wants them or can use their skills, but not for reasons of incompatibility. And finally, workers have the opportunity to shape and modify both the base pay plan as well as a bonus plan based on each individual team's production output that can pay out more than $1,300 every six months. (Workers earn 10 percent of all sales beyond the plant's break-even point.)

The decision by Super Sack to empower employees has had a remarkable impact on the $25 million company. Between 1990 and 1995, scrap/rework has gone from 2.1 percent of sales to 0.2 percent of sales. Total inventory is down 23 percent, and work-in-process inventory is down 50 percent. What's more, in the latest 12 months, sales volume at the two plants is up 55 percent, with sales per worker-hour up 21 percent. In addition, a mandatory muscle-stretching program at the start of each shift has helped reduce the number of lost-day injuries at the two northern Texas plants by 80 percent since 1990.

There's also a strong environmental commitment. ◎ **All products can be reprocessed, and all scrap is recycled.** Super Sack reconditions used bags, pioneered the use of water-based inks for printing on woven plastic bags, and built a treatment system to remove all ink solids from its wastewater.

But the transformation from departmentalized batch manufacturing, where buffer stock was built up and each bag traveled throughout the entire plant, did not come easily. Super Sack first experimented with a self-directed work team in order to make a specialized product at Savoy in 1989, then moved the concept to its start-up plant in Bonham in 1990. But when Super Sack first tried to switch the Savoy plant to self-directed work teams, the plan failed miserably.

"We had a random drawing of names to set up the first team of 8 to 10 people," says Gott, and then set out to teach them "interpersonal skills so that they could coexist and address the conflicts that were inevitable." But "we did not give them all the information they needed," continues Kellenberger. "One of our greatest mistakes at the beginning was our failure to recognize how important training was. ◎ **You need to make a huge commitment of time and resources for training people in communication, goal-setting, and general team-building skills to make a successful transition. . . .**

"We had told employees that they were empowered, but they didn't know what it was that we wanted them to do, and they didn't understand that empowerment meant responsibility and accountability."

Despite those early struggles, the new way of working began to succeed, says Kellenberger, for two reasons. First, Super Sack had faith enough in its concept to switch the entire plant to self-directed teams over Labor Day weekend in 1991. "We never wavered, even though we did agonize," says Gott. Second, Super Sack finally convinced its managers to let teams make mistakes. "We had to learn to let them come to us for information and not just make the decisions ourselves. We had to learn to let the teams make mistakes so they could learn to manage themselves," remembers Kellenberger.

—MICHAEL A. VERESPEJ

Super Sack Manufacturing Corp.
Contact: Janet Gott, General Manager
P.O. Box 245, Savoy, TX 75479
Telephone: (903) 965-7713
Employees: 325
Major products: Flexible intermediate bulk containers

SYMBIOSIS CORPORATION

Miami, Florida

People don't get into trouble around here for doing things that they think need to get done.

—BILL BOX, PRESIDENT

EMPLOYEE INVOLVEMENT
CUSTOMER FOCUS

1995 Best Plant award winner Symbiosis Corporation was created in 1988 when two engineers—Kevin Smith, its first CEO, and Charlie Slater—left their jobs with another medical device company to form a syringe manufacturing business. Started in a garage on a shoestring, the company has grown into a $50 million business, which today commands about 72 percent of the world market for disposable gastrointestinal biopsy forceps. It also makes disposable laparoscopic scissors and other instruments used for abdominal surgery.

A major turning point occurred in 1989, when the fledgling company's syringe sales were drying up. "That's when Boston Scientific threw out a challenge to us to find a way to make a low-cost biopsy forceps," recalls President Bill Box, another founder and former vice president of manufacturing. "We took the challenge, mostly because we were starved." At the time, surgeons were using reusable forceps—priced at $300 to $400—which are difficult to sterilize. Boston Scientific Corporation, one of several Symbiosis partners that market the Miami firm's devices to the medical community, saw an opportunity to reinvent the market if a way could be found to produce inexpensive instruments that would sell as disposables. "We worked like crazy to develop a low-cost product," Box says.

One of the keys to the new product was the refinement of investment-casting techniques, traditionally used in jewelry making, to produce very small, very precise metal parts that traditionally had to be machined. One advantage that microcasting offers is flexibility in making design changes. "Once you've got a stamping die, if you want to change a design, it will cost you another $50,000 or $60,000 to change the dies," says Scott Jahrmarkt, vice president of manufacturing. "But with investment casting, you can do multiple iterations of a design before you get to production."

Oddly enough, the idea of using investment casting sprang from Box's fascination with model railroading. To support his hobby, he had learned the technique and owned some rudimentary equipment. ◎ **Many of the people who've been instrumental in Symbiosis' success, Box notes, have drawn upon skills developed in the pursuit of hobbies—including watch repair, auto racing, and model airplanes.** "We all use our hobbies as a vehicle to develop certain skill sets," he says. "You try something new, you perfect a skill, and then you put it on the shelf with the idea that some day you may need that skill."

That's just one aspect of the agility that has become a forte of the Miami firm, which is known for its ability to quickly design highly engineered new products and move them smoothly into manufacturing. One wall in the second-floor lobby of its office/manufacturing complex features a display of plaques commemorating many of the 150-plus patents awarded to the young firm's creative engineering staff.

Many of Symbiosis' new-product designs are created on a Pro/Engineer CAD system with 3-D solid-modeling capability, enabling designers to generate realistic views of products before physical prototypes are built. A complementary software system—Autodesk 3D Studio—creates photo-realistic print renderings of parts and finished products. ◎ **"This lets us finalize the aesthetic properties of a part and show it to the customer, who will know exactly what the product looks like before we ever make one," says Peter Kratsch, senior product designer.** Thus, fewer prototypes are needed, and product development time is reduced.

Acquired in 1992 by American Home Products Corporation, Symbiosis has also adopted a focused factory structure, integrating subassembly and assembly operations into customer-focused teams. "That enabled us to drive total manufacturing throughput time down because there wasn't a handoff anymore," Jahrmarkt says. Since 1990, manufacturing cycle time has been sliced by 50 percent. Meanwhile, spurred by volume increases and process improvements, overall productivity has soared 169 percent.

The emphasis on agility has been important in enabling Symbiosis to cope with dramatic growth spurts. "We always have second and third sets of tooling built when we build the first one," the manufacturing vice president explains. "We don't want to be in a situation where we have to wait to start up a second [assembly] line because of the lead time on tooling. To meet customer needs, I want to be able to turn on a line overnight."

◎ **The 850-employee firm has implemented a voluntary mentoring program in which experienced personnel guide relative newcomers, helping to acclimate them to the company environment.** And special provisions have been made to support the firm's 16 deaf and hearing-impaired workers. Marianne Stokes, a sign-language expert, was hired as an employee-relations administrator in the human resources department. She serves as an interpreter when needed and also teaches sign-language classes for supervisors and other employees who want to communicate better with their deaf coworkers.

In March 1996, Boston Scientific acquired Symbiosis.

—JOHN H. SHERIDAN

Symbiosis Corporation

Contact: Tom Elwood, Director of Manufacturing

8600 NW 41st St., Miami, FL 33166

Telephone: (305) 597-4000

Employees: 850

Major products: Disposable medical devices

TENNESSEE EASTMAN COMPANY

Kingsport, Tennessee

Employee Involvement
Environment and Safety
Quality

A 1 percent improvement in equipment reliability in this plant is worth \$8 million to the bottom line.

—Bill Maggard, TPM Manager

At first glance, what seems most striking about Tennessee Eastman Company's (TEC) Kingsport facility is its size. The 825-acre chemical complex, straddling the Holston River in northeastern Tennessee, is a sprawling maze of interdependent production and support operations occupying 376 buildings. Stretching about 1.5 miles from end to end, the 1991 Best Plant award-winning site hosts seven product divisions, each a large business in itself. Support facilities include a 170-megawatt power plant that generates enough electricity for a city of 80,000 homes. The plant's employees produce 350 different chemical, plastic, and fiber products, creating a revenue stream of nearly \$2 billion a year for its parent firm, Eastman Chemical Company.

But on closer inspection, what seems truly impressive about Tennessee Eastman is its ability to learn—to nurture a culture in which change has become a way of life. That is no easy task in such a large, diverse organization. Just ask Bill Garwood, TEC's president. He led the charge as Tennessee Eastman embarked on a multifaceted improvement program. Unlike many companies that have launched similar efforts, TEC was doing well financially. "Our managers wondered why we should change something that had been successful," Garwood says. "But I kept saying, 'There are companies out there that are ready to take our business away.' " Even today, he adds, one of his toughest challenges is communicating the need for change. He spends about 25 percent of his time working with managers on change issues.

◎ **Supporting its push to world-class performance, Tennessee Eastman has put a heavy accent on training.** The typical TEC employee spends about 80 hours a year in training. And, in parts of the company that have implemented advanced work-system concepts, employees spend up to 200 hours a year in training.

The investment has paid off. Between 1981 and 1991, productivity (measured as sales per employee) increased by nearly 70 percent. A team-based approach to quality improvement—supported by formal training in process evaluation control and improvement—has maintained product quality at "best-in-industry" levels. First-pass yields in 1991 averaged 95 percent for all products; in some operations, they were nearly 99 percent. Further, customer satisfaction levels were on the rise. Surveys, conducted as part of an extensive customer satisfaction process administered by the marketing arm of Eastman Chemical, show that 75 percent of the firm's customers rated it their number one supplier, up from about 50 percent in just four years.

◎ **Tennessee Eastman's management game plan is based upon seven "visions," including an environmental vision that the company act as "an exemplary steward" of the land, water, air, and raw materials it uses.** In 1987, it built an $80 million treatment plant to handle all wastewater from the site. The company also forged a partnership with Waste Management Incorporated to build a facility on TEC property to recycle up to 50 million lb/yr of cardboard, paper, glass, aluminum, and plastic.

Not surprisingly, technical innovation is important at TEC, which employs 175 PhDs in its research center. A major technological achievement: the establishment of the world's only commercial-scale "chemicals from coal" gasification facility. Daily, it converts 10 to 12 railcar loads of high-sulfur coal into hydrogen, carbon monoxide, and sulfur. The recovered sulfur is sold and the two gases become raw materials for other TEC divisions.

But perhaps the real hallmark of Tennessee Eastman is its conviction that employee empowerment is the key to the future. The latest thrust has been the creation of self-regulating teams and new work systems in some departments. "The new work system is designed so that employees feel like they own the business and act accordingly," says Nick Grabar, an internal consultant in organizational design.

Within TEC, operating divisions are given leeway to design their own versions of new work systems. In the Filter Products Division, for example, self-regulating teams are creating a six-level pay-for-skills plan. Each team has a coordinator for production, maintenance, quality, safety, training, labor utilization, and environment. "We took the shift supervisor's job and broke it down and gave parts of the job to the coordinators," says Karen Rowell, a department manager.

The former shift supervisors were given extensive training and now serve as team facilitators. Ray Gentry, a supervisor-turned-coach, notes that his role now is to ensure that the coordinators are doing their jobs. "We'll stand back and let them make mistakes—let them learn from their mistakes," he says. "Then we'll coach them on what they should have done. But we'll step in if we see a serious safety or quality mistake being made."

◎ **An advanced total productive maintenance (TPM) effort has dramatically improved equipment uptime.** Launched in 1986, the program has resulted in cumulative savings of more than $24 million. The primary reason for TEC's strong emphasis on TPM is to improve equipment reliability. TPM manager Bill Maggard says the key to implementing TPM is "interface management"—breaking down organizational barriers between operators and maintenance specialists and overcoming the "that's not my job" syndrome.

Using a zone concept, maintenance tasks are divided into those that can be done by operators, those requiring a maintenance mechanic, and those that can be done by either. Operators perform some 160,000 lower-level maintenance tasks yearly.

—John H. Sheridan

Tennessee Eastman Company

Contact: Katherine Watkins, Coordinator, Corporate Information Center

P.O. Box 511, Kingsport, TN 37662

Telephone: (423) 229-3078

Employees: 8,000

Major products: Chemicals, fibers, plastics

TEXAS INSTRUMENTS, INCORPORATED

Lubbock, Texas

AGILE MANUFACTURING
NEW PRODUCT DEVELOPMENT
TECHNOLOGY

Just 20 minutes after raw plastic pellets are placed into the hopper of the injection-molding machine, finished calculators come off the end of line—tested, packaged, and ready to ship.

—TOM BARRICK, MANAGER

Paul Greenwood has a stubborn streak, at least when it comes to manufacturing competitiveness. As a Texas Instruments (TI) engineering manager in the early 1980s, Greenwood was involved in the production of pocket calculators. At the time, TI was the only company still making low-priced calculators in the United States. When competitive pressures forced a decision in 1982 to shift production to the Far East, he took it as something of a personal defeat.

"Ever since, I wanted to figure out a way to bring the calculator business back to the United States," says Greenwood, an operations manager at TI's Lubbock plant. "I wasn't going to give up." He didn't, and neither did the members of his 1990 Best Plants award-winning team. So they conceived a highly automated, continuous-flow production system that could manufacture solar-powered scientific pocket calculators at a competitive cost.

At that point, all Greenwood needed was a customer. And, in May 1989, he found one in the Wal-Mart chain, which was putting heavy emphasis on its "Made in the U.S.A." products. ◎ **Based on Wal-Mart's enthusiastic response to the proposal, Greenwood assembled a cross-functional team to design the final product and the manufacturing process concurrently.** They accomplished the feat in record time—just five months.

And in November 1989, low-end calculator manufacturing returned to U.S. soil as the Lubbock plant began producing its line of TI-25 calculators in five models, priced from $5 to $10. More than two million were made in 1990.

Among the keys to making the calculators at a competitive cost, Greenwood says, were the selection of top-quality suppliers and the determination to eliminate waste activity, such as inspecting plastic cases after the injection-molding process and then packing them for temporary storage. "Our line is one continuous process and there are not a lot of non-value-added operations," Greenwood points out.

◎ Designing the product for ease of assembly was another important factor in cost reduction. With only seven components, the TI-25 has the fewest parts of any scientific calculator in the world.

But the continuous-flow system would not have been possible without the assurance of high-quality components, particularly the plant's plastics injection-molding equipment. In the past, because yields tended to be unpredictable, it was difficult to synchronize the production of plastic cases with the pace of the assembly process.

That problem was licked with homegrown technology. At the front of the calculator line, three 250-ton Cincinnati Milacron injection-molding presses are governed by TI-530 programmable logic controllers (PLCs) equipped with TI's Turbomold module for plastic presses. The controller senses both the velocity and pressure of the equipment and "constantly makes adjustments at a very fast and precise rate so that the parts come out perfect every time," explains Tom Barrick, manager of plastic controls engineering. A similar application in the plant's educational-toy line has reduced defects from 5,800 ppm to just 162 ppm. **◎ The PLCs are linked to TI's software-based "Cell/View" system, which provides on-line monitoring and real-time statistical process control.**

With high-quality output assured, it became possible to feed plastic cases directly to the automated assembly line, eliminating inspection, WIP storage, and all the handling and time delays associated with those activities.

The most critical stations of the assembly process have been automated. Robots equipped with suction devices remove plastic parts from the presses and place them on a transfer device. In the assembly area, robots apply decals, heat-seal components that are inserted manually, and join the top and bottom halves of the calculator prior to ultrasonic welding.

After tests for functionality at the end of the line, the 2 × 4 inch calculators are blister-packed into red, white, and blue "Made in U.S.A." packaging.

One quality goal set for the calculator line was to keep scrap to 0.5 percent, and that goal has been exceeded, says Greenwood. Other achievements include limiting WIP to one shift's worth and keeping customer lead time to just five days.

The Made-in-America label has been a big plus, Greenwood asserts. "The calculators are selling better than we anticipated."

In April 1995, TI consolidated the Lubbock operation with its other contract manufacturing units. In the spring of 1996, all were sold to Solectron.

—JOHN H. SHERIDAN

THE TIMKEN COMPANY

Bucyrus, Ohio

Total associate involvement is the most powerful machine any company can have.

—Tom Strouble, Marketing Director

Agile Manufacturing
Employee Involvement

At first glance, the industrial complex that interrupts the pastoral landscape where U.S. Route 30 glides into Bucyrus could easily be mistaken for any number of aging Rust Belt manufacturing facilities. But, within The Timken Company's 42-year-old, high-volume tapered-roller-bearing plant—with its long rows of cutting, grinding, heat-treating, and polishing equipment—a remarkable transformation has been taking place.

Some of the changes have been physical, such as the installation of a tool setup station in the cup "green-machining" area—requested by ingenious equipment operators who wanted to do their own setups to speed changeovers. But more dramatic has been the culture change that has occurred since 1985, when the plant's workforce held its collective breath in anticipation of a shutdown order that never came.

During a mid-1980s restructuring, Timken management decided to close one of the company's two high-volume bearings plants. And even after the Bucyrus plant was spared, many of the workers feared that they had been granted only a temporary reprieve. "We had a demoralized workforce," recalls Jim Benson, who arrived in Bucyrus in 1986 as general manager of the bearing plant. "The workers told me, 'There's no future for us. It's only a matter of time.' "

But they were wrong. And they have worked hard at a wide array of continuous-improvement projects to prove themselves wrong. By 1992, the year they earned their Best Plants award, the empowered Bucyrus workers were making a significant contribution to Timken's ambitious companywide "Vision 2000" campaign—a quest to become "the best manufacturing company in the world" by the turn of the century.

◎ **When Vision 2000 was conceived in 1989, heavy emphasis was placed on developing "centers of expertise" within the company's network of plants, notes Tom Strouble, formerly director of manufacturing for Timken's North American and South American bearings business.** The idea is to have different facilities master various "tools for accelerating the change process," explains Strouble. Each center is required to develop in-depth knowledge in its specialties, keep abreast of the latest developments, and spread that knowledge to other company units. The Bucyrus plant, producing two million tapered roller bearings a week, has become Timken's center of expertise in four areas.

The first is self-directed work teams, which have boosted productivity, elevated an already-superb quality record, reduced maintenance costs, and improved the plant's ability to meet production schedules. The teams were introduced with a pilot project in December 1990, following a year and a half of intensive research, design, and preparation. In addition to making day-to-day operating decisions once reserved for management, the teams reach across functional boundaries and perform many tasks, such as setup and maintenance, previously handled by support functions. To prepare a manufacturing cell or department for conversion to self-directed status, a design team, including hourly workers, does extensive groundwork. "They ask, 'How do we get the job done today? What is the best, the simplest, way to get it done?' " explains Benson. Within a self-directed team, members rotate six key-role positions in which they assume responsibilities formerly handled by supervisors—for equipment, quality, cost performance, safety and housekeeping, training, and human resources. The key-role people serve on one of six plantwide Key Role Councils, which provide a vehicle for sharing information and improvement ideas.

◎ **A second center of expertise revolves around the CEDAC method—a total-employee-involvement activity that gives everybody a crack at diagnosing and solving designated problems.** The acronym CEDAC stands for "cause-and-effect diagramming with the addition of cards." The plant's first CEDAC project, which zeroed in on quality/downtime problems with an outside-diameter grinding machine, saved $285,000. In applying CEDAC, a project team, often headed by an hourly worker, establishes a goal, such as a 50 percent reduction in defects, and a time frame for meeting it. The team then installs a bulletin board near the problem area and invites workers and managers throughout the plant to contribute ideas on blue "fact" cards or yellow "solution" cards. The project team evaluates the ideas and decides which to adopt.

A third center is the use of part-time employment—to handle swings in demand without resorting to layoffs of full-time workers. One benefit of the use of part-timers for less-skilled tasks is that full-time workers now have greater flexibility in setting their own work schedules. The plant's final center is a world-class suggestion system, known as the Timken Idea Proposal Program (TIPP). **◎ TIPP places decision-making responsibility for ideas at the frontline level.** "Fewer signoffs are required," explains Mike Neff, manager of production operations. "Now, an associate who has an idea works it out with his supervisor and implements it without going through several layers of management for approval." The chief benefit is faster implementation of good ideas.

It all adds up to a total commitment to quality. The plant's customer-complaint records document an impressive seven-sigma performance in shipped product—a mere 0.31 defects per million bearings in 1991, despite very tight tolerances. And quality achievements are only one reason the employees at Bucyrus feel a sense of pride. "We have an 'alive' workforce today," Benson smiles. "They believe we have a future. . . . We believe we can compete. And we believe we are creating a competitive advantage."

—John H. Sheridan

The Timken Company

Contact: R. Dean Hunter, Manager, Production Control and Logistics

P.O. Box 391, Bucyrus, OH 44820

Telephone: (419) 563-2200

Employees: 1,000

Major products: Tapered roller bearings

THE TIMKEN COMPANY

Canton, Ohio

If you put the best people on the job, then treat them like partners, the rest almost takes care of itself.

—Lee Sholley, Plant Manager

Technology
Agile Manufacturing
Supplier Relations

Timken conceived the Faircrest Steel Plant (FSP) amidst bleak forecasts for the U.S. steel industry in the early 1980s, rolling the dice with a $450 million investment—nearly two-thirds of the company's net worth. The 900,000-square-foot facility, spread over 450 acres, opened its doors in 1985, only to mature as one of the most advanced integrated alloy-steel manufacturers in the world.

From the beginning, Timken outfitted the 1994 Best Plant award winner with the latest and greatest toys of steelmaking, including a 160-ton, ultra-high-powered, electric-arc furnace; a Hydris analyzer for measuring precise amounts of hydrogen in liquid steel; and a unique Elkem inspection unit for rapid inspection of bar surfaces.

◎ **High-tech aficionados would appreciate Faircrest's Hierarchical Computer Control System (HCCS)—typifying the influence of Silicon Valley on the Rust Belt.** The first fully integrated computer network in the steel business, the HCCS operates on both the plant and process level, managing information flow and also directing specific equipment. Alloy bin computers, for example, calculate the amount and type of alloy additives released into molten steel. Performance is tracked on-line against goals, then conscientiously displayed for employees to see at terminals and monitors throughout the plant. Programmers can log into the system from their homes if necessary. And field sales engineers deliver customer feedback in real time, fueling the customer-driven approach.

To truly appreciate how advanced the processes are at Faircrest, it is not enough merely to walk through the plant. "You really have to go and visit another plant just to see how good we are," says Mark Piatt, a scrap loader and enthusiastic tour guide. Even Shoichiro Toyoda, chairman of Toyota Motor Corporation, called Faircrest the cleanest steel mill he had ever seen.

Complementing the impressive Timken technology is the culture of FSP. A special contract with union affiliates has allowed management to define its workforce in terms of teams; hourly workers and staff are all considered associates. This partnership led to a 100 percent conversion of the plant's workforce to self-directed work teams.

FSP's transition to a totally team-driven plant stems from a four-year quest to overcome a simple issue of supply and demand. As product specialist Dave Thompson remarks, "If we could have produced 50,000 or 100,000 more tons [of alloy bars] it could have all gone to the marketplace." Capitalizing on that demand was another story. "Sure," says Sholley, "for another $350 million you could buy your way to additional capacity." But instead, the plant took the process improvement path, riding a wave of change on the strength of its teams. All told, the plant has dramatically maximized production capacity (originally designed for 550,000 tons of steel per year) to 770,000 tons per year—a 38 percent capacity boost without any major capital investment.

◎ **Faircrest's management team engineered a three-tiered strategy to fit under the umbrella of "accelerated continuous improvement," which is Timken's corporate vision.** The first principle is to run the mill faster. Running computerized plant simulations enabled efficient refinements in the synchronous-flow philosophy. And techniques such as single-minute exchange of dies have reduced routine work significantly. Roll changes that used to take half an hour to complete, for example, now take less than 10 minutes.

The second strategic element, which amplifies the speed principle, involves running the mill longer. An early-relief system, easing shift transitions into nearly perpetual cycles, coupled with a total productive maintenance program, has reduced downtime from 600 hours to 175 hours per year. "Much of what we do, such as scheduling, is structured through our computer systems," Sholley adds.

Furthermore, stringent quality measures have shaved rework time and salvaged alloy bars from scrap. Devices such as the diameter vision gauge, which measures bar diameter in just 10 seconds, allow time for adjustments by rolling-mill operators like Tim Morris before the steel is rendered unworkable. "This way, I don't make any scrap," he says, it's "just quality out." Overall, Faircrest has lowered its customer rejects to just 22.9 tons per million tons produced, a 62 percent reduction.

The final ingredient in the formula involves managing the entire plant as a team rather than as individuals performing independent functions. This is true not only for employees, but suppliers and customers as well. ◎ **Across the street stands "Supplier City," a 10-acre complex currently housing six sup-**

pliers, with more to be added by year's end. Electronic data interchange enables deliveries within 15 minutes of placing an order.

The plant's suppliers are each awarded 20-year agreements targeting a 10 percent value improvement every year. "That's something you can't get unless you're in a long-term relationship," says Scot Bowman, procurement specialist. Timken has already saved $3 million in inventory costs, with greater savings expected during the next few years.

Impressive results spew from Faircrest like sparks from its monstrous steel cooker. Between 1989 and 1994, manufacturing costs dropped by 25 percent, and maintenance costs were cut by 37 percent. Faircrest's output is nearly four times that of the U.S. or Japanese average (in worker-hours per ton). Cycle time has been compressed by 67 percent. Through just-in-time and synchronous-flow techniques, lot sizes have been reduced by 50 percent, and product inventory has been cut 70 percent ahead of finishing operations.

—David Bottoms

The Timken Company, Faircrest Steel Plant

Contact: Robert W. Merrell, Principal Quality Engineer

P.O. Box 6931, Canton, OH 44706

Telephone: (330) 471-7462

Employees: 458

Major products: Alloy steel bars

TRW VEHICLE SAFETY SYSTEMS, INCORPORATED

Cookeville, Tennessee

QUALITY
EMPLOYEE INVOLVEMENT

We're trying to run the business on the basis of facts and data, and have an ever-increasing number of people dealing with that data and participating in obtaining resolution.

—DAN SKIBA, PLANT MANAGER

Moving like astronauts on the moon—trailing umbilical cords that pump air into full protective suits and wearing foil-hooded face masks—technicians drop disks of sodium azide propellant into fixtures that facilitate loading them into foot-long metal canisters. Vacuum-equipped workstations remove any potentially explosive propellant dust, and rows of angled nozzles hang poised to deliver 300 gallons of water per minute at the slightest sign of heat or bright light.

Are we at a NASA rocket-test center or weapons depot? No, we are in the propellant load room, where the charges that inflate automotive air bags are inserted, at TRW Vehicle Safety Systems Incorporated's (VSSI) plant in Cookeville, a 1995 Best Plant award winner. As a relatively new facility, with first production in July 1991, TRW VSSI had the opportunity to build a world-class facility right off the bat. The challenge has been the 20-fold ramp-up of volume during the last three years, and an employment growth from 7 to 800 people to meet demand that outpaces supply of passenger-side airbag modules and inflators.

Producing some six million airbag modules in 1995, TRW VSSI is organized into customer-focused manufacturing work cells, each producing specific products for specific customers such as Ford, General Motors, Chrysler, BMW, Saturn, Toyota, Volkswagen, and Jaguar. Cells are further grouped by customers or market segments into business teams. Each team is led by a business-team manager who coordinates the activities of his or her cross-functional team, the individual members of which have dotted-line responsibility to the business-team manager and direct responsibility to their functional manager. This approach gives any team the feel of operating in a small company, yet with big-company support.

◎ **The principal avenue for the communication of customer focus and quality values at TRW VSSI is the Cookeville Operating System (COS).** "This is a comprehensive set of living documentation which monitors 15 to 20 key plant indexes," says industrial engineering manager Walter Marcum. Indexes include cost/unit, waste elimination, customer rejects, on-time delivery, and cycle time, as well as more subjective factors such as quality of work life, customer satisfaction, and a communication index. Reviewed monthly, each factor is assigned a goal and a champion who has the responsibility to spread initiatives throughout the plant to help meet that goal. The COS goals are spread pyramid-style down the organizational structure to individuals in the work cells.

At the production-cell level, the COS is reflected in targets set for the Six Fundamentals of Manufacturing, which are represented in a logo resembling a house. Good housekeeping is the foundation of the house, with pillars of safety, productivity, delivery performance, and waste elimination supporting a roof of quality. A lasting impression of the TRW VSSI facility is the housekeeping—the plant glows with cleanliness. This and the other five fundamentals are tracked and reported daily, with a manufacturing cell awarded a medallion at month's end for each target achieved. If six medallions are gained, the line is awarded a Six-Fundamentals flag, which is displayed at the "public" end of the line. "The employees take a lot of pride in that flag," says Cowan.

TRW VSSI is 100 percent team-based. ◎ **It is a union-free, all-salaried shop, with a deliberate effort made "to remove as many of the perceived and real barriers as possible that divide groups, and focus on the things that are most important," says human resources manager Lanny Knight.** There are no time clocks or assigned parking. When it came time to expand cafeteria food service, the employees voted for a McDonald's on-site. Although the golden arches declined, the company did attract the first Subway franchise operating in a U.S. plant.

Team training sessions are conducted at the TRW VSSI Development Center. This dedicated learning facility is loaded with tapes, books, and self-study programs on everything from spreadsheets and word processing to foreign languages, self-help, and parenting. Full-time trainers teach classes sponsored by the center, including Team Boot Camp, where virtually every employee was trained in a two-and-a-half day course on the basics of working together.

◎ **The development center also sponsors a unique internal internship program.** Here an employee proposes an internship in any department, typically with career exploration and advancement in mind. The employee determines his or her own learning objectives. Over a period of 12 weeks, not exceeding four hours per week, the employee completes the program on his or her own time and earns a certificate based on achieving the objectives.

When TRW VSSI attacks targets for waste and cost reduction, quality improvement, or better resource utilization, or is challenged by increased market demand, it turns to other structured initiatives. For instance, the Continuous Improvement Process Plus (AKA CIP+) is a kaizenlike, highly focused 10-step process designed to encompass all functions in the operation. The results of its application have been stunning. Implementation of 48 new ideas on a Chrysler line yielded a capacity increase from 252,000 modules a year to 324,000 at the same workforce level, while effectively reducing cost of labor 34 percent.

Less tangible, yet equally important, says Marcum, is that "technicians feel a part of the decision-making process since they all participate in brainstorming sessions, and a few volunteer to sit on the core improvement team. It encourages everyone on the shop floor to think in terms of improvement."

—TIM STEVENS

TRW Vehicle Safety Systems, Inc.

Contact: Walter Marcum, Industrial Engineering Section Head

P.O. Box 3027, Cookeville, TN 38502

Telephone: (615) 528-3611

Employees: 750

Major products: Passenger airbag modules and inflators

UNISYS CORPORATION

Pueblo, Colorado

. . . work with the customer to solve the problem up front, rather than fight about it later.

—Rolf Anderson, Manager-Business Development

Supplier Relations
Technology
Employee Involvement
New Product Development

The Pueblo Operations plant of Unisys Corporation's Government Systems Group, a manufacturer of printed circuit card assemblies, computers, and information-processing systems for the Department of Defense (DOD) and, increasingly, other government agencies, opened in 1985 with an extremely young workforce. The new employees "were fired with the enthusiasm of youth and were given some basic training through Pueblo Community College," recalls Darell Grass, a 27-year Unisys engineer who has been at the plant since its opening. "But they were basically inexperienced. That created problems at first. It took until about 1990 for this plant really to jell."

The jelling came at a fortunate time—just before the plant received a contract from the Air Force to build the Weasel Attack Signal Processor (WASP), the computer aboard the F-4 Wild Weasel aircraft that knocks out surface-to-air missile launchers. Delivery was scheduled in 18 months. But as the Persian Gulf War loomed, the Air Force wanted that shaved to 12 months.

"We weren't sure we could do it," confesses Melvin Murray, director of operations. ◎ **But by sending teams to vendors' facilities to help them shorten their lead times, forming special teams to cut assembly time in half (with employees voluntarily working extra hours), and setting up still other teams to trim test time, the plant went the Air Force one better.** It delivered the WASPs in only nine months—a significant factor in the smashing Desert Storm victory.

In the 12 months prior to winning its 1993 Best Plant status, Pueblo Operations also cut the cycle time for its printed circuit card assemblies by a spectacular 75 percent, from as much as four weeks to only six or seven days. The plant had been content with its previous cycle time. (Even though the facility builds more than 1,000 different types of circuit boards and must meet certain specifications and perform contractual testing not required by commercial customers, it still bettered industry standards.) Nevertheless, in its relentless search for improvement, Pueblo Operations found that it had built significant amounts of dead time into its system. And, as with the WASP, it zealously sought to identify operations that could be performed concurrently instead of sequentially. Total Quality Process (TQP) teams did the work in identifying opportunities and implementing improvements.

Between 1991 and 1993, the plant reduced setup time for its printed circuit assemblies by a remarkable 80 percent. It also trimmed its work-in-process inventory by 71 percent, total inventory by 60 percent, and total costs by 50 percent. Productivity soared 55 percent. Pueblo Operations also boasts a 100 percent on-time delivery record that stretched for nearly five years and a 100 percent customer acceptance rate.

From 1988 to 1993, Unisys invested $4 million in the plant, a unit of the Government Systems Group's Electronic Systems Division, for a dazzling array of new equipment. The facility's robotics technology is state of the art. ◎ **Its award-winning Material Management Center, located next door to the manufacturing facility to receive and test components, is equipped with a robotic material-handling system that stores and retrieves some 320,000 different parts with 99 percent accuracy.**

Throughout the plant, all operations are linked by a local area network of 350 personal computers, providing on-line, real-time management and planning information. Besides that, an electronic communications system, featuring 11 giant television screens on the shop floor, provides an instant, plantwide supplement to normal communications media.

◎ **"We go overboard on communications," says Murray.** He personally keeps an open door to all employees, conducts an "all-hands chat" with all employees 10 times a year, and every other week sits down with groups of 15 to 20 employees for informal breakfast "information exchanges." He also has instituted "skip-level meetings" in which employees can talk directly to their bosses' superiors.

Such communication contributes to the feeling of family that visitors to the plant quickly sense, a feeling that helps explain the plant's low turnover rate of 10.8 percent last year (8 percentage points below the industry average) and why only six employees chose to take advantage of a lucrative early-retirement offer. Indeed, employee involvement is the hallmark of the TQP program launched by Murray upon his arrival in January 1990. The program, which expands on Unisys' corporatewide Total Quality Management Process, has included a flattening of the organization from five layers to three and a proliferation of teams. Along with the empowerment has come a heavy emphasis on training, which increased from 717 hours in 1991 to 19,725 hours in 1993.

The TQP effort extends far beyond the shop floor. ◎ **Included is a stepped-up focus on customers, who now write requirement specifications, partner with the plant in the design process, and conduct design reviews.** An on-site videoconferencing center enables teams to meet with customers electronically. Suppliers also are a part of Pueblo Operations' TQP push. Trying to develop stronger relationships with fewer, but higher-performing, vendors, the plant has reduced the number of suppliers from 2,400 to about 1,500 and hopes to pare the figure to 1,000.

In environmental protection, the plant has eliminated CFC emissions two years ahead of the schedule called for by the Montreal Protocol environmental agreement. And it is a model community neighbor, having helped persuade 19 other companies, providing 3,600 jobs, to locate in Pueblo.

In March 1995, Unisys sold its defense operations, including the Pueblo plant to Loral Corporation. Four months later, the plant was closed.

—William H. Miller

VARIAN ASSOCIATES, INCORPORATED, NMR INSTRUMENTS

Palo Alto, California

QUALITY
TECHNOLOGY
CUSTOMER FOCUS

We intentionally try to move people around. It fosters teamwork. And before you know it, the different functions are talking more to each other.

—RAY SHAW, DIVISION GENERAL MANAGER

At his desk, surrounded by three computer terminals, Dr. Steven Patt, a Ph.D. chemist with 25 years of software engineering experience, is discussing an electronic newsletter that he dispatches weekly over the Internet to hundreds of customers who use the sophisticated nuclear magnetic resonance (NMR) spectrometers built by his company, the NMR Instruments unit of Varian Associates in Palo Alto, a 1994 Best Plants award winner.

Recipients of the newsletter are scientists who spend most of their work day using NMR spectrometers to analyze molecular structures by manipulating their magnetic characteristics. In their eyes, Patt says, information that helps them take full advantage of the expensive equipment ($150,000 to $2.5 million per machine) is "like gold."

If it seems a bit unusual to find a software engineer publishing a newsletter, how about a purchasing/materials executive heading up a new-product development project. Meet Debra Wollesen, NMR Instruments' materials manager, who is doing just that. More accustomed to working with suppliers, she exudes enthusiasm about her current challenge. "It provides me with a broader scope of the business—the customer side," Wollesen says. "In the materials field, you don't often get a lot of customer interaction." Call it cross-pollination, cross-functional management, or whatever you like. But in Varian's NMR Instruments business, heavy emphasis is placed on broadening people's horizons to give them a balanced view of the business.

After slipping to a distant second in the world market in the late 1970s behind a German competitor, NMR Instruments committed itself to retaking the technological lead in its industry and to more closely matching its offerings to customer needs. Between 1990 and 1993, its global market share climbed from 30 to 44 percent. (More than half of its products are sold outside the United States.)

If there is a simple explanation for the comeback, suggests Allen Lauer, Varian executive vice president, it is one word: *perseverance.* "The people here really believed that they could gain market share in a very competitive, technology-driven market," he says. "They never gave up. They never stopped looking for continuous improvement."

◎ **On a companywide basis, Varian's top management team emphasizes an "operational excellence" strategy that seeks continuous improvement in five areas: customer focus, unbending commitment to quality, highly flexible and responsive factories, speeding time to market, and organizational excellence characterized by outstanding communication, teamwork, and training.** It is "a very powerful strategy," Lauer insists. "And we put customer satisfaction at the top of the list."

While improving market share in the 1990–1993 period, NMR Instruments boosted sales per employee by 30 percent, reduced scrap and rework by 23 percent, reduced system cycle time by 35 percent, and trimmed warranty costs by 41 percent. Meanwhile, it earned ISO 9001 certification and got a leg up on its competitors by becoming the first company to offer powerful 750-MHz spectrometers which represent a technological breakthrough in its market.

The NMR Instruments plant has improved productivity an impressive 64 percent through a combination of initiatives, including design for manufacturability, investment in networked computers, and conversion to demand-flow manufacturing using kanban signals. Practices such as concurrent engineering have helped to shrink the product development cycle. For example, use of advanced tools—such as computer simulation in the design of complex, 12-layer printed circuit boards—has contributed to a 50 percent reduction in time to market.

A modular design-for-manufacturability approach provides much greater flexibility in offering customized instruments. The plant's configure-to-order system enables customers to get the most value for their dollar. ◎ **On the plant floor, a video-enhanced computer system delivers illustrated step-by-step assembly instructions to employees' computer terminals.** Customized builds are guided by a "customer option sheet," highlighting information from the sales order, that pops up on the terminal screens. Moreover, during production, a computerized configuration-management system creates a record of the components going into each product as it is built for a specific customer. The data is stored in the Varian system indefinitely in order to facilitate subsequent field support and upgrades.

The business unit also trimmed raw materials inventories, in part by forging strong partnership links with key suppliers in a value-managed relationship program. In effect, says manufacturing manager Gary Jow, it amounts to "making our suppliers an extension of our factory." Kanban arrangements with suppliers, which now cover some 75 percent of the volume of purchased materials, have significantly shortened parts lead times.

◎ **Insights into customer needs and concerns emerge from the stream of interactive communication over its Internet-based "virtual customer support network."** In addition to distributing Dr. Patt's electronic newsletter, the network stimulates E-mail communication with customers and between customers. "And because of the global nature of the Internet," Patt notes, "that person doesn't have to be in this building, or even in this country. One of our primary software support people is a fellow who works out of his home in Switzerland. . . . So we are able to pull in the resources of our company from around the world to solve customers' problems."

—JOHN H. SHERIDAN

Varian, Nuclear Magnetic Resonance Instruments
Contact: Gary Jow, Manufacturing Manager
3120 Hansen Way, Palo Alto, CA 94304
Telephone: (415) 493-4000
Employees: 340
Major products: NMR spectrometers

VARIAN ASSOCIATES, INCORPORATED, ONCOLOGY SYSTEMS

Palo Alto, California

CUSTOMER FOCUS
SUPPLIER RELATIONS

The sheer speed of change in our businesses makes it impractical for one person to make all the decisions needed for us to be competitive. Individuals must step forward in situations that call for their specific mix of skills and judgments.

—JIM YOUNKIN, MANUFACTURING MANAGER

High performance and constant improvement are goals everywhere in Jim Younkin's life. It's even apparent as he tools along El Camino Real to his job as manufacturing manager at Varian Associates Incorporated's Oncology Systems unit (formerly Medical Equipment) in Palo Alto. His conveyance of choice is a performance pickup, a GMC Syclone, 285 horsepower in factory trim, but which he has boosted to 350 horsepower in his quest for ever more spectacular performance. That kind of effort is also evident at Varian, where his determined efforts continue to refine what originally was a powerful manufacturing machine.

The accomplishments of Younkin's manufacturing team at the Oncology Systems unit are definitely not an example of a U.S. organization trying to regain dominance in its product niche. In accelerators, the 1992 Best Plant award winner is the established world leader among a distinguished group of competitors that includes Siemens, Philips, and General Electric. Instead, Younkin's efforts are directed at making this winner even better. To stay at the top, the Oncology Systems group is focusing its efforts on delivering ever increasing quality, product effectiveness, customer satisfaction, and production flexibility.

The knowledge that their product is used to treat cancers is a strong motivator. Product lines include the Clinac medical radiotherapy accelerator, the Ximatron radiotherapy simulator, and related products such as imaging systems, information-management systems, and accessories. In addition, for industrial applications, the plant makes the Linatron high-energy linear accelerator for nondestructive testing. More than 2,100 medical accelerators and 373 simulators are in place around the world, treating nearly 73,500 patients daily.

The Oncology Systems group's pursuit of excellence fits neatly into the overall corporate revitalization strategy of CEO J. Tracy O'Rourke. Replacing retiring chairman and CEO Thomas Sege in February 1990, O'Rourke inherited a $1.37 billion company that, like its neighbor Hewlett-Packard, is considered an entrepreneurial pioneer in Silicon Valley. Credited with the 1937 invention of the klystron tube, brothers Russell and Sigurd Varian founded the company in 1948 with $22,000 in capital. (The klystron played a vital role in World War II as a microwave source for the radar that helped Britain defend its shores.) In addition to microwave tubes, the company established patent rights for nuclear magnetic resonance (NMR) and began building its formidable reputation in the analytical equipment market. Its position was further strengthened in the 1950s with the invention of the electronic vacuum pump. That led to a role in the semiconductor industry. Skills in manipulating and depositing materials on semiconductor wafers evolved to ion implanters and sputtering systems, where Varian is a market leader.

By the time O'Rourke arrived, the company had diversified beyond its core businesses, was beginning to lose focus, and critics saw its profit potential slipping. And while Varian was already a multinational organization, no analyst described it as world class. Varian, however, was still a leader in technology. "It was technology-driven and -dominated," says O'Rourke. On the other hand, "Varian had trouble with strategic focus. Like a hummingbird, we were going from opportunity to opportunity, only to abandon them when the competition got too hot. We were simply spreading ourselves too thin."

In 1992, Varian was about midway through a three-phase master plan that O'Rourke hoped would restore greater profitability. Phase two of the program is a drive to achieve operational excellence. So Oncology Systems and the other three core businesses (analytical instruments, semiconductor production equipment, and electron devices) are focusing on such internal improvements as quality, customer satisfaction, shorter time to market, and flexible factories. All of these categories are increasing profitability.

The Oncology Systems unit takes operational excellence beyond the shipping dock. After all, a customer is not satisfied until the product is installed and operating. ◎ **In this instance, Oncology Systems created a task force that identified several hundred problems delaying setup and certifications of its Clinac medical linear accelerators.** It found that several hours could be trimmed if spare-component boards were set up to machine parameters in the factory rather than in the field. Shipping cables in plastic bags rather than in foam "popcorn" filler saved 30 minutes of cleanup time. Total installation time savings after the first round of improvements: 95 hours.

Supplier integration is a key element in the success of Oncology Systems' plant. While the facility performs system integration, final assembly, and testing, fully 70 percent of its manufacturing cost is in purchased components and subassemblies. Oncology Systems' goal is to develop supplier/partners capable of providing 100 percent defect-free parts, just in time, directly to the point of use. ◎ **Its supplier certification and partnership program is called Value Managed Relations (VMR).** To get VMR contract, the supplier must become a virtual business partner. Those who do enjoy long-term exclusive contracts (three to five years); access to training, tooling, and technical assistance; streamlined administrative procedures; and better scheduling visibility.

O'Rourke emphasizes that phase two, the drive to achieve operational excellence, is not a culture change, but a means of identifying bad habits and empowering people to effectively change them. "For example, people had fallen into the habit that it was OK to be late, and, as a result, customer shipments from some of our core businesses would lag, sometimes by several months." In 1992, Oncology Systems was meeting production schedules 92 percent of the time and customer delivery dates 100 percent of the time.

—John Teresko

Varian Associates, Oncology Systems

Contact: Jim Younkin, Manufacturing Manager

911 Hansen Way, Palo Alto, CA 94304

Telephone: (415) 424-5060

Employees: 1,200

Major products: Accelerators, simulators

WILSON SPORTING GOODS COMPANY

Humboldt, Tennessee

The people here take a proactive approach—and they are persistent. When they identify an objective, they pursue it until they achieve it.

—BILL SCHLEGEL, DIRECTOR OF OPERATIONS

EMPLOYEE INVOLVEMENT
QUALITY
SUPPLIER RELATIONS

Drop in on a Monday morning staff meeting at Wilson Sporting Goods Company's Humboldt plant and you are bound to notice the casual attire. Not a necktie in the room. All of the managers huddled with plant manager Al Scott are wearing golf shirts. And why not? After all, they are in the golf business—manufacturing, as Scott puts it, "the world's finest golf balls."

But there is another, more significant reason why neckties are taboo in the plant. "We're trying to break down barriers and create a team environment," says Scott, who has presided over an impressive turnaround at the plant since his arrival in 1985. Indeed, teams and teamwork are a permeating theme in the 1992 Best Plant award-winning facility, which turns out some 96 million golf balls a year—including the Ultra, Ultra Competition, and ProStaff models. Although participation is voluntary, some 66 percent of the workers have enlisted in formal teams that tackle problems and exploit opportunities for improvement. Many of the teams are headed by hourly associates, and they're authorized to spend up to $500 on projects without management approval. Minutes of weekly team meetings are posted on the large "Team Wilson" bulletin board for all to review.

In sports, successful teams have good coaching. ◎ **And at Humboldt, the managers and supervisors tend to think of themselves as coaches rather than bosses.** "We've tried to find supervisors who have the ability to coach people," says John Thoma, manufacturing superintendent in the compression-molding department.

The team atmosphere has been a major force in transforming the plant from one of Wilson's least efficient into one of its finest. Perhaps the most telling statistic: Between 1985, when the company decided to consolidate all of its golf ball production in Humboldt, and 1992, its market share has climbed from 2 percent to a robust 17 percent. Despite the eightfold increase in production during that period, total inventory has been slashed by 67 percent. And plantwide productivity is up 121 percent.

One element that made the comeback possible is a dedication to the five "guiding principles" that underpin Al Scott's formula for world-class competitiveness—continuous improvement, associate involvement, total quality management, just-in-time techniques to eliminate waste, and a focus on lowest total cost manufacturing. "We try to make every decision consistent with those five philosophies," Scott asserts.

The folks in Humboldt established a goal of becoming a leader in the "special order" golf ball business, producing products with customized logos as well as personalized golf balls imprinted with golfers' names. Thanks to an emphasis on quality, quick response, and customer service, that business has mushroomed from 185,000 dozen in 1985 to nearly two million dozen in 1992. At the same time, "We reduced the queue time for plates from six weeks to one day," says Bill Johnson, who manages the special-order unit. Consequently, Humboldt is capable of shipping a rush order within one day after receiving the artwork for a logo. Normal lead time, however, is 10 days compared with a six-week industry standard. The plant has accepted special orders as small as one dozen balls and as large as 85,000 dozen. The special-order unit, essentially a job shop that ships direct to its customers, is justifiably proud of its 99.9 percent on-time delivery performance.

◎ **Another goal that has been pursued relentlessly is the reduction of manufacturing losses. With associate involvement and use of the Demos Control Chart System, which integrates and pyramids plantwide quality data, the plant has achieved a 67 percent reduction in scrap and rework, along with savings of $9.5 million in quality costs between 1985 and 1992.** The Demos system involves "operator-participative record keeping" and making daily comparisons of quality data against targets for each operation. "We continually look at the exceptions where we failed to meet the target," notes Mike Kramer, manager of quality and materials engineering. "That enables us to focus on the elements where the greatest opportunities for improvement exist. The targets become tighter all the time. It is a continuous-improvement effort." One result: in the tricky injection-molding process, where DuPont Surlyn covers are applied in production of two-piece balls, the defect rate has been slashed from about 15 percent in 1985 to a mere 1.4 percent in 1992.

◎ **A strong supplier-partnership program, coordinated by materials manager Mark Loscudo, has also paid off handsomely, reducing the cost of hold-**

ing inventory and contributing to a dramatic gain in annual inventory turns from 6.5 in 1985 to 85 by 1992. "We didn't do what the auto companies typically have done—that is, tell the suppliers that they have to carry our inventory," stresses Scott. "Instead, we've given them our forecasts and asked them to try to realign their operations to more of a just-in-time basis." Consequently, lead times from key suppliers have been slashed, and inventories held by suppliers have actually been decreased by more than $1 million.

Internally, communication plays an important role in developing teamwork. ◎ **At quarterly "State of the Plant" meetings, the associates get an update on the financial picture, as well as a review of their performance on quality and other measures.** Upcoming events and future production plans, including anticipated volumes and workdays for each of the next 12 months, are posted on a "near-term planning board" for all to see. "The associates get the same rolling forecasts that our suppliers get," Scott observes.

—JOHN H. SHERIDAN

Wilson Sporting Goods Company
Contact: Ted Harpole, Training Manager
2330 Ultra Dr., Humboldt, TN 38343
Telephone: (901) 784-5335
Employees: 680
Major products: Golf balls

XEL COMMUNICATIONS, INCORPORATED

Aurora, Colorado

Without the team structure, we wouldn't be nearly as good on quality.

—John Gilpin, Vice President for Quality

Customer Focus
Employee Involvement

When GTE Corporation established an electronics manufacturing division in Aurora in 1980, it was considered an "intrepreneurial experiment" that the telecommunications giant hoped would develop the quick response capability to compete with smaller, more agile competitors. The experiment proved a success—although many of its most significant achievements occurred after GTE spun off the business in 1984.

GTE planned to close the facility as part of a strategic decision to divest manufacturing operations. But Bill Sanko, who had been division vice president and general manager, persuaded a handful of his comanagers to pool their resources—taking second mortgages on their homes—to buy the business, which became 1995 Best Plants award winner XEL Communications, Incorporated.

The bold move not only saved 70 jobs, but it helped to establish risk taking as one of the core values, which has since driven the firm to new heights. A maker of electronic products for the telecommunications industry, including the in-flight Airfone marketed by GTE, it now boasts 305 employees, and in 1994 the company recorded a 122 percent leap in revenues.

Not long after the buyout, the firm's senior managers asked themselves—and their employees—what kind of company they wanted XEL to become. As part of the exercise, they reflected on their commonly held values. Risk taking and innovation were high on the list. Key executives also shared a belief in the power of self-management long before it became one of the buzzwords of the empowerment movement. "We said it would be great if we were a company of self-managers, where everybody accepted responsibility for the quality of their work and for customer satisfaction," recalls Sanko, who is now XEL's president and CEO.

Self-management is now a reality. XEL's plant teams include process-oriented teams (which oversee key circuit board manufacturing operations such as autoinsertion and wave soldering) and product manufacturing teams (which handle manual assembly tasks, do touch-up work, and take responsibility for getting customer orders shipped on time). Line supervisor positions were eliminated on the plant floor. ◎ **"The teams make the commitment to the customer," says Sanko.**

It was fluctuation—and the unpredictability—of demand that initially drove XEL to explore ways to improve its manufacturing flexibility, thus enabling it to conserve cash by keeping finished-goods inventory at a minimum. "We have a marketplace that doesn't forecast very well," explains Bill Sanko. "But when they want products, they want them *now.*" Developing a culture of innovation and team decision making, coupled with the adoption of just-in-time (JIT) methods with cellular manufacturing and kanban replenishment systems, has been instrumental in providing the quick response capability on which XEL now prides itself.

Aware that product development is an area where quick response can be critical, XEL has sliced its development cycle by a remarkable 91 percent since 1986. It streamlined the relationship between sales and engineering design, and provided facilities where customer personnel and XEL engineers could work together on new products and incorporate modifications into prototypes on the fly.

In XEL's early days, it had a leg up on the competition, with a 42-day manufacturing cycle. Yet the company has since managed to compress that down to just three days and hopes to reduce it even further, thanks to improvements that will be realized from its recent relocation of its previous 53,000-square-foot home to a new 115,000-square-foot facility. The new building, which has several conference rooms for team meetings and a computer-based-skills training center, features greater emphasis on point-of-use stocking.

Manufacturing vice president John Puckett, who championed the need for cycle-time reduction, recognized that teams, JIT, and closer links with key suppliers would all play important roles. ◎ **He attributes much of the improvement to the plant's emphasis on cross training employees.** "Cross training in a team environment really helps us to respond to fluctuations in demand," he says, "because you are able to move resources where you need them very quickly, including engineering resources."

Nearly any discussion of the firm's manufacturing achievements, including a 92.5 percent productivity gain from 1990 to 1995 and a 58 percent drop in warranty claims, quickly turns to the role of teams. Under the plant's Total Quality

Control program, operators inspect all components installed at the previous workstation. ◎ **Each team sets its own quarterly goals for assembly defects (in parts per million), audited defects, on-time delivery, rework, productivity, cycle time, and cost per component.** Because production schedules may change with little notice, one of the toughest issues the teams must deal with is "keeping our priorities updated—what ships today, what needs to be built first," notes Fred Arent, a member of the Red Team, one of several product teams. "You have to make sure everyone is working on the most important jobs. We hold two meetings a day just to keep ourselves organized."

A rating system, based on the teams' success against their own goals, figures heavily in payouts made under the company's merit-pay system. The top-ranked team might, for example, get a 7 percent bonus while the bottom-performing team gets much less—or perhaps nothing at all. In fact, a team with a very poor performance rating could lose the right, at least temporarily, to be a self-managed team. Within a given team, adjustments can be made, based on peer appraisals, to increase the merit-pay amount awarded to a top-performing employee.

In October 1995, XEL Communications was acquired by Reading, Pennsylvania–based Gilbert Associates, a holding company.

—John H. Sheridan

XEL Communications, Inc.

Contact: John Puckett, VP, Manufacturing

17101 East Ohio Dr., Aurora, CO 80017

Telephone: (303) 369-7000

Employees: 305

Major products: Telecommunications equipment

XEROX CORPORATION

Webster, New York

QUALITY
AGILE MANUFACTURING

[Hourly employees] have had a large degree of input into the type of equipment we put on the floor.

—CHARLES BRAUN, ENGINEERING AND QUALITY MANAGER

The Malcolm Baldrige National Quality Award is a highly visible symbol of what a resourceful, purpose-driven company can accomplish. But Xerox's now-famous quality thrust of the 1980s gave "The Document Company" more than a fancy trophy. As Chairman David T. Kearns noted in his 1989 annual letter to shareholders: "We may be the only American company in an industry targeted by the Japanese to actually regain market share without the aid of tariff protection or other government help."

That may be the real reason for the large banner hanging at the entrance to one of the buildings in the firm's Webster manufacturing complex. It's a tribute to the troops for winning the Baldrige Award. The last line says, simply: "Thank you, Team Xerox."

The 1990 Best Plants award-winning Webster site, which includes the Components Manufacturing Operations (CMO) and New Build Operations (NBO) plants, typifies the company's commitment to continuous improvement. The CMO plant does sheet-metal work (including frame welding), machining, plating, harness assembly, and plastic injection molding. It supplies about 15 percent of the components that are assembled into finished products in the NBO plant across the street.

Thanks to improved designs and an extensive effort to upgrade the quality of both internal and external suppliers, the number of defective parts caught on the line at Webster plummeted from 10,000 parts per million (ppm) in 1980 to just 360 ppm by 1990. In that same period the number of defects found in finished products dropped 78 percent, from 36 per 100 machines to just 8. What makes that so remarkable is the complexity of a Xerox copier: The top-of-the-line 5090 duplicator contains no fewer than 30,000 parts.

Webster's quality gains can be traced to Xerox's "Leadership Through Quality" effort, inaugurated in 1984, which relies on an extensive employee-involvement program. Webster production workers participate in problem-solving teams known as Business Area Work Groups.

Despite the successes to date, Xerox has no intention of resting on its laurels. "The Leadership Through Quality program is evolving," explains Jim Horn, vice president of manufacturing for the Webster site. "And the tools are evolving."

◎ **One of the newest techniques, dubbed "A-delta-T," involves the use of a ratio to measure the gap between actual performance and a "theoretical best" level.** "If you come up with a ratio of 20, that means there is lots of room for improvement," notes David Knauss, materials manager. "A-delta-T is a tool that a team can use to set goals."

In the components manufacturing operation, each of the plant's five "business centers" has an A-delta-T team working on pilot projects. And each business center has a full-time quality coordinator.

Quality also has been enhanced by state-of-the-art equipment. Due to the consistency of robotic welds in the 5090 frame cell, first-pass yields on completed frames rose from 85 percent in 1989 to better than 99 percent in 1990. In the NBO assembly plant, an Adept robotic vision system in the customer-replaceable-unit department helps to assure quality by detecting missing or flawed components.

At the end of the semiautomatic assembly line for low-volume 5028 copiers there is a two-employee "major repair" station, which corrects any problems detected in the manufacturing process. But thanks to Xerox's broad-based quality effort, they're not very busy these days. Nodding toward one of the two workers, a tour guide quips: "She has a lot of time for reading now."

On the flagship 5090 duplicator line, a dedicated effort to lick quality problems with the delicate document-handler systems led to the recent elimination of a final test station for document handlers. By helping operators devise ways to eliminate defects, the 20 people in the final test area "worked themselves out of a job," says Fran Shippers, one of the 20 who now coordinates tours of the NBO plant.

But quality is not the only important thrust at Webster these days. JIT methods, electronic kanban links to suppliers, setup reduction, cross-trained workers, self-scheduling teams, and flexible equipment are all getting heavy emphasis. "Just in time is probably our number one strategy right now," says Horn.

One JIT showcase is the CMO's 5090 frame-welding cell, developed by a joint union/management team. The cell includes a 200-ton stamping press, a numerically controlled vertical milling center, robotic subassembly welding, a manual final welding station, and a high-speed coordinate-measuring machine.

◎ **The use of a pull system and smaller lot sizes in the frame cell has reduced work in process (WIP) and thereby improved quality.** For example, the number of units between the cell's machining and welding operations was reduced from 300 sets to 10. "Now if there's a problem, we don't have to rework 80 or 100 frames," says Al Gallina, an engineer.

Quick setup methods also reduced WIP in the CMO's wire-harness center. Setup time for the flexible, automated Condor machine, which cuts wires for harness subassemblies, was reduced 75 percent. Hourly workers recommended the purchase of the Condor machine, which led to an 85 percent productivity improvement in the area.

—John H. Sheridan

Xerox Corporation, Webster Manufacturing
800 Long Ridge Rd., Stamford, CT 06904
Telephone: (203) 968-3000
Employees: 2,829
Major products: Copy machines

ZYTEC CORPORATION

Redwood Falls, Minnesota

Employee Involvement

We've been here a long time, and we haven't changed our tune. If you change your tune, your people won't take you seriously, and they won't trust you.

—Ron Schmidt, CEO

Ron Schmidt, chairman, president, and CEO of Zytec Corporation, is different. He does not have the gift of gab, nor that regal, starched air that people of position, it is assumed, must have. Ron Schmidt is an unpolished, unassuming introvert who talks straight. In his short-sleeved button-down shirt—no tie, no jacket with crisply folded pocket square—he is indistinguishable from the engineers who constitute the majority of employees at Zytec's headquarters building.

Still, more than attitude and attire sets Ron Schmidt and Zytec, a manufacturer of electronic power supplies, apart from the pack. The company's unwavering dedication to the late Dr. W. Edwards Deming's quality principles has won management the trust of employees, produced 28 percent revenue growth during the 1991 recession, and won the Zytec team the Malcolm Baldrige National Quality Award, the Minnesota Quality Award, and a 1992 Best Plants award.

Most companies embrace a good portion of Deming's philosophies but find quite a few of the quality guru's tonics too tough to swallow. These often relate to the Deming directives on performance reviews, compensation and rewards, quotas, and even some of his most basic quality improvement methodologies. Schmidt admits that, like other companies, Zytec struggled with Deming's approaches, but the fact that management has not compromised or sidestepped the Deming principles helped forge a high degree of trust among management and nonmanagement employees.

Deming made a point of emphasizing the importance of this action/reaction relationship. In the "14 points" that form the crux of his management philosophy, point 1 is "Create constancy of purpose toward improvement of product and service. . . ." ◎ **Constancy of purpose is essential, Schmidt has found, to**

gaining the commitment of employees. "Employees are used to the program of the month, and they have been conditioned to wait it out until the program runs out of steam and the next new program arrives or the next new manager comes in and announces another companywide drive."

Schmidt says that to foster trust, managers must be up front and honest in their communications with employees at all times, but especially when business conditions sour. "We've told our people that we will try to keep all of our permanent employees employed," he says. "But back in the fourth quarter of 1984 we were running at a rate of $80 million a year revenue, and six months later we were running at a rate of $24 million a year—and we were a leveraged buyout company. We had to shut down our factory for about six weeks, and we cut back everyone's salaries temporarily once we brought them back to work.

"Those were difficult times, but I think the employees understood that we had to do it, that it was a matter of survival. It was scary for them, and they didn't like what they were hearing, but we were honest from the start, and they trusted us. We survived that crisis, and now everyone's benefiting."

Zytec's quality and productivity achievements highlight the success of the management approach. By 1992, Zytec's on-time delivery performance for all of its products has improved 22 percent from 1990 levels to 99.4 percent. Between 1987 and 1992, scrap and rework were reduced 66 to 0.23 percent of sales, and warranty costs had fallen 72 percent. Productivity improved 75 percent during the same period because of extensive use of autoinsertion rather than hand-insertion techniques and automatic testing rather than operator-assisted testing.

Schmidt says that adopting Deming's views on compensation policies and performance reviews was hard for Zytec. Deming maintained (and statistics support his claim) that the variance in performance among the large majority of employees within any one job classification is small—too small to accurately draw distinctions and determine rewards. Rewarding employees differently based on minor performance differences can also be counterproductive because it creates envy and disappointments and thwarts the formation of strong teams. ◎ **So, at Zytec every employee within a specific job classification receives the same pay, and there are no individual bonuses paid out at the end of the year.** "We feel that it is the quality of our system and processes that determines the quality of our products," explains Schmidt. "We feel that we can improve these processes best through team efforts, and that to improve teamwork, team members within job classifications should be equally compensated."

In January 1992, Zytec even eliminated individual performance incentives for its sales force. Salespeople are compensated entirely through their salaries. Although the sales force was reluctant to adopt the approach, one of the benefits is that past variances in their compensation that arose from differences in the size of the territory they covered or variation in regional business conditions have been wiped out.

◎ **In line with Deming's beliefs, Zytec also axed traditional performance reviews.** We replaced performance reviews with a meeting between supervisor and employee," says Schmidt. "Unlike performance reviews, these aren't geared to linking performance and pay. Instead, we talk about obstacles to doing the best possible work, training that might be needed to improve job performance, and how the supervisor/employee relationship can be improved. It's a goal-setting and development tool."

Schmidt has emphasized that true quality comes from pride, not just incentives. "What we are working toward is getting people to want to pursue quality because they take pride in their work, not just because they're told to do it. We are moving from a controlled environment to a committed environment."

"I've observed," concludes Schmidt, "that just about everyone who has really had success in their total quality programs tells me how great their employees are. I believe it's all a reflection of the culture that management creates rather than where the employees come from. In the end, managements get the quality of employees that they deserve."

—DAVE ALTANY

Zytec Corporation

Contact: Max Davis, Vice President of Manufacturing

1425 E. Bridge St., Redwood Falls, MN 56283

Telephone: (612) 941-1100

Employees: 800

Major products: Electronic power supplies

STATISTICAL PROFILE

The America's Best Plants Statistical Profile presents the quantitative indicators of world-class manufacturing. This is a unique numerical portrait of excellence, and they are included here as a resource and a challenge. As a resource, these statistics offer a rare look into the operating parameters of a select group of highly successful manufacturers. This is hard information that provides benchmarks for your own work. As a challenge, the statistical achievements of the Best Plants represent the highest standards in manufacturing. If you take up the challenge and successfully match the results of the Best Plants, your operation will stand among the best—the most efficient, effective, and profitable in the world.

The sections of the Statistical Profile largely correspond to the Assessment Survey that directly follows this section. After completing the Survey, readers can compare their own statistical results to those of the Best Plants.

In order to give the database greater depth and reliability, statistics drawn from entries of all 25 annual Best Plants finalists are included in the profile. When no specific year is indicated for a statistic, it was calculated using the 1995 finalists, the latest year available prior to this book's publication.

Before we turn you loose, it is only fair to warn that numbers can lie, although in this case it is certainly not intentional. Results vary widely with the type of operation. And, as *IndustryWeek* senior editor John Sheridan notes, ". . . readers should bear in mind that performance levels in one industry may not be realistic goals for another. For, example, while a high-volume repetitive manufacturing operation may surpass 200 WIP (work-in-process) turns annually, a much lower figure conceivably could represent world-class inventory management for a low-volume maker of complex, highly customized products." Recognizing that reality, the Profile also offers a series of break-out statistics by business type. The overall measures and the break-out figures should be used whenever possible.

AMERICA'S BEST PLANTS STATISTICAL PROFILE

Comprehensiveness of Effort

Three-Year Comparisons (*percent of plants indicating areas of major emphasis*)

	1993	1994	1995
Total Quality Management	96	96	92
Management by Policy (breakthroughs)	20	44	52
Cycle-time reduction	92	96	96
JIT/continuous-flow production	96	96	100
Cellular manufacturing	72	80	92
Focused-factory production concepts	84	88	84
Supplier partnerships	96	96	96
Customer satisfaction programs	96	96	100
EDI links to customers/suppliers	72	80	88
Competitive benchmarking	76	68	96
Continuous improvement	100	100	100
Visibility systems	—	68	—
Employee empowerment	100	100	100
Accelerated worker training	80	68	68
Apprenticeship programs	40	44	56
Use of work teams	100	100	96
Employee cross training	96	100	100
Use of cross-functional teams	96	100	96
Performance measurement and reward systems	84	100	92
Total Productive Maintenance (TPM)	64	72	68
Advanced process technology	84	84	76
Advanced material technologies	56	72	48
Inventory reduction	96	100	92
Concurrent engineering	92	92	84
Design for assembly/manufacturability	84	96	88
Design for quality	88	88	92
MRP II system enhancement	72	72	52
Delivery dependability	96	100	100
Reducing order-to-shipment lead time	88	100	96
Flexible manufacturing methods	88	92	84
Agile manufacturing strategies	—	80	88
Streamlined production flow	88	80	92

Computer-integrated manufacturing (CIM)	76	88	80
Enterprise integration*	64	60	68
Improving union/management cooperation	40	40	20

* Described in 1993 as "cross-functional computer integration."

Quality

Quality Indicators (*in percent unless otherwise noted*)

Products and Components	High	Low	Avg.	Median
Finished-product first-pass yield*	100.0	90.1	97.2	98.3
Finished-product yield improvement, 5 years†	90.0	0.0	31.6	16.0
Typical finished-product yield for industry‡	98.9	41.0	86.8	95.0
Finished-product pass-through yield	99.6	75.0	93.6	95.9
Major component first-pass yield	99.9	88.3	97.2	98.0
Component yield improvement, 5 years†	94.0	0.0	31.9	19.7
Typical component yield for industry‡	99.6	65.0	92.7	95.0
Component process capability (*Cpk*)§	6.42	1.17	2.47	1.90

* In calculating yield, 23 of the plants deduct for items requiring rework in the immediate upstream stage of the process. One plant considered itself to be only a producer of components and thus did not report for this category. A second plant did not calculate quality based on this indicator.
† "Percent" yield improvement was defined to mean percent reduction in rejects.
‡ Not all finalists reporting product and component first-pass yields reported industry averages.
§ Twenty plants reported that they calculate Cpk to gauge process capabilities.

Across All Processes	High	Low	Avg.	Median
Average *Cpk value* across all manufacturing processes	4.09	1.08	1.96	1.60
First-pass yield for *all* finished products (weighted average)	99.99	76.0	95.7	97.9
Total scrap/rework as a *% of sales*	9.0	0.0	1.50	0.8
Scrap/rework reduction in 5 years	91.2	0.0	51.3	50.0
Customer reject rate on finished products (*parts per million*)*	16,000.0	0.0	3,082.5	492.0
Reject-rate reduction in last 5 years	91.0	35.0	64.6	63.6
Reduction in warranty costs within last 5 years	84.0	6.0	41.9	29.5
Number of benchmarking studies conducted in last 3 years	35.0	0.0	10.8	7.0

* Parts per million = ppm.

Three-Year Comparisons of Medians (*in percent unless otherwise noted*)

	1993	1994	1995
Finished-product first-pass yield	98.7	98.9	98.3
Yield improvement, last 5 years*	25.0	10.0	16.0
Finished-product pass-through yield	—	—	95.9
Major component first-pass yield	98.1	98.1	98.0
Component yield improvement*	20.0	7.5	19.7
Total scrap/rework as a *% of sales*	0.7	0.9	0.8
Scrap/rework reduction in 5 years	32.0	48.5	50.0
Finished product customer reject rate (*ppm*)	184.0	250.0	492.0
Reject-rate reduction in 5 years	57.0	60.0	63.6
Warranty cost reduction in 5 years	27.2	50.0	29.5
Number of benchmarking studies in last 3 years	7.0	7.0	7.0

* Data reporting is somewhat inconsistent: "percentage point increase" in yield was asked for in past years, but several plants reported *percent reduction in defects,* since that is how they track yield gains.

Top Quality Performances
Finished Product First-Pass Yields

Product	*Percent*
1995	
Diffusion pumps	100.0
Water pumps	99.9
Box-level electronics	99.6
Passenger airbag modules	99.5
Drawn housings	99.5
Hydraulic valve lifters	99.1
Bulk bags	99.1
Handheld electric power tools	99.0
Ethyleneamines	98.9
Cold-rolled sheet steel	98.9
1994	
Spectrometers	100.0
Electronic weapons systems	100.0
Filled solution bags	99.9

Rotary actuator	99.8
Electronic products (box level)	99.6
Hydraulic valve lifters	99.5
A/C compressors	99.5
Radial tires	99.4
Power supplies	99.1
Feminine sanitary products	99.0
1993	
Polyisobutylene	100.0
Tank gunner's control	100.0
Medical equipment	100.0
Bellows assembly	99.9
Surgical drape elements	99.8
Process control cabinet	99.7
Radial tires	99.3
Gasoline fuel injectors	99.3
Electrical products	99.3
Imaging equipment	99.0

Major Component First-Pass Yields

Product	*Percent*
1995	
Engines/short blocks	99.99
Custom metal stampings	99.9
Hydrocarbons	99.9
Output shaft for electric drill	99.8
Bag webbing	99.7
Central processing unit	99.4
Valve body	99.2
Gear case	99.1
Add-on electronic components	99.0
Automotive axle	98.6
1994	
Surgical equipment	100.0
Actuator gear	99.9

TV picture tubes	99.7
Leadless chip carriers	99.6
Machinery components	99.0
Medical devices	99.0
Camera components	99.0
Acoustical ceiling products	98.9
Webbing for containers	98.2
Empty solution bags	98.1
1993	
Tank control	100.0
Electrical cabinets	99.9
Surgical gowns	99.9
Electrical receptacles	99.8
Ethylene	99.4
Polypropylene granules	99.0
Input/output board	99.0
Rubber	98.5
Valve group	98.1
Ice cream mix	98.0

Lowest Scrap Rework

Product	*Percent of sales*
1995	
Multiuser computer systems	0.00
Computer workstations	0.04
Electronic products	0.07
Axles	0.15
Escalators	0.23
Plated stampings	0.26
Portable electric power tools	0.30
Diffusion pumps	0.50
Handheld electric power tools	0.50
Chemicals	0.60

1994	
Electronic products	0.01
Steelmaking	0.01
Service station equipment	0.07
Surgical sutures, needles	0.30
Bulk containers, filling equipment	0.49
Laser components	0.50
Pharmaceuticals, medical devices	0.66
NMR spectrometers	0.70
Aerospace flight control systems	0.76
Pharmaceutical solutions	0.80
1993	
Fuel-dispensing equipment	0.06
Nuclear power station*	0.07
Surgical supplies	0.10
Process control equipment	0.20
Chemicals†	0.26
Ice cream	0.40
Injection-molding machinery	0.40
Electrical equipment	0.50
Military equipment	0.51
Medical equipment	0.54

* Figure represents radioactive waste disposal cost as a percent of revenues.
† Figure represents "flaring factor."

Other Quality Issues

Three-Year Comparisons	*Percent of plants*		
	1993	1994	1995
Has plant applied for ISO 9000 certification?	44	52	—
Has ISO 9000 certification been issued?	32	28	56
Does plant extensively use computerized SPC?	80	72	56
Does SPC provide real-time process feedback?	96	80	68
Does plant calculate Cpk values?	—	—	80

Quality Methods Used	Percent of plants		
	1993	1994	1995
Manual SPC	84	84	60
Computerized SPC	88	72	56
Operators inspect own work	92	100	100
Quality Function Deployment (QFD)	32	52	60
Total Quality Management (TQM)	88	88	92
Design of experiments/Taguchi methods	44	60	60
Poka-yoke methods	48	64	56
Design for quality	68	80	88
Quality circles	44	24	32
Employee problem-solving teams	100	100	100
Other*	—	36	40

* Includes: corrective action teams; process improvement teams; computerized TQC points; activity-based measurement performance reports for office personnel; process validation; gain-sharing tied to quality; internal quality audits; supplier development program; expanded training; failure effect and mode analysis (FEMA), cause-and-effect diagramming with cards (CEDAC); five-step problem-resolution process; concurrent engineering; demand-flow manufacturing.

Employee Empowerment/Involvement

Empowerment Indicators (*in percent unless otherwise noted*)

	High	Low	Avg.	Median
Percent of production workforce now participating in self-directed work teams*	100.0	0.0	72.2	95.0
Percent of *total* workforce participating in self-directed work teams	100.0	0.0	67.3	83.0
Formal training per employee annually (*days*)	15.0	1.0	7.5	7.5
Current employees to supervisors (*ratio*)	325:1	7.5:1	53:1	28.5:1
Current annual labor turnover rate	30.6	0.01	5.7	2.5

* Eleven of the plants reported 100% of production workforce participates in self-directed work teams.

Three-Year Comparisons of Medians

	1993	1994	1995
Percent of production workforce now participating in self-directed work teams	75.0	68.0	95.0
Percent of *total* workforce participating in self-directed work teams	—	66.0	83.0
Formal training per employee annually (*days*)	7.0	8.0	7.5
Current employees to supervisors (*ratio*)	20:1	31:1	28.5:1
Current annual labor turnover rate	—	2.8	2.5

Three-Year Comparisons

	Percent of plants answering yes		
	1993	1994	1995
Do self-directed work teams make day-to-day decisions on production operations?	92	80	88
Does plant use work teams other than self-directed teams?	100	88	92
Has supervisor/employee ratio been reduced?	96	88	72
Do production workers participate in cross-functional teams for special projects?	100	100	—
Do production workers participate in process development efforts?	100	96	100
Have plant personnel participated in concurrent engineering to reduce product development cycle?	92	96	100
Has plant adopted a *kaizen* approach to continuous improvement?	—	80	96
Do all employees receive frequent training?	96	96	—
Does plant emphasize cross training of production employees?	100	100	100
Are employees paid for skills acquired?	60	68	68
Are plant employees involved in competitive benchmarking efforts?	84	84	80
Does plant provide financial rewards for *team-based* performance/results?	—	76	84

Manufacturing Operations and Flexibility

Manufacturing Practices
Three-Year Comparison

	Percent of plants answering yes		
	1993	1994	1995
Does plant manufacture or assemble components in small lot sizes?	84	92	96
Has plant adopted focused factory production systems?	84	84	92
Does plant practice Total Productive Maintenance (TPM)?	68	80	72
Has plant adopted JIT/continuous-flow production methods?	96	100	96
If yes, are JIT methods predominant?	96	88	76
Does plant employ a "pull" system with kanban signals?	—	88	72
Does plant offer JIT delivery to customers?	92	88	96
If JIT delivery is offered to customers, how is it achieved?*			
Shipments from finished goods	20	4	24
Primarily build/produce to order	72	84	68
In improving repetitive operations, is takt time considered and analyzed?	—	52	—
Have standard order-to-shipment lead times been significantly reduced?	88	96	80
Does lead time create competitive advantage?†	88	68	—
Does plant build primarily to order rather than to stock?	80	84	88
Have quick changeover methods been widely adopted?	96	84	—

* Of the 24 plants offering JIT delivery, only 23 specified how it was achieved.

† Seven of the 1994 finalists—28 percent—indicated they were "not sure" whether order lead time represented a significant competitive advantage.

Responsiveness and Measurements (*in percent unless otherwise noted*)

	High	Low	Avg.	Median
Standard order-to-shipment lead time for major products (*days*)	120.0	1.0	24.0	11.0
Reduction in order-to-shipment lead time	97.0	14.0	55.9	55.5
Shortest lead time on rush orders for above major product (*days*)	28.0	0.08	4.6	1.0
Reduction in component lot sizes	99.0	0.0	58.5	69.0
Manufacturing cycle time for a typical product (*days*)	35.0	0.001*	6.3	3.0
Percent reduction in manufacturing cycle time within last 5 years	99.0	10.0	58.7	59.5
On-time delivery rate	100.0	70.0	95.2	98.0
Based on customer request date	60% of plants			
Based on date promised	40% of plants			
Production schedules met (*% of time*)	100.0	70.0	95.2	98.0
Average machine availability rate†	100.0	80.0	94.9	95.0

* Steel fittings and components: Reported cycle time was two minutes.
† Some plants use the term *machine uptime*.

Three-Year Comparisons of Medians (*in percent unless otherwise noted*)

	1993	1994	1995
Standard order-to-shipment lead time for major products (*days*)	14.0	14.0	11.0
Reduction in order-to-shipment lead times (%)	50.0	50.0	55.5
Shortest order-to-shipment lead time (*days*)	—	—	1.0

Shortest Standard Order-to-Shipment Lead Times

Product	*Lead time (in days)*
1995	
Vacuum valves and flanges	1.0
Axles	1.5
Fuel injectors	2.0
Handheld electric power tools	2.0
Stamped motor housings	3.0
Biopsy forceps	3.0
Steel products	7.0
Engines	7.0
Computer workstations	10.0
Symmetric multiprocessors	10.0
Complete motherboards	10.0
Aluminum termination lugs	11.0
System-level electronic components	12.0
Roller hydraulic valve lifters	13.0
Service valves	14.0
1994	
Single-use cameras	1.0
Feminine sanitary napkins	1.0
Pharmaceuticals	1.0
Power supplies	2.0
Magnets	2.0
Biopsy forceps	3.0
Medical sutures	3.0
Pharmaceutical solutions	3.2
Wiring harnesses	3.5
Commercial electronic products	5.0
Vinyl ceiling panels	6.5
Air-conditioning compressors	12.5
Hydraulic valve lifters	13.0
Radial tires	14.0
Plastics injection-molding equipment	14.0

	1993
Ethylene	Almost immediate (pipeline)
Propylene	0.5
Polypropylene	1.0
Electrical receptacles	1.5
Electronic engine control	2.0
Ice cream	3.0
Yogurt	3.0
Surgical gowns	3.0
Electrical switch	3.0
Electrical drive	3.0
Electronic products	4.0
Surgical drapes	5.0
Camera components	5.0
Antilock braking systems	5.0
Bellows assembly	7.0

Manufacturing Operations
Three-Year Comparisons of Medians

	1993	1994	1995
Reduction in component lot sizes (%)	75.0	75.0	69.0
Manufacturing cycle time for a typical product (*days*)	7.0	3.0	3.0
Percent reduction in manufacturing cycle time within last 5 years (%)	62.0	50.0	59.5
Manufacturing cycle time as a percent of order-to-ship lead time (%)	—	20.5	30.0
On-time delivery rate* (*% of time*)	98.5	99.0	98.0
Production schedules met (*% of time*)	98.0	99.0	98.0
Average machine availability rate (*% of time*)	96.9	95.0	95.0

* On-time delivery calculations differ. Of the 1993 finalists, 48 percent based it on date requested by customer; 44 percent based it on promised delivery date; one plant used both measurements. Of the 1994 finalists, 16—or 64 percent—use customer request date as basis for calculation. Of the 1995 finalists, 60 percent use customer request date.

Environmental/Safety Issues

	Percent of plants answering yes		
	1993	1994	1995
Has plant adopted a pollution prevention strategy to eliminate pollutants at an upstream stage of the process rather than at the end of the pipe?	96	96	100
Has plant implemented pollution control measures that exceed current regulatory requirements?	96	80	92
Does plant have a strong worker-safety program?	100	100	—
Has plant been cited by federal or state Environmental Protection Agency authorities for serious violations of environmental laws in last 5 years?	—	0	0

Shortest Manufacturing Cycle Times and 5-Year Reductions

Product	*Cycle time*	*% reduced*
1995		
Steel fittings and components	2.0 minutes	50.0
Stamped motor housings	15.0 minutes	99.0
Automotive axles	89.8 minutes	22.0
Handheld electric power tools	2.6 hours	14.0
Flexible intermediate bulk containers	5.0 hours	90.0
Engine	20.5 hours	10.0
Aluminum termination lugs	1.0 day	90.0
Vacuum valves and flanges	1.0 day	90.0
Service valves	2.0 days	99.0
Handheld electric power tools	2.0 days	97.0
Desktop computers	3.0 days	40.0
Biopsy forceps	3.0 days	50.0
Channel cards	3.5 days	35.0
Symmetric multiprocessors	4.5 days	55.0
Circuit card assemblies	5.0 days	64.0
Telecommunications equipment	6.0 days	80.0

1994		
Acoustical ceiling panels	5.0 minutes	20.0
Color TVs	2.0 hours	50.0
High-volume computer product	2.0 hours	—
Printed circuit boards	4.0 hours	98.0
Pharmaceuticals	4.5 hours	25.0
Aluminum alloy machining stock	7.0 hours	14.0
Bulk bags	8.0 hours	83.0
Hot-rolled steel	12.0 hours	—
A/C compressors	1.33 days	55.0
Controller consoles	1.5 days	50.0
Wiring harnesses	2.0 days	93.0
Radial tires	2.6 days	15.2
Microsurgical instruments	3.0 days	—
Pharmaceutical solutions	3.0 days	66.0
Gasoline dispensing systems	3.0 days	50.0
1993		
Surgical gowns	4.0 hours	50.0
Low-voltage controls	28.5 hours	82.0
Electronic engine control	1.7 days	85.0
Rubber products	1.7 days	89.0
Electrical safety switch	3.0 days	95.0
Packaged ice cream	5.0 days	67.0
Electrical receptacle	5.0 days	50.0
Bellows assembly	7.0 days	75.0
Printed circuit assembly	7.0 days	75.0
Magnetic resonance spectrometers	7.0 days	53.0
Pressure transmitters	8.0 days	62.0
Receiver	9.0 days	75.0
Thermoelectric coolers	10.0 days	25.0
Process control systems	14.0 days	89.0

Inventory Management

	High	Low	Avg.	Median
Annual inventory turns	98.3	2.8	17.0	12.0
Annual work-in-process (WIP) turns	656.0	8.5	80.3	39.9
Total inventory reduction in 5 years (%)*	73.0	0.0	36.4	35.0
WIP inventory reduction in 5 years (%)	90.0	0.0	42.5	42.0

* In a few cases, inventories were not reduced on an absolute basis; however, due to increases in production, inventory was reduced as a percent of production volume.

Three-Year Comparisons of Medians

	1993	1994	1995
Annual inventory turns	7.6	7.0	12.0
Annual work-in-process (WIP) turns	41.3	42.6	39.9
Total inventory reduction in 5 years (%)	51.0	41.0	35.0
WIP inventory reduction in 5 years (%)	49.7	50.0	42.0

Highest Annual WIP Turns and 5-Year WIP Reductions

Product	*Annual WIP turns*	*% reduced*
1995		
Flexible intermediate bulk containers	656.0	50.0
Passenger airbag modules and inflators	240.0	80.0
Steel fittings and components	180.0	50.0
Remanufactured diesel engines	122.0	69.0
Computer workstations and servers	79.4	NA*
Plated stampings and machined parts	71.5	48.0
Electronic products	63.3	19.0
Multiuser computer systems	57.0	0.0
Hydraulic valve lifters	50.1	42.0
Commodity and specialty chemicals	50.0	33.0
Silicon wafers	45.5	55.0
Fuel-injection components	40.1	12.7
Fluid conveying products	39.6	67.0
Disposable medical devices	35.0	90.0
Automotive axles	34.9	34.9

1994		
Acoustical ceiling panels	817.0	50.0
Bulk containers	263.0	75.0
Plastic molded components	200.0	NA*
Consumer electronics	193.0	32.0
Pharmaceutical solutions	173.0	64.0
Pharmaceuticals and medical devices	148.0	40.0
Laser assemblies	140.0	96.0
Electronic products	110.0	75.0
Air-conditioning compressors	55.0	90.0
Radial tires	52.0	NM†
Wiring harnesses	45.0	70.0
Hydraulic valve lifters	43.0	50.0
Service station equipment	42.2	40.3
Piston rings	40.8	27.0
Surgical instruments	35.0	90.0
1993		
Petrochemicals	203.0	41.0
Chemicals	194.2	38.5
Rubber products	150.0	60.0
Automotive electronics	148.0	80.0
Passenger vehicles	120.0	NM†
Automotive equipment	46.2	21.0
Electrical supplies	43.8	55.0
Pressure transmitters	41.3	49.6

* NA = no answer from plant.
† NM = not meaningful.

Technology

Technology Investments

	Percent of total plant budget			
	High	Low	Avg.	Median
Approximate annual investment in information systems and technology (includes hardware, software, and training and support costs)	30.0	1.0	7.9	5.0

Use of Advanced Technology
Three-Year Comparisons

	Percent of plants answering yes		
	1993	1994	1995
Does plant make *extensive* use of computer-integrated manufacturing (CIM) technologies?	76	76	84
Prior to implementing CIM and/or automation, was the production process streamlined to eliminate non-value-added activities?	84	72	—
Has plant implemented an enterprise-integration strategy, using information technology to link manufacturing/engineering with other functional areas of the business?	—	—	76
Has plant begun to use Internet/online services to support strategic business objectives?	—	—	52

Percent of plants using the following technologies *extensively*

	1993	1994	1995
Computer-aided design (CAD)	96	88	96
CAD with solid modeling	52	40	40
CAD integrated with CAM	60	60	52
Finite element analysis	52	44	60
Computerized process simulation	52	48	72
CIM plantwide network	56	64	64
Flexible manufacturing systems	76	72	72
Flexible assembly systems	48	64	76
Flexible machining centers	32	48	48
Robotics	40	52	64
Computerized SPC with real-time feedback	72	64	48
Automated guided vehicles	16	24	8
Automated storage/retrieval systems	—	24	32
Voice recognition/response	4	12	4
Computer-aided process planning (CAPP)	52	20	68
Rapid prototyping (e.g., stereolithography)	48	76	24
Advanced MRP II systems	76	80	68
Finite-capacity scheduling	44	56	48

	1993	1994	1995
Manufacturing execution systems (MES)	20	24	32
Product data management (PDM) systems	—	—	52
EDI links to customers	52	64	76
EDI links to suppliers	60	64	72
Bar coding	68	88	88
Shop-floor data-collection systems	88	88	84
Machine vision	28	32	48
Expert systems (artificial intelligence)	24	24	20
Radio frequency identification	—	—	24

Product Development Cycle

Three-Year Comparisons

	1993	1994	1995
Has plant significantly reduced product development cycle time? (*% of plants*)	72.0%	76.0%	96.0%
If yes, reduction in product development cycle (*median*)	50.0%	45.0%	50.0%

	High	Low	Avg.	Median
Percent reduction in product development cycle	91.4%	7.0%	46.1%	50.0%
Percent of production represented by new or redesigned products introduced in last 12 months	100.0%	4.0%	36.0%	25.4%

Practices used by product design teams:

Design for manufacturability	100% of plants
Design for quality	100% of plants
Design for assembly	64% of plants
Design for procurement	64% of plants

Methods Used to Shorten Product Development Cycles

Sampling of 1995 Responses

- Early supplier involvement (*cited by many plants*)
- Employee empowerment
- Extensive emphasis on training
- Simplification of manufacturing processes and automated assembly, calibration, and testing
- Manufacturing simulation models (computer) for flow-time optimization
- Parametric and life-cycle cost models
- CAD/CAM design tools
- Design of experiments
- Utilization of some existing stock materials for components
- Communicating process capabilities to design engineers
- Systematic design review procedure
- Early customer involvement
- CAD solid modeling
- Flexible CNC (computerized numerical control) processes
- User-group product focus meetings
- Salvage development*
- Design for testability
- Extensive LAN (local area network) for improved communication flow
- Integrated product and processes development (IPPD)
- Concurrent engineering
- Application of detailed marketing and technical specifications
- Well-planned field trials
- Computer-based on-line documentation that provides a near-paperless release system for new and modified products
- Simulation tools
- Customer-designed test sites
- Electronic exchange of design data
- Coresidency of manufacturing engineers with product design engineers

* Development of methods to remanufacture products deemed nonsalvageable.

- Long-term supplier agreements
- Dedicated, cross-functional product teams
- Customer input during product definition and development
- Computer-aided engineering
- Soft- and hard-tool development
- Applications engineering
- Implementation of cross-functional product teams
- Project team structures
- Concurrent engineering of product and process design (cited by many finalists)
- Statistical design of experiments
- Concurrent manufacturing with up-front engineering, CAD/CAM/CIM, and parallel work flow
- Working with customer to develop standardization programs
- Cross-functional strategic team establishes product requirements
- Cross-functional teams including customers and suppliers
- Early involvement of regulatory agency (FDA)
- Supplier engineering and laboratory support from concept inception
- Rapid prototyping
- Use of Quality Function Deployment (QFD) matrix
- DFM review in product definition stage
- Finite element analysis
- Failure mode effect analysis
- Involvement of purchasing, engineering, management, and labor employees during prototype stage
- Flexibility in manufacturing processes to allow for design changes
- Acceleration of bench testing to reduce total testing cycle
- Integrated product development approach involving multifunctional disciplines from design to final delivery
- Early analytical evaluation techniques for design analysis prior to fabrication of prototypes
- Solid modeling to evaluate product geometry
- Stereolithography to provide early review prototypes
- Concurrent engineering teams empowered to make decisions, allocate resources, and spend funds with minimal senior management review
- New CAD systems with advanced software and new CNC manufacturing equipment
- Multifunctional collocated design teams using integrated computer systems and shared databases

- Involvement of factory assembly operators and test technicians in the design process
- Supplier involvement in new-product design processes
- Use of engineering modeling simulations in the design process
- Use of manufacturing software simulations to improve material flow through the plant
- Empowerment of office cross-functional team that develops specifications
- Establishing state-of-the-art tool room
- Close relationships with key vendors for prototyping and production
- Use of an engineering project-tracking system
- Use of quick turnaround injection molds
- Using a design-of-experiments approach to develop necessary production procedures
- Emphasis on modular design
- Outsourcing of commodity technology
- Increased integration of CAD, CAM, and CAE tools
- Close customer interface throughout the development process

Manufacturing Cost Reduction

Three-Year Comparisons

	Percent of plants answering yes		
	1993	1994	1995
Has plant significantly reduced manufacturing costs in the last 5 years?	96.0	92.0	92.0
Have manufacturing unit-cost reductions translated into market-price reductions?	84.0	88.0	80.0
Medians	*Percent reductions*		
Approximate manufacturing cost reduction in last five years*	30.0	22.0	20.0
Approximate overall cost reduction in last 5 years, including materials	—	15.0	12.2

* Year-to-year comparisons may be misleading. Of the 1993 finalists, 11 did include purchased materials costs in calculations; for 12 others, purchased materials costs were excluded; two did not specify. In 1994, candidates were asked to provide cost calculations both ways. For the three-year comparison, the 1994 and 1995 data is that which *excludes* materials costs.

	Percent reduction			
	High	Low	Avg.	Median
Approximate manufacturing cost reduction (excluding materials costs)	77.0%	0.3%	24.5%	20.0%
Approximate overall cost reduction in last 5 years, including materials	40.0%	2.0%	16.1%	12.2%
Approximate cost reduction per unit of product shipped†	60.0%	1.0%	22.0%	16.0%

† *Note:* Cost per unit may be affected by increase or decrease in volumes.

Evidence of Improved Competitiveness

Productivity

	High	Low	Avg.	Median
% productivity improvement in 5 years—increase in value-added per employee based on employment, not just direct labor	169.0	1.0	59.1	48.7

Three-Year Comparison of Medians

	1993	1994	1995
Productivity improvement in last 5 years	38.4%	40.0%	48.7%

Largest Productivity Increases

Product/process	*Percent*
1995	
Disposable medical devices	169.0
Escalators	132.0
Plated stampings and machined parts	115.8
Diesel engines and components	100.0
Telecommunication products	92.6
Fuel-injection components	89.0
Electronic products for military and avionics	73.0

Intermediate flexible bulk containers	72.0
Handheld electric power tools	65.0
Electronic products	65.0
Airbag modules and inflators	60.1
1994	
Medical instruments (surgical)	100.0
Consumer electronics	85.5
Bulk containers	78.5
Service station equipment	69.0
Contract electronics manufacturing	65.0
Magnetic resonance spectrometers	64.0
Military electronics systems	64.0
Laser assemblies	60.0
Aluminum alloy stock	60.0
Air-conditioning compressors	57.0
Aircraft flight-control systems	40.0
1993	
Undersea weapons	363.0
Nuclear power	225.0
Electrical equipment	156.0
Medical equipment	89.0
Thermoelectric coolers	73.0
Military equipment	72.2
Welded bellows assemblies	63.0
Ice cream	57.0
High-reliability computer systems	55.0
Automotive equipment	52.0
Rubber products	46.0
Acoustical ceiling products	38.4

Major Reasons for Productivity Improvement

Sampling of 1995 Finalists Reporting Gains of 40% or More

Cited by Multiple Finalists:

- Implementation of cellular manufacturing
- Improved and/or increased training procedures

- Investment in or improved use of state-of-the-art equipment and technology: machine tools, CAD/CAM, modeling systems, processing technology, etc.
- Employee empowerment
- Supplier partnerships
- Implementation of JIT demand-flow manufacturing
- Increased automation
- Implementation of design for manufacturability
- Concurrent engineering
- Design of experiments

Other:

- Development of managerial and supervisory staff from within
- Computerized business management system
- Advancement in product knowledge for general plant population
- Reduction in internal cycle time
- Restricting overhead expansion during periods of rapid sales growth
- Rapid product prototyping, development, and implementation
- Focused meetings with manufacturing engineers to reduce scrap and increase productivity
- Core product groups consisting of customers, design engineers, quality engineers, manufacturing engineers, buyers, planners, supervisors, and key production personnel
- Supplier management as a core competency
- Quality improvements resulting in less waste
- Yield improvements result in less handling, less inspections, less retest, and less engineering changes
- Philosophy of "pulling" production through the operation
- Production equipment improvements that increase operations speed and accuracy, improve flexibility, and reduce downtime
- Better R&D
- Organization restructuring resulting in fewer layers of management and improved communication flow
- Computer information and communication systems
- Continuous improvement teams
- Focused business units
- Internal certification programs
- Identifying "cost-out" activities as projects

- Implementation of plantwide pull-board inventory control system
- Implementation of workplace organization and visual controls
- Improved first-time quality in all product lines
- Eliminating inspectors—operators responsible for their own work
- Seventy-five percent reduction in janitors—operations and maintenance personnel responsible for housekeeping
- Upgrade of in-plant lighting
- Implementation of planned maintenance plantwide
- Installation of new heat-treat technology
- Implementation of on-site commodity management
- Cell management
- Process analysis
- Open, honest, and respectful communications
- Strong support from corporate management for plant efforts
- Improved information management
- Streamlining of process to eliminate waste
- Monitoring of continuous improvement process
- Integrated product/process teams (IPTs)
- Use of TQM tools (including Pareto, run, and control charts) and cause-and-effect, flow, and scatter diagrams
- Factorywide tracking system
- Goal-oriented toward six-sigma quality
- Process verification/proofing phase for each product prior to start-up or production runs
- Design of experiments
- Manufacturing processes are flow-charted to identify impediments to flow
- Minimization of inventories using JIT and kanban techniques
- High-performance work systems (HPWS)
- Concurrent engineering
- Producibility engineering tied to manufacturing with design drawing sign-off approval authorization
- Improved manufacturing cost visibility through the MRP II system
- Statistical process control
- Flexible manufacturing methods
- Employee training in specific skills
- Safe working environment
- Product and process quality improvements
- High-volume assembly cells
- Overhaul of prehiring training program

- Reward systems, for both individual employees and for teams
- Extensive use of benchmarking of European plants
- Component cycle-time reduction and first-pass yield
- Scrap reduction and first-pass yield
- In-kind replication of processes to support volume growth
- Administrative process reengineering and office automation
- Engineering design cost reductions
- Build to order
- Information technology investments

Sampling of 1994 Finalists Reporting Gains of 40% or More

- Improvements in product design for manufacturability
- Employee involvement in continuous improvement: kaizen, quick-step, team-building
- Manufacturing process improvements (reducing nonproductive time, scrap, rework)
- Employee training in specific job skill requirements, including equipment maintenance
- "Get fast or go broke" attitude
- Requiring machinists to be responsible for checking their own work, not inspection
- Manufacturing and manufacturing engineering worked together to bring processes under control
- Implementing a cost-reduction team to evaluate the largest product line
- Engineering and management rotational training program
- Discipline and cleanliness
- Design of experiments
- Extensive use of CAD/CAM
- Use of robots
- Manufacturability studies with customers
- Total Productive Maintenance
- Use of Design for Manufacturability and Assembly (DFMA) techniques that make new products easier to build and improve quality
- Older products redesigned, standardizing parts and components, and using DFMA
- Setup times and lot sizes were reduced, thus reducing cycle times
- Cellular manufacturing techniques resulting in less inventory, better quality, reduced manufacturing space, and reduced cycle times

- A kanban pull system resulting in less inventory and reduced wastes in terms of indirect activities
- Suppliers are provided training at no cost, treated as partners, and given long-term contracts, which improves delivery performance and quality
- Improved relationship with union—grievances reduced 88 percent since 1986
- Adoption of TQM process leading to continuous improvement in many aspects of production; formal inspection nearly eliminated
- Improved business systems (hardware and software) to eliminate waste and reduce cycle times in administrative functions—from order entry to scheduling
- Concepts of teamwork, self-directed work teams, and employee empowerment; emphasis on importance of people, treating people with dignity and respect; and ongoing training in both hard and soft skills has led to decreasing absenteeism, improved morale, and belief in the concept of employee ownership of quality
- Expanded employee training
- Investment in state-of-the-art production equipment
- Work-flow analysis and process simplification
- Real-time process and operation feedback
- Bar-code implementation of product flow
- Workforce's focus on the customers.
- Operations were added to meet the needs of our customers as we became aware of their special requirements
- Self-directed work teams; technicians work together to continuously improve the process
- Variable compensation bonus plan, based on performance indices that include recovery, a major component of overall productivity, and conversion cost; with these as focus items, our teams are anxious to improve productivity
- High-performance work systems
- Concurrent engineering efforts that include design for manufacturability
- Improved manufacturing cost visibility (MRP II)—actual costs versus contract bid cost comparisons
- Statistical Process Control
- Flexible manufacturing methods
- Continuous product redesign

- Installation of computer-controlled jigs working in concert with a computer bus system that has enabled faster line rates, higher quality, and quicker troubleshooting
- Component standardization across multiple models
- Employee involvement in work redesign
- Reduced employee turnover, contributing to a more experienced, highly trained workforce
- Steady capital investment in automation
- Use of self-directed work teams, [which has] transformed us into an organization where everyone is encouraged, helped, and productive
- Information sharing about product sales prices, previous day's output, and performance of members and all teams
- Team members who are flexible enough to help in other jobs
- SDWT members set goals for the day every morning
- Plantwide incentive system is designed to reward members for increasing productivity ("fear of change has been replaced by the excitement of new challenges")
- Across-the-board implementation of "lean manufacturing" techniques, such as JIT, dynamic inventory modeling, focused factories, U-shaped lines, cross-functional teams, and TQM
- Computerized business management system
- Rapid product prototyping, development, and implementation
- State-of-the-art machine tools, CAD/CAM, and modeling systems. Products are designed on 3-D modeling systems, toolpaths are created, and prototypes are machined—all paperless
- Excellent training and education for all employees
- Conversion to demand-flow manufacturing, use of kanbans and backflushing [inventory] techniques
- Strong "design for manufacturability" policy, coupled with concurrent engineering
- Seamless organizational structure to encourage cross-functional activities
- Employee involvement with expanded scope of responsibilities
- Investment in high-tech tools, networking, and computers to aid in productivity
- JIT implementation
- Skill-based pay and training
- Worker empowerment and self-directed work teams

Customer Focus

Three-Year Comparisons

	Percent of plants answering yes		
	1993	1994	1995
Does company have a formal "customer satisfaction" program?	—	96	100
Are customer satisfaction surveys conducted regularly?	92	76	92
Are results of customer satisfaction surveys shared with all employees?	92	72	96
Does management create opportunities for employee contact with customers?	100	96	100
Do customers participate in product development efforts?	88	96	92

Supplier Partnerships

Three-Year Comparisons

	Percent of plants answering yes		
	1993	1994	1995
Does company/plant emphasize *early* supplier involvement in product development?	96	92	100
Do major suppliers receive long-term purchase agreements?	92	92	92
Do high-volume suppliers deliver to point use in plant?	64	68	—
Does plant have EDI links to major suppliers?	60	68	72
Do on-site "resident suppliers" manage and replenish inventories of items purchased from their firms?	—	52	46
Are key suppliers asked to evaluate *your* performance as a customer?	—	72	80
Do major suppliers contribute to cost-reduction and/or quality improvement efforts?	100	100	100
Does plant/company work closely with suppliers improve their quality, cost, or delivery performance?	—	100	100

	High	Low	Avg.	Median
Approximate number of suppliers who account for 80% of the dollar volume of purchased material (1995 only)	588.0	2.0	72.4	20.0

Market Results

Three-Year Comparisons

	Percent of plants answering yes		
	1993	1994	1995
Has plant increased *domestic* market share in primary product lines?*	92.0	80.0	88.0
Has plant increased *world* market share in primary product lines*†	84.0	68.0	88.0
Does plant measure customer-retention rate for major customers?	—	64.0	92.0
Is plant currently profitable?‡	100.0	88.0	96.0
If yes, has level of profitability been increasing?§	92.0	95.2	92.0
Medians			
Approximate rank in domestic market	1st	1st	1st
Approximate percentage-point gain in domestic market share since 1990*	9.0	12.1	9.0
Approximate rank in world market	—	1.0	—
Approximate percentage of plant sales/shipments to export markets	15.0	12.0	15.5
Customer retention rate¶	—	100.0	100.0
Plant-level return on assets**	—	25.0	25.6

* For 1993 finalists, increase was reported from base year 1985; for 1995 and 1994 respondents, base year was 1990.

† World market share not applicable to some defense contractors. Several finalists in all three years did not respond to this question; their nonresponse was interpreted as a no.

‡ In 1994, two plants did not answer the question; 22 of the other 23 (or 96 percent) said they were profitable.

§ One of the 1993 finalists did not respond to this question.

¶ Thirteen plants in 1995 reported 100 percent retention over four years; eleven in 1994 reported 100 percent retention over three years (since 1990).

** Formula for calculating plant-level return on assets varies among finalists.

Market Findings

	High	Low	Avg.	Median
Approximate rank in domestic market*	1st	15th	2.61	1st
Approximate *percentage-point* gain in domestic market since 1990†	90 pts	−1.1 pts	17.47pts	9 pts
Approximate percentage of plant sales/ shipments to export markets‡	51.0%	2.0%	18.7%	15.5%
Approximate increase in export volume since 1990	143.0%	0.0%	46.6%	30.0%
Major customer-retention rate over four years§	100.0%	82.0%	97.2%	100.0%
Plant-level return on assets	105.0%	6.1%	32.5%	25.6%
Approximate increase in unit sales volume for specific major products since 1990	1441.1%	9.2%	207.7%	100.0%

* Fourteen plants reported a number one market ranking.

† A few plants may have reported percentage gain in market share instead of percentage-point gain.

‡ Three plants did not report.

§ Thirteen plants reported retention rates of 100 percent.

ASSESSMENT SURVEY

Any committed benchmarker will quickly point out that studying the best practices of other companies is ultimately useless without a clear understanding of their own organization's current position. Without that internal portrait, there is no logical way to identify the knowledge that will most benefit the company. Further, without full self-knowledge, benchmarkers cannot sensibly adopt and implement the practices they uncover in their studies. This applies as well to Best Plants status.

The following survey, which is an adaptation of the application form that was used to begin judging the finalists for 1996's Best Plant awards, will help create your operational self-portrait. Completing it in detail will reveal the strengths, weaknesses, and perhaps the altogether-missing pieces in your company's performance. Comparing it to the three previous sections in this book—components of Best Plants performance revealed in the chapters of Section 1; the individual profiles of the Best Plants; and the Best Plants Statistical Profile—will yield a treasure trove of useful insights.

The Assessment Survey is best utilized as a customized improvement plan. Identify those areas in which your facility is overachieving and celebrate your accomplishments. Find those areas where your operation is underachieving and target them for improvement. Use the Assessment Survey as a guide to the critical measures of operational excellence. For example, if you cannot provide cycle-time measurements, chances are good that your operation's cycle time is not getting any faster. The Survey will also force you to confront the components of world-class manufacturing. If your facility cannot offer examples of environmental initiatives or community action, the time has come to address these areas. Collect the ideas offered in this book, contact the Best Plants for more ideas, and create an action plan. This book and this Survey are meant to initiate improvement.

One caution: Readers who plan to formally enter the America's Best Plants contest can use this survey as a guide, but not as a formal application. The idea of continuous improvement is not the sole property of manufacturers, and each year the contest's entry form is slightly modified to support the current state of the art. A new entry is available each January and can be obtained by calling (216) 696-7000, faxing your request to 216-696-7670, or writing to *Industry-Week,* America's Best Plants, 1100 Superior Ave., Cleveland, OH 44114-2543.

I. Comprehensiveness of Effort

Please check all activities which have been an area of *major emphasis* at this facility. (Use "NA" to indicate activities not applicable at this plant):

___ Total Quality Management
___ Management by Policy (MBP)
___ Cycle-time reduction
___ JIT/continuous-flow production
___ Cellular manufacturing
___ Focused-factory production concepts
___ Supplier-partnership programs
___ Customer satisfaction programs
___ EDI links to customers and/or suppliers
___ Competitive benchmarking
___ Continuous improvement
___ Hoshin/breakthrough strategies
___ Employee empowerment
___ Accelerated worker training
___ Apprenticeship programs
___ Use of work teams
___ Use of cross-functional teams
___ Employee cross training
___ Performance measurement and reward systems
___ Visibility systems (visual management)
___ Predictive/preventive maintenance
___ Total Productive Maintenance (TPM)
___ Advanced process technology
___ Advanced material technologies
___ Inventory reduction
___ Concurrent engineering
___ Design for assembly or manufacturability
___ Design for quality
___ Design for procurement
___ MRP II system enhancement
___ Enterprise Resource Planning (ERP) systems
___ Delivery dependability
___ Reducing order-to-shipment lead times
___ Flexible manufacturing methods
___ Agile manufacturing strategies
___ Streamlined production flow
___ Computer-integrated manufacturing (CIM)
___ Enterprise integration
___ Improving union/management cooperation

II. Measurement

In gauging plant performance improvements, against which of the following does management measure current performance?

_____Past plant performance

______Performance of "best practices" companies
______The ultimate possible performance

Number of benchmarking studies conducted in last three years: ______

How did plant obtain benchmarking data? Explain:

If possible, cite benchmarking data comparing recent plant performance with that of other leading operations in the *same* industry:

III. Management Practices

Please describe *two* of the most significant management practices and/or policies that have contributed to the competitiveness of this plant. (Examples may include methods for sharing information with employees, practices designed to cultivate a team culture, development of formal customer satisfaction programs, adoption of employee recognition or gainsharing programs, or other practices.) Explain how these policies were implemented and how obstacles (if any) were overcome.

IV. Process Flow

Please describe the sequence of manufacturing operations in the production process for a typical product. (Attach process flowchart to supporting statement.) Briefly describe major changes/improvements to product flow.

V. Quality Achievements

Briefly describe major quality initiatives and the most significant results:

Has plant sought ISO 9000 or QS 9000 certification? _____Yes _____No
Has plant received ISO 9000 or QS 9000 certification? _____Yes _____No
Does plant make extensive use of computerized SPC? _____Yes _____No
Does SPC provide real-time feedback on process variables? _____Yes _____No

Does plant calculate Cpk values to gauge process capability? _____Yes ____No

Are Cpk measurements used *extensively* throughout the plant? _____Yes ____No

If no, how does plant gauge process capability?

Quality Indicators for a Typical Finished Product and Major Component

A. Finished Product ______________________________

Current first-pass yield ______%

Yield improvement within last five years ______%

Typical yields for your industry ______%

Pass-through yield ______%

B. Major Component ______________________________

Current first-pass yield ______%

Yield improvement within last five years ______%

Typical yield for your industry ______%

Cpk value for *this* component manufacturing process ______

Quality Indicators for All products (Averages)

First-pass yield for all finished products (use weighted average)* ______%

Average Cpk value—or comparable measure—across all processes where Cpk measurements are applicable (weighted average)* ______

If measurement is other than Cpk value, explain:

Percent of production (units shipped) represented by new products introduced within the last 12 months ______%

* Weighted average should take into account differences in product volumes or value added.

Scrap/rework *as a percent of sales* ______%
Percent reduction in scrap/rework as a percent of sales in last five years ______%
Customer reject rate on finished products (ppm) ______
Percent reduction in customer reject rate in last five years ______%
Warranty costs *as a percent of sales* ______%
Percent reduction in warranty costs as a percent of sales within last five years ______%

Quality Methods

Which of the following quality methods are used *extensively*?

____ Manual SPC
____ Computerized SPC
____ Operators inspect own work
____ Quality Function Deployment
____ Total Quality Management

____ OTHER:

____ Design of Experiments (e.g., Taguchi methods)
____ Poka-yoke (fail-safing) methods
____ Design for Quality
____ Quality circles
____ Employee problem-solving teams

Comment:

VI. Employee Involvement/Empowerment

Briefly describe the plant's most significant employee involvement/empowerment practices, including employee education and development:

List the most impressive *achievements* attributable to involvement/empowerment practices—and explain how these have contributed to marketplace success:

Do self-directed or empowered work teams make daily decisions on production operations? ______Yes ______No

Indicate which of the following responsibilities/decisions are handled by teams rather than supervisors:

____ Production scheduling
____ Training
____ Quality assurance
____ Skills certification
____ Vacation/work scheduling
____ Daily job assignments
____ Safety review and compliance
____ Environmental compliance
____ Interteam communications
____ Hiring/firing of team members
____ Disciplinary actions
____ Performance reviews (peer evaluation)
____ Materials management

____ OTHER:

Percent of production workforce now participating in *self-directed* work teams: ______%

Percent of production workforce now participating in *empowered* work teams: ______%

Percent of total plant workforce participating in *self-directed* work teams: ______%

Percent of total plant workforce participating in *empowered* work teams: ______%

Are other types of work teams or cross-functional teams used? ______Yes ______No

If yes, describe the role of the teams:

If "team leader" position has been created, does it rotate among team members? ______Yes ______No

Please list team leader's responsibilities:

Has plant-floor supervisor/employee ratio been reduced? ______Yes ______No

Current ratio (no. of employees per supervisor) ______

Previous ratio ______ Year: ______

Do production workers participate in process development efforts? ______Yes ______No

Have production personnel participated in concurrent engineering activities to help improve manufacturability or reduce the product development cycle? ______Yes ______No

Has plant adopted a kaizen approach to continuous improvement? ______Yes ______No

What is the plant's current annual labor turnover rate? ______%

What was labor turnover rate five years ago? ______%

Percentage point reduction in annual labor turnover rate over last five years: ______

Average annual days of formal training per employee: ______

Has plant established a training curriculum with local college or other provider? ______Yes ______No

Does plant emphasize cross training of production employees? ______Yes ______No

Are employees paid for skills or knowledge acquired? ______Yes ______No

If yes, did workforce participate in the design of pay-for-skills/knowledge plan? ______Yes ______No

Are plant employees involved in benchmarking activities? ______Yes ______No

Does plant provide financial rewards for team-based performance/results? ______Yes ______No

Does plant provide financial rewards for individual employee performance? ______Yes ______No

If yes to either of the last two questions briefly describe reward system:

Average hourly wage (without overtime) of production employees: $ ______/hr
How does average hourly wage compare with production employee wages in other plants in the region—and with other plants in your industry?

VII. Customer Focus

Briefly describe major *continuing* efforts to understand and anticipate customers' needs and wants:

Does company have a formal "customer satisfaction" program in place? ______Yes ______No
If yes, briefly describe the elements of the program:

Are customer satisfaction surveys conducted regularly? ______Yes ______No
Are results of customer satisfaction surveys shared with *all* employees? ______Yes ______No
Does management create opportunities for employee contact with customers? ______Yes ______No

If yes, give examples:

Do customers participate in product development efforts? _____Yes _____No
If yes, describe how they participate:

VIII. Supplier Partnership

Describe plant's most important "supplier partnership" strategies and practices—and how plant has benefited from these partnerships:

Describe the type of information plant shares with key suppliers:

Does company/plant emphasize *early* supplier involvement in product development? ____Yes ____No
If yes, what is supplier's role?

Do major suppliers receive long-term purchase agreements? ____Yes ____No
Approximate number of production-material suppliers who account for 80 percent of the dollar volume of purchased material:
Currently ______
Three years ago ______
Does plant use point-of-use delivery/storage systems? ____Yes ____No
Do on-site resident suppliers (representing key vendors) manage and replenish inventories of items purchased from their firms? ____Yes ____No

Are key suppliers asked to evaluate *your* performance as a customer? ____Yes ____No

Do major suppliers contribute to cost-reduction and/or quality improvement efforts? ____Yes ____No

If yes, cite measurable results:

Percent of purchased material (dollar volume) that no longer requires incoming inspection: ______%

Does plant work closely with suppliers to improve *their* cost, quality, or delivery performance? ____Yes ____No

IX. Technology

Does plant extensively use computer-integrated manufacturing (CIM) technologies? ____Yes ____No

If yes, briefly describe CIM applications and results:

Has plant implemented an *enterprise integration* strategy, using information technology to link manufacturing/engineering with other functional departments or geographic locations of the business? ____Yes ____No

If yes, briefly describe enterprise integration approach and results:

Do enterprise integration links extend to customers and suppliers? ____Yes ____No

Has plant begun to use Internet/on-line services to support strategic business objectives? ____Yes ____No

If yes, briefly describe Internet/on-line activities:

Has plant implemented Enterprise Resource Planning (ERP) systems in a client/server environment? ____Yes ____No

If yes, which manufacturing software package was used? ______________

Was extensive customization required? ____Yes ____No

How have investments in information technology improved plant's competitive position or enhanced customer support and satisfaction?

Please check any of the following technologies that are used extensively in this plant. (Use "NA" to designate technologies that are not applicable.)

____ Computer-aided design
____ CAD with solid modeling
____ CAD integrated with CAM
____ Plantwide CIM network
____ Finite Element Analysis (FEA)
____ Computerized process simulation
____ Flexible manufacturing systems
____ Flexible assembly systems
____ Flexible machining centers
____ Robotics
____ Computerized SPC with real-time feedback
____ Automated guided vehicles
____ Automated storage/retrieval systems
____ Voice recognition/response
____ Computer-aided process planning
____ Rapid prototyping (e.g., stereolithography)
____ Advanced MRP II systems
____ Enterprise Resource Planning system
____ Finite-capacity scheduling
____ Manufacturing execution systems (MES)
____ Product data management (PDM) systems
____ EDI links to suppliers
____ EDI links to customers
____ Bar coding
____ Shop-floor data-collection systems
____ Machine vision
____ Expert systems (artificial intelligence)
____ Radio frequency (RF) identification

Describe any particularly innovative or state-of-the-art applications of *production* technology:

Describe any particularly innovative or state-of-the-art applications of *information* technology:

X. Manufacturing Operations and Flexibility

What practices, methods, or technologies have been most effective in improving plant's flexibility and agility?

Does plant manufacture or assemble components in small lot sizes? ____Yes ____No

By approximately what percentage have lot sizes been reduced? ______%

What are the manufacturing cycle times for *typical* products?

Product A ____________________ Cycle time ______

Product B ____________________ Cycle time ______

By what percent has cycle time been reduced within the last five years?

Product A ______%

Product B ______%

Has plant adopted focused factory production systems? ____Yes ____No

Has plant adopted JIT/continuous-flow production methods? ____Yes ____No

If yes, are JIT methods predominant? ____Yes ____No

Briefly describe major benefits achieved by adopting JIT/continuous-flow manufacturing methods:

Does plant employ a pull system with kanban signals? ____Yes ____No

Does plant offer JIT delivery to customers? ____Yes ____No

If yes, does plant *primarily* (*select one*):

______ Rely on shipments from finished-goods inventory?

______ Rely on quick completion of subassemblies into finished goods?

______ Build/produce entirely to order using a JIT/pull system?

For *major* products, what is your current *standard* order-to-shipment lead time?

Product A ____________________ Lead time ______

Product B ____________________ Lead time ______

Has order-to-shipment lead time been significantly reduced? ____Yes ____No

If yes, by what percentage?

Product A ______%

Product B ______%

What is the shortest lead time on rush orders?

Product A ______

Product B ______

For product A, what is the *typical* order-to-shipment lead time in *your industry*? ______

Do you build primarily to order rather than to stock? ____Yes ____No

On-time delivery (% of time) ______%

Is the above on-time rate based on (*select one*):

______ Date customer requested

______ Date promised

How often are production schedules met (% of time)? ______%

Inventory Management

Annual inventory turns*: ______

Annual work-in-process (WIP) turns†: ______

Percent reduction in total inventory since 1991: ______%

Percent reduction in WIP inventory since 1991: ______%

Percent reduction in finished-goods inventory since 1991: ______%

XI. Maintenance

Describe key elements of maintenance programs and practices:

Does plant practice Total Productive Maintenance? ____Yes ____No

Does plant employ predictive/preventive maintenance techniques? ____Yes ____No

What is the average machine availability rate? ______%

* Inventory turns = value of annual shipments ÷ current inventory value. (Use "plant cost" valuation.)

† WIP turns = value of annual shipments ÷ current WIP value.

XII. New-Product Development

Has plant significantly reduced product development cycle time?* ____Yes ____No

If yes, by what approximate percentage has cycle time been reduced? ______%

What percentage of plant's current production (based on dollar value of shipments) is represented by new or redesigned products introduced within the previous 12 months? ______%

Do plant personnel participate in concurrent engineering activities to reduce product development cycle? ____Yes ____No

Which of the following practices have been adopted by the product development team?

______ Design for manufacturability ______ Design for quality
______ Design for assembly ______ Design for procurement

How have the above practices significantly reduced manufacturing costs or time to market?

XIII. Environment and Safety Programs

Has plant adopted a pollution prevention strategy to eliminate pollutants at an upstream stage of the process rather than at the end of the pipe? ____Yes ____No

Has plant implemented pollution control measures that exceed current regulatory requirements? ____Yes ____No

If yes, give one or two examples:

* Assume product development cycle begins with initial product concept and ends with production of salable product.

List plant's most significant accomplishments in environmental protection.

Has plant been cited by federal or state Environmental Protection Agency authorities for violations of environmental laws within the last five years?* ____Yes ____No

If yes, explain:

List plant's most significant *worker safety* achievements (provide statistical results where possible):

For most recent 12-month period, what was plant's lost-workday rate per 1,000 employees? ______

For the same period, what was plant's OSHA-reportable injury rate per 1,000 employees?* ______

If above rates have improved, provide specifics:

* If plant is located in Mexico or Canada, use comparable authoritative agencies for those countries.

XIV. Community Involvement

Briefly describe this plant's most significant community involvement activities:

XV. Evidence of Improved Competitiveness

Productivity

Using a value-added-per-employee yardstick, based on the total employment at this plant (not just direct labor), how much has productivity improved in the last five years? ______%

Also calculate five-year productivity increase based on total sales per employee: ______%

Describe major reasons for productivity improvement:

Cost Reduction

Has this plant significantly reduced manufacturing costs in the last five years? ____Yes ____No

Approximate manufacturing cost reduction per unit of product shipped, *excluding* purchased-materials costs* ______%

Approximate cost reduction per unit, *including* purchased-materials costs: ______%

Have unit cost reductions translated into market-price reductions? ____Yes ____No

* For manufacturing-cost calculation here, include: quality-related costs, direct and indirect labor, manufacturing support and overhead, equipment repair and maintenance, and other costs directly associated with manufacturing operations. Do not include purchased-materials costs here.

Market Results

Has plant increased *domestic* market share in primary product lines since 1991? ____Yes ____No

Approximate *percentage-point* gain in domestic market share since 1991: ______

Approximate *rank* in domestic market: ______

Has plant increased *world* market share in primary product lines since 1991? ____Yes ____No

Approximate rank in *world* market: ______

Approximate percentage of plant sales/shipments to export markets: ______%

Approximate increase in export volume since 1991: ______%

Does plant measure customer-retention rate for major customers? ____Yes ____No

If yes, what has been the retention rate for the last five years?* ______%

Is plant currently profitable? ____Yes ____No

If yes, has plant-level profitability been increasing? ____Yes ____No

Comment:

What is plant-level *return on assets?* ______%

Approximate increase in unit sales volume for major products since 1991:

Product A ________________________________ Increase: ______%

Product B ________________________________ Increase: ______%

Other Marketplace Achievements

Please describe market-related achievements not otherwise delineated in this section:

* Percent of significant customers in 1991 who are still customers.

AFTERWORD

Ford Motor Company was started with the idea and determination of one man and quickly grew to employ thousands. It was Henry Ford's innovative philosophies regarding quality and efficiency in manufacturing that gave Ford its firm foundation on which to build. Mr. Ford made the following statement decades ago, "Save ten steps a day for each of twelve thousand employees and you will have saved fifty miles of wasted motion and misspent energy." It is a very basic concept but it rings true even for today's manufacturers.

Customers, of course, don't buy processes, they buy products. And products aren't built by processes, they're built by people. When it comes to providing a well-built product, our strength lies in our people. They represent a vast fund of knowledge, experience, and creative ideas—resources that meet or create demand in the market.

The world is getting smaller and the competition is getting tougher. Only the most efficient and quality-oriented manufacturers will survive. Those businesses that are constantly improving, constantly seeking to meet or create a demand, constantly finding lower-cost ways to bring high-quality goods to the greatest number of potential customers will be the survivors.

Alex Trotman
Chairman and Chief Executive Officer
Ford Motor Company

INDUSTRYWEEK EDITORIAL STAFF CONTACTS

IndustryWeek magazine and its staff can be contacted by mail, telephone, fax, or on-line via CompuServe Information Service and the World Wide Web. Mail should be sent to the magazine at 1100 Superior Ave., Cleveland, OH 44114-2543. Telephone: (216) 696-7000. Fax: 216-696-7670.

E-mail queries can be addressed to the staff at the listed CompuServe addresses or through *IndustryWeek Interactive*, a CompuServe forum that features an on-line version of the magazine. *IndustryWeek*'s World Wide Web site (http://www.industryweek.com) is currently under construction and will also offer contact information for the staff. Further information will be posted as available at Penton Publishing's Web site located at: http://www.penton.com.

David Bottoms, Associate Editor	74774.2323
John R. Brandt, Editor-in-Chief	74774.3150
Miriam Carey, Conference Editor	74774.2316
Shari Caudron, Contributing Editor	74777.3672
Tanya Clark, Contributing Editor	74777.3673
Daniel Dombey, Contributing Editor	74052.2234
Peter Fletcher, Contributing Editor	74777.3663
Theodore B. Kinni, Director, *IW* Books	74774.2215
John S. McClenahen, Senior Editor	74774.2217
William H. Miller, Senior Editor	74774.2321
Brian S. Moskal, Senior Editor	74774.2330
Robert Patton, Contributing Editor	74777.3665
John H. Sheridan, Senior Editor	74774.2325
Tim Stevens, Associate Editor	74774.2322
George Taninecz, Associate Editor	74777.3670
John Teresko, Senior Technology Editor	74774.2324
Michael A. Verespej, Senior Editor	74774.2320
Alfred Vollmer, Contributing Editor	75162.1246
Janet Walter, Manager, Editorial Information Systems	74774.2327

INDEX

Where to Get More Information About *IndustryWeek*

IndustryWeek is highly respected by industry management because it champions the importance of manufacturing to a healthy, global economy, and recognizes the distinct and unique challenges its executives face.

IW's mission is to help executives create more effective, more humane, and more profitable organizations. *IW* is committed to:

1. Providing its executive customers with the most relevant, current, and reliable information on management best practices.
2. Providing the information in the media and forums of our customers'choice.